**NATO Science for Peace and Security Series C: Environmental Security**

NATO Science for Peace and Security Series

Sub-Series C. Environmental Security

Volumes in the Environmental Security Sub-Series deal with scientific and technological aspects of innovative and promising projects related to environmental security, including regional environmental problems and natural and man-made disasters. Floods, droughts, landslides, desertification, earthquakes and industrial accidents are examples of topics covered, as are regional environmental problems related to transboundary water resources or transfrontier air pollution. Other areas covered relate to climate change, environmental chemistry, disaster prevention and forecast, remote sensing and detection, and decision support systems/risk assessment.

Special offer: For all clients with a print standing order we offer free access to the electronic volumes of the Series published in the current year.

Antonella Angelini · Luigi di Bitonto ·
Carlo Pastore
Editors

# Security Enhancement for Climate Change Impacting Urban Resources - SECCURe

Springer

*Editors*
Antonella Angelini
Water Research Institute (IRSA)
National Research Council (CNR)
Bari, Italy

Luigi di Bitonto
Water Research Institute (IRSA)
National Research Council (CNR)
Bari, Italy

Carlo Pastore
Water Research Institute (IRSA)
National Research Council (CNR)
Bari, Italy

ISSN 1874-6519 ISSN 1874-6543 (electronic)
NATO Science for Peace and Security Series C: Environmental Security
ISBN 978-94-024-2344-0 ISBN 978-94-024-2345-7 (eBook)
https://doi.org/10.1007/978-94-024-2345-7

Proceedings of the Advanced Research Workshop on Security Enhancement for Climate Change Impacting Urban Resources - SECCURe, 9–12 July 2024, Montelibretti (Rome), Italy



This Springer imprint is published by the registered company Springer Nature B.V.
The registered company address is: Van Godewijckstraat 30, 3311 GX Dordrecht, The Netherlands

# Organization

## Program Committee Chairs

Daniela D'Agostino, CIHEAM Bari, Valenzano (Bari), Italy
Carlo Pastore, National Research Council, Water Research Institute, Bari, Italy

## Program Committee Members

Daniela D'Agostino, CIHEAM Bari, Valenzano (Bari), Italy
Carlo Pastore, National Research Council, Water Research Institute, Bari, Italy

## Reviewers

Mostafa Abdelgeliel, Arab Academy for Science, Technology and Maritime Transport (AASTMT), Energy Research Unit, and Electrical and Control Engineering, Alexandrias, Egypt
Ahmed Hamza H. Ali, Assiut University, Department of Mechanical Power Engineering, Assiut, Egypt
Antonella Angelini, National Research Council, Water Research Institute, Bari, Italy
Georgia Antonoupolu, University of Patras, Department of Sustainable Agriculture, Agrinio, Greece
Zoughaib Assaad, Mines Paris PSL, CEEP, Paris, France
Camilla Braguglia, Maria Water Research Institute of National Research Council, Monterotondo (Rome), Italy
Daniela D'Agostino, CIHEAM Bari, Valenzano (Bari), Italy
Massimiliano Errico, University of Southern Denmark, Department of Green Technology, Odense, Denmark

Gerasimos Lyberatos, National Technical University of Athens, Chemical Engineering, Zografou, Greece
Mariano Martín, Universidad de Salamanca, Chemical Engineering, Salamanca, Spain
Henrique Matos, Instituto Superior Técnico/Universidade Lisboa, Chemical Engineering Department, Lisboa, Portugal
Rabi Mohtar Texas A&M University, Energy Institute, College Station, USA
Ioanna Ntaikou University of Patras, Civil Engineering, Patras, Greece
Karina Angelica Ojeda Delgado, Universidad de Cartagena, Chemical Engineering, Cartagena, Colombia
Alessandro Pagano, Department of Civil, Environmental, Land, Construction and Chemistry, Polytechnic University of Bari, Bari, Italy
Konstantinos Pappas, Texas A&M University, Energy Institute, College Station, USA
Stefania Tronci, Università di Cagliari, Cagliari, Italy
Luigi di Bitonto, National Research Council, Water Research Institute, Bari, Italy

# Preface

The Water-Energy-Food (WEF) Nexus is a critical area of research aimed at building a more sustainable society. As climate change intensifies, the pressures of water and food scarcity, along with rising energy demands, call for careful planning, international cooperation, and solutions that address environmental security challenges. Beyond resource management, the WEF Nexus must also consider the social security issues tied to environmental and resource limitations, such as health risks. In today's rapidly changing and unpredictable climate landscape, a multidisciplinary approach to the WEF Nexus is essential. This requires collaboration and continuous dialogue between experts across various fields. While no single solution can address the global environmental crisis, every effort to reduce greenhouse gas (GHG) emissions represents a step toward mitigating climate change. Different geographical conditions necessitate tailored solutions, highlighting the importance of a multi-faceted approach to climate resilience.

New sustainable strategies are needed to improve resource management. Applying Circular Economy principles to the water-energy-food interconnections across diverse regions will be crucial in identifying alternative, beneficial pathways. The use of renewable resources, advancements in wastewater treatment, smart cultivation practices, and biomass valorization can bolster societal resilience, benefiting the entire community. Additionally, integrating innovative technologies and process designs can lead to biorefineries with lower GHG emissions, aligning with the goals of sustainable growth and human well-being. This book addresses these pressing issues, providing a comprehensive overview of the challenges posed by climate change and their impacts on society. It also presents a series of innovative actions aimed at enhancing the resilience of modern society.

Bari, Italy

Antonella Angelini
Luigi di Bitonto
Carlo Pastore

# Contents

# Part I
# Energy/Food/Water Nexus: From Worldwide Contexts to the Mediterranean Area

# Chapter 1
# Energy-Food-Health Nexus: The Role of Water

**Iqbal M. Mujtaba**

**Abstract** The 17 UN Sustainable Development Goals (UN SDGs) are to create a smart, resilient, healthy and cohesive society. Water (SDG 6—Clean water) is a key contributing factor to several of these SDGs, for example, 2-zero hunger, 3-good health & wellbeing, 6—(clean water and) sanitation, 7- affordable and clean energy, 8—decent work & economic growth, 16—peace and justice. Climate action (SDG 13) while impacting society's health & wellbeing (SDG 3), impacts the surface water level too (SDG 6) and thus contributes to SDGs 2, 7, 8, 16 indirectly. This article highlights the link between water and several of these SDGs and discusses the role society should play in mitigating the challenges in energy, food and health.

**Keywords** Energy · Food · Water · Health · Climate

## 1.1 Introduction

Improved standards of living require increasing amounts of freshwater. Consequently, in the last 50 years, a sharp increase in the volume of industrial effluent being generated and disposed of into rivers and oceans is witnessed, causing significant harm to our ecosystem and health. Although the Ancient Mariners' rime: "Water, water everywhere/Not any drop to drink" was referring to 97% of the planet's water (seawater), vast amount polluted river streams around the world would fit equally to this rime.

Climate change also results in river water levels falling in many parts of the world to a dangerous level impacting on agriculture, transportation of essential commodities, etc. (see further details in Sect. 1.5). Currently, 11% of the world population do

I. M. Mujtaba (✉)
Chemical Engineering Department, Faculty of Management, Sciences & Engineering, University of Bradford, Bradford, UK
e-mail: I.M.Mujtaba@bradford.ac.uk



A. Angelini et al. (eds.), *Security Enhancement for Climate Change Impacting Urban Resources - SECCURe*, NATO Science for Peace and Security Series C: Environmental Security, https://doi.org/10.1007/978-94-024-2345-7_1

not have clean water close to home. Globally, about 1.7 billion people use a drinking water source contaminated with faeces (https://www.wateraid.org/facts-and-statis tics). Water has been a device used for religious conflict and regional and local battles leading to waves of migration to other countries. The disputes over water will inevitably become more common, as 280 river basins globally are shared by two or more countries (see Sect. 1.7 for further details).

Interestingly, the UN report estimated that three out of four jobs that make up the global workforce are either heavily or moderately dependent on water (https:// www.unesco.org/en/wwap/wwdr/2016 accessed on 01/03/2025). This means that water shortages and problems of access to water and sanitation could limit economic growth and job creation in the coming decades. The report also highlights that water, and jobs are inextricably linked to economic, environmental, or social perspective.

Finally, the worldwide food wastage contributes to huge amounts of usage of water and land and impacts environment via $CO_2$ emissions. This paper will highlight these global issues linked to water and will reflect on potential solutions.

## 1.2 Water-Energy Nexus

Sustainable supply of energy is essential to have a healthy and prosperous society. However, energy production requires large amounts of water. Mekonnen et al. [1] reported consumption of 378 billion $m^3$ of water per year is required for producing electricity using coal, lignite, natural gas, oil, uranium or biomass as well as wind, solar, geothermal energy and hydropower. In Africa alone, according to Gerbens-Leenes et al. [2], total water consumption (WC) was reported to be 23.822 billion $m^3$/ year for an annual electricity production of 1,050, 674 GWh based on 2534 individual power plants (including 1447 fossil (coal, oil and natural gas), 1071 renewable (wind, sun, biomass, geothermal and hydropower) and 16 other (waste heat and nuclear) power plants.

High oil prices around the globe make the production of biofuel very attractive in recent years. However, environmental performance, economic feasibility and social desirability are of concern for biofuel production. For example, the impact of biofuel on water usage is hardly mentioned. Table 1.1 shows the water consumption for different crops used for bioenergy [3]. Note, many of these crops are food crops. Therefore, biomass production required for biofuel production may compete with food crops. However, it is assumed that not only crop yields, but total biomass yields are used for the generation of electricity.

de Fraiture et al. [4] mentioned that biofuel production will require to utilize about 3% of crop water use by 2030 and many countries (such as India and China) may decide not to pursue biofuel options based on traditional food crops.

**Table 1.1** Water consumption of different crops providing electricity

| Crops | Water consumption ($m^3$/GJ of electricity) |
|---|---|
| Sugar beet | 46 |
| Maize | 50 |
| Sugar Cane | 50 |
| Barley | 70 |
| Rye | 77 |
| Paddy rice | 85 |
| Wheat | 93 |
| Potato | 105 |
| Cassava | 148 |
| Soybean | 173 |
| Sorghum | 180 |
| Rapeseed | 383 |
| Jatropha | 396 |

*Source* Adapted from Gerbens-Leenes et al. [3]

## 1.3 Water and Food Nexus

A vast amount of water is required for sustainable production and processing of foods. In many parts of the world 35–90% water goes into agriculture (Fig. 1.1), south Asia using the most while Europe and Central Asia using the least.

Table 1.2 shows the amount of water needed to process different types of food. This clearly demonstrates the requirement of large amounts of water being used in

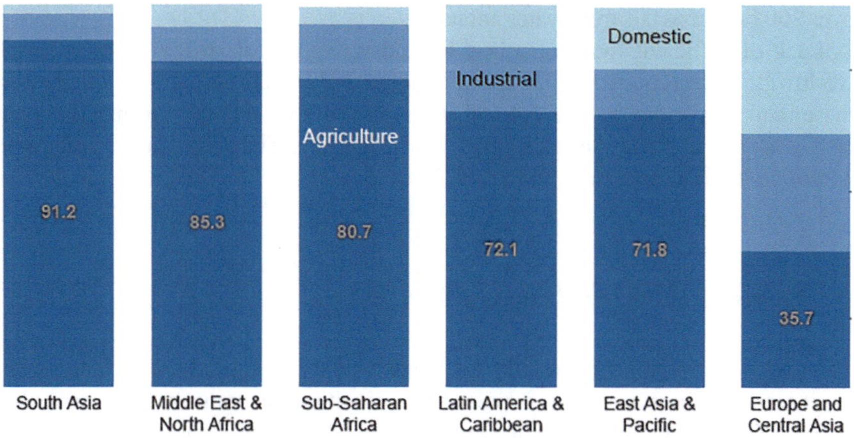


**Fig. 1.1** Water use in the world (https://blogs.worldbank.org/opendata/chart-globally-70-freshwater-used-agriculture, accessed on 25/02/2025 and adapted)

**Table 1.2** Water consumption for producing different types of foods

| Foodstuff | Quantity | Water consumption, litres |
|---|---|---|
| Chocolate | 1 kg | 17,196 |
| Beef | 1 kg | 15,415 |
| Sheep Meat | 1 kg | 10,412 |
| Pork | 1 kg | 5988 |
| Butter | 1 kg | 5553 |
| Chicken meat | 1 kg | 4325 |
| Cheese | 1 kg | 3178 |
| Rice | 1 kg | 2497 |
| Cotton | 1 @ 250g | 2495 |
| Pasta (dry) | 1 kg | 1849 |
| Bread | 1 kg | 1608 |
| Pizza | 1 unit | 1239 |
| Potatoes | 1 kg | 287 |
| Milk | 1 × 250ml glass | 255 |
| Tomato | 1 kg | 214 |
| Egg | 1 | 196 |

*Source* IMechE, adapted from https://www.theguardian.com/news/datablog/2013/jan/10/how-much-water-food-production-waste#data accessed on 25/02/2025

the food sector. Dairy industries worldwide are considered to be the most consuming and polluting plants in food manufacturing industries with a significant number of pollutants and value-added compounds in the water streams. For example, Buabeng-Baidoo et al. [5] metntioned that the dairy industry in India is anticipated to consume over 200 billion $m^3$ of freshwater annually.

Lack of sufficient food for a healthy and active life of several million people worldwide were reported by FAO and IFAD in 2015. Addressing food security for these individuals places an additional strain on water resources over the 70 percent of available water being already used for agriculture. Transboundary water issues contribute to food insecurity further [6, 7].

### *1.3.1 Water-Precision Agriculture*

To combat the significant amount of water usage for agriculture, scientists and engineers have been working on precision agricultural techniques. Note, according to FAO (Food and Agricultural Organisation), 250,000 $m^3$ of water is used to produce food that is lost or wasted worldwide. Precision agriculture (PA), satellite farming or site specific crop management (SSCM) system uses real-time sensors that measure everything from chlorophyll levels to plant water status. The goal of

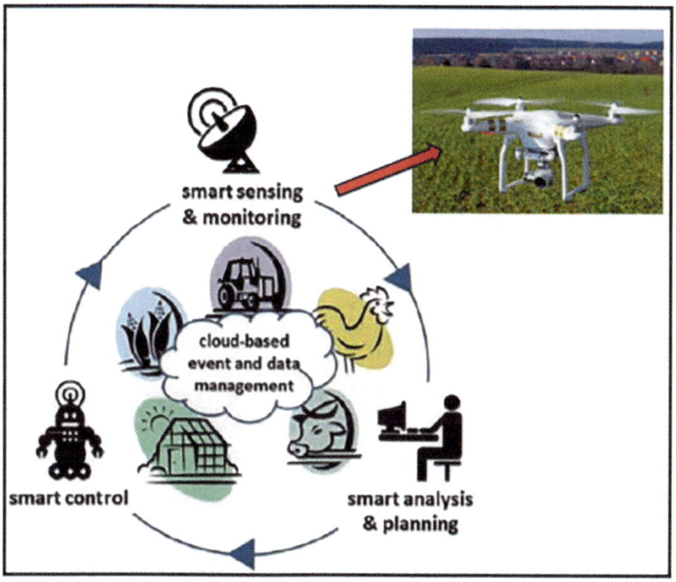


**Fig. 1.2** Smart farming (adapted from [8, 9])

precision agriculture research is to define a decision support system (DSS) for whole farm management with the goal of optimizing returns on inputs while preserving resources such as water, pesticides. Figures 1.2 and 1.3 shows the concept of future smart farms. These figures show that field data are recorded using drone (or IoT/cloud based systems) which is then processed using data analytics resulting in planning and control for supplying appropriate amounts of water and pesticides.

## 1.4 Water and Health Nexus

Clean water, decent toilets and good hygiene are essential elements of a dignified and healthy life. Globally, more than 10% of the population do not have clean water close to home, 1.5 billion people do not have a decent toilet of their own, 30% of the schools do not have a basic water service (https://www.wateraid.org/facts-and-statistics). Globally, almost 1.7 billion people use drinking water sources contaminated with faeces with significant risk of microbial contamination (https://www.who.int/news-room/fact-sheets/detail/drinking-water). Contaminated water and poor sanitation are responsible for transmitting diseases such as cholera, diarrhea, dysentery, typhoid, polio and many others resulting in over half a million death a year and huge health expenditure at both personal and at state levels.

However, the impact of water on improving child health has mainly focused on reducing child mortality and morbidity associated with microbiological pathogens

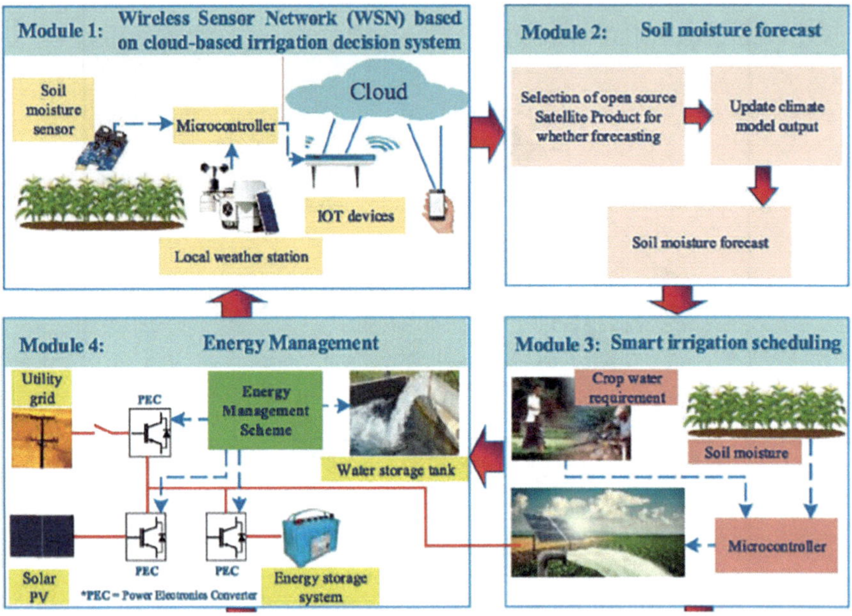


**Fig. 1.3** Smart farming water-food-energy nexus (adapted from [10])

and diarrheal disease. The review by Rhue et al. [11] discussed many alternative pathways for improving health and wellbeing throughout childhood by ensuring improved household access to safe and sustainable water supply. In addition, contaminated water can impact fetal health of pregnant mothers [12].

Furthermore, contaminated or unprotected water for bathing leads to high proportion of bacterial vaginosis (BV), the most common vaginal infection for women [13]. The estimated annual global economic burden of treating symptomatic BV is about US $4.8 billion [14].

### *1.4.1 Water-Economic Growth-Health*

According to UN Report 3 out of 4 jobs that make up the global workforce are either heavily or moderately dependent on water. Water shortages and lack of access to water and sanitation could limit economic growth and job creation in the coming decades leading to mental health (to start with and then impacting physical health). It is mentioned in the report that water and jobs are inextricably linked to economic, health or social perspective (https://www.unesco.org/en/wwap/wwdr/2016 accessed on 01/03/2025).

Roggenburg et al. [15] noted that the US-Mexico border was experiencing a decline in the availability of this critical resource (water), due to mismanagement and climate change. The natural water in the area would continue to deplete and the region would not be able to support local societies and industry without proper planning such as considering the use of desalination techniques combined with renewable energy sources.

### 1.4.2 Wastewater and Health Nexus

The exponential population growth and the growth in industrialization are continuously increasing the amount of wastewater containing many harmful synthetic and geogenic organic compounds often being disposed of into rivers, lakes, and groundwater and severely affect human health and the natural ecosystem [16]. For example, some well-known endocrine-disrupting chemicals (EDCs) often found in water sources include pesticides, fungicides, and dioxins, butyl benzyl phthalate and diethyl phthalate, bisphenol A, tamoxifen, natural sex hormones estrone can cause adverse effects on humans and aquatic life [17].

The quality of wastewater depends on the type of industry it serves and also on the standards and regulations of the country in question. Also, around 150 million tonnes of fertilizers and several million tonnes of pesticides are used globally each year contributing to huge and various types of contaminants in the water sources. As par Schwarzenbach et al. [18] report, 300 million tonnes of synthetic compounds per year were being disposed of into natural waters and this has grown many folds since then.

Bolong et al. [19] mentioned 564 toxic chemicals in wastewater impacting health and well-being and the environment. Bueno et al. [20] reported presence of a group of 100 organic compounds in wastewater treatment plants (WWTPs) belonging to several chemical groups (pharmaceuticals, personal care products, pesticides and metabolites). Fijalkowski et al. [21] has mentioned that over 1500 organic compounds being registered with vast part of them being found in wastewater and later in sewage sludge. In addition, microplastics in water streams has been a major public concern in recent years. Production of plastics has increased by more than 20 times over the last half century and improper disposal has contributed to the existence of plastic debris in wastewater, marine water, freshwater, etc. leading to increased ecological risks, and chronic toxicity to living organisms globally [22].

Health agencies around the world set tight targets for removing these harmful pollutants from industrial effluents before being discharged into the surface water. For example, the maximum allowable limit of dimethylphenol concentration in surface water is 0.05 ppm as set by the Agency of Toxic Substances and Disease Registry (ATSDR). The World Health Organization (WHO) has set the phenol concentration to 1 μg/l in drinking water. Note, phenolic compounds can be found in several industrial effluents (refineries 6–500 ppm, coal 9–6800 ppm, petrochemical 28–1220 ppm, pharmaceutical production 0.1–1600 ppm). Another common toxic contaminant found in wastewater systems is N-nitrosamine (especially NDMA, N-nitrosodimethylamine-D6). Many water authorities around the globe strictly regulates N-nitrosamine concentration level in drinking water and recycled water. The International Agency for Research on Cancer and the US Environmental Protection Agency have classified N-nitrosamine as a cancer risk and have restricted the concentration of N-nitrosamine to maximum 0.7 ng/L concentration in recycled water [16].

### *1.4.3 Water—Sanitation-Health*

As mentioned earlier, 1.5 billion people do not have a decent toilet of their own. UN SDGs 6.1 and 6.2 aim to have access to safe drinking water, access to adequate and equitable sanitation and hygiene for all, and an end to open defecation by 2030. Bisung and Elliott [23] argued that the availability of safe water and adequate sanitation determine the psychosocial well-being of individuals and households. Note, a very large amount of water is required for sanitation and personal hygiene. Sultana et al. [24] reported that on average (male, Female. Adults, children) 39 L of water is consumed per capita per day in lower-income urban families. In the event of humanitarian emergencies, the first priority is to provide an adequate quantity of water, even if its safety cannot be guaranteed, and to protect water sources from contamination. WHO has recommended that a minimum of 15 L per person per day should be provided (https://www.who.int/teams/environment-climate-change-and-health/water-sanitation-and-health/environmental-health-in-emergencies/humanitarian-emergencies, accessed on 28 February 2022).

### *1.4.4 Wastewater and Reuse*

Water recovery and reuse from industrial and municipal wastewater are increasing even in countries with little or no water shortage [14] (Fig. 1.4). The wastewater from different sources are collected and treated using advanced treatment methods. This in turn has reduced the quantity of wastewater being dumped into water sources significantly. Jones et al. [23] estimated a global wastewater production of 359.4 $\times 10^9$ $m^3$/year, of which 63% (225.6 $\times 10^9$ $m^3$/year) is collected and 52% (188.1 $\times 10^9$ $m^3$/year) is treated. Table 1.3 highlights wastewater production, treatment and reuse by region [25]. More than 5000 municipal WWTPs operates in China, with a daily treatment capacity of nearly 200 million $m^3$/d [26]. Nyamayedenga [27] reported around 50% (net) reduction in freshwater consumption and wastewater generation in an industrial plant. Bhavsar et al. [28] considered recovery and recycle of wastewater in a multi-batch textile industry and reported 59% reduction in freshwater consumption in the plant.

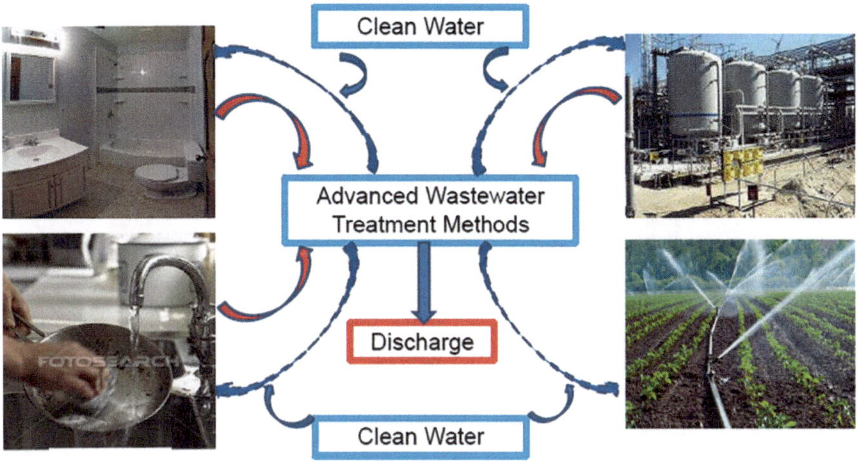


**Fig. 1.4** Water sources → treatment → clean water → wastewater → wastewater treatment and reuse [33]

**Table 1.3** Wastewater production, treatment and reuse by region

| Region | Production ($10^9$ $m^3$/year) | Treatment ($10^9$ $m^3$/year) | Reuse ($10^9$ $m^3$/year) |
|---|---|---|---|
| North America | 74.7 | 50.6 | 9.1 |
| Latin America and Caribbean | 42.1 | 15,4 | 2.1 |
| Western Europe | 38.5 | 33.0 | 6.7 |
| Middel East and North Africa | 21.9 | 11.4 | 6.1 |
| Sub-Saharan Africa | 11.0 | 1.8 | 1.6 |
| South Asia | 25.6 | 4.0 | 0.5 |
| Eastern Europe and Central Asia | 28.2 | 14.9 | 2.6 |
| East Asia and Pacific | 117.6 | 57 | 11.9 |

*Source*: Adapted from Jones et al. [25]

The yuck factor, the term *toilet-to-tap* used by media gives significant negative images preventing the use of reclaimed wastewater, especially for drinking and agricultural production purposes [29, 30]. According to Ravishankar et al. [31], 28% of the population in Bengaluru would not use the reclaimed water for drinking (Fig. 1.5). Recent study in Oman shows over 74% of the population would not drink treated wastewater ([32], Fig. 1.5).

CNN news on 19 March 2019 revealed severe water shortages that could plunge England into 'jaws of death' in 25 years and according to BBC news on 10 May 2013 revealed that Londoners were not ready to drink treated wastewater (Fig. 1.6).

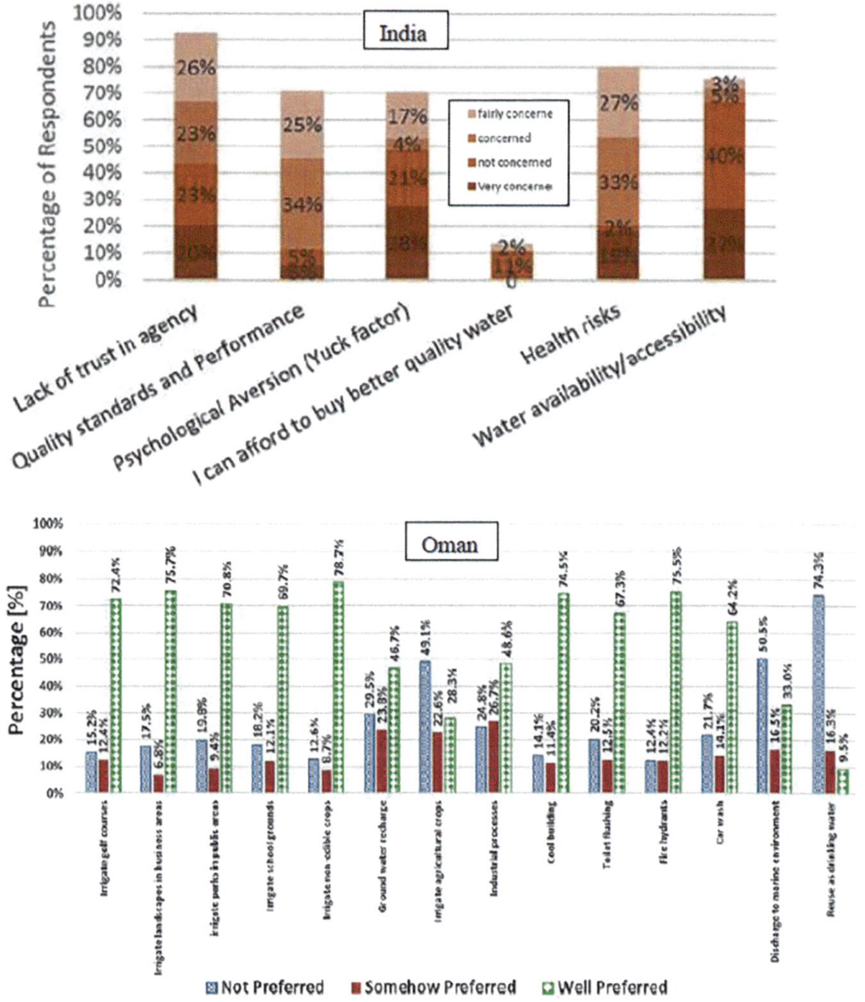


**Fig. 1.5** People's perception on the use of treated wastewater (adapted from [31, 32])

## 1.5 Climate and Water Nexus

Existing pressures on water resources will continue to be seen not only due to an increase in population but also by climate change. Gosling and Arling [34] carried out global-scale assessment of the impact of climate change on water scarcity. Under the current projected climate change, people living under conditions of severe water stress will continue to rise for the years to come. The current accelerated warming being observed in deserts compared to other terrestrial areas in recent decades is predicted to continue in the future making it difficult and challenging to maintain water security in the drylands [35]. Climate change is also projected to impact river,

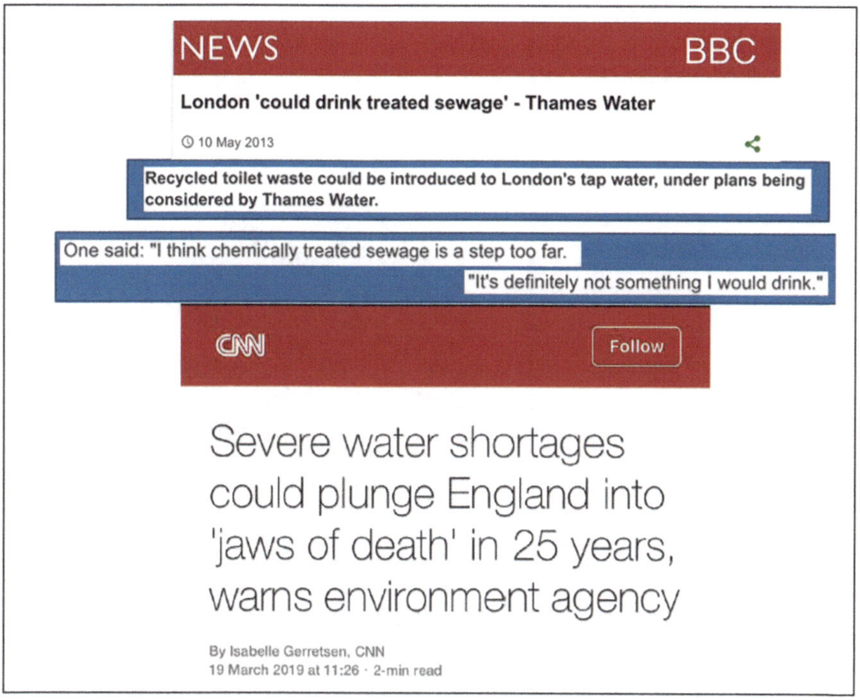


**Fig. 1.6** BBC News (10 May 2013) and CNN News (19 March 2019)

lake, and wetland hydrology and thus agriculture and supply chain [36]. In 2022, Spain has observed her water reservoirs at 40% capacity; Italy declared state of emergency due to low water level in River Po impacting agriculture severely; Germany observing serious disruption in supply chain due to dangerous low water level in river Rhine (Fig. 1.7).

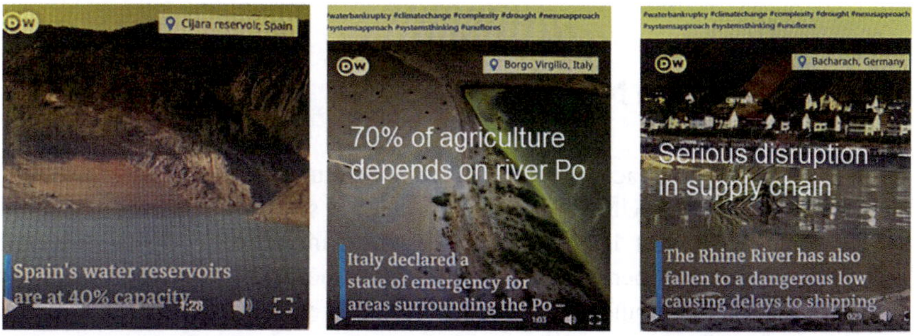


**Fig. 1.7** Climate change impacting surface water level and the consequences (DW Tv, 22 August 2022)

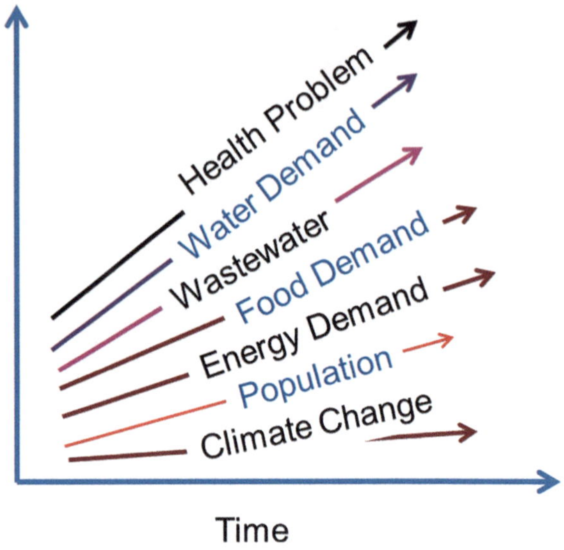


**Fig. 1.8** Population growth, climate change and associated challenges

## 1.6 Water and Associated Challenges

The discussions presented so far clearly highlight that water impacts significantly on our life. Figure 1.8 clearly captures (not to scale) that increase in population increases the demand in water, food & energy which then results in increasing amount of wastewater leading to increasing health challenges. In addition, continuously changing climate (rising temperature) impacts on the available and usable surface water. Figure 1.9 highlights list of available water resources and technologies which can be used for freshwater production (from seawater, brackish water, rivers, lakes) and wastewater treatment. The question remains whether technical solutions alone can combat the global challenge associated with water [6, 16, 37, 38].

## 1.7 Water—Peace Nexus

UN SDG 16 calls for peace, justice and strong institutions. In this section, we will look at the link between UN SDG 6 (Clean water and sanitation) and 16. Myers [39] quoted "During the next 10 years, many countries important to the United States will almost certainly experience water problems—shortages, poor water quality, or floods—that will contribute to the risk of instability and state failure, and increase regional tensions."

The link between environmental problems and conflict is already well established and given the extreme water shortages, the Middle East is particularly vulnerable to

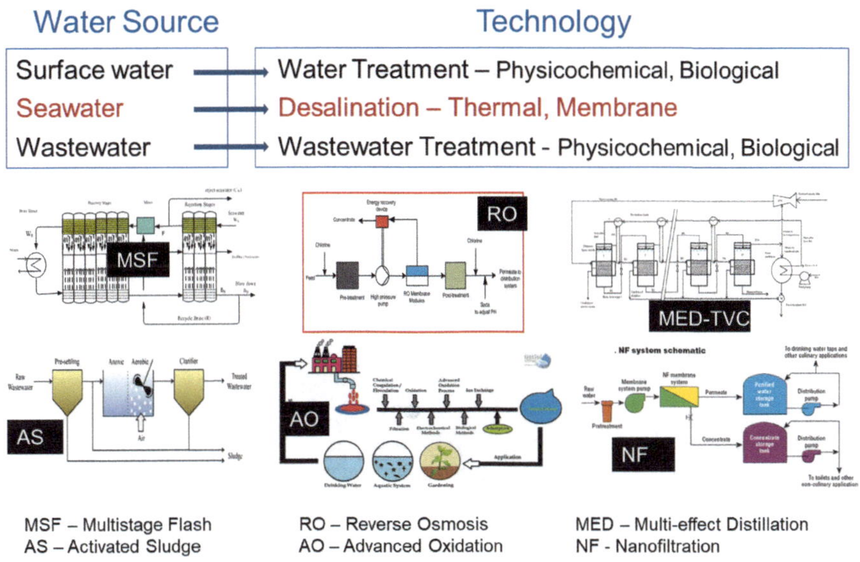


**Fig. 1.9** Available water resources and technologies for freshwater production and wastewater treatment

environmentally induced instabilities [40]. More than 280 river basins are shared by two or more countries in the world leading to frequent disputes between upstream and downstream countries (https://climate-diplomacy.org/water-diplomacy accessed on 3 March 2025). For example, the Nile River and the Red Sea, have long defined the politics and rivalries of this region [41]. The war between Ukraine and Russia starting in February 2022 will have long-standing environmental footprint, especially regarding water resources and management [42]. Climate change and manmade interventions (e.g., Farakka barrage built in 1975) in the Ganga–Padma River (shared between India and Bangladesh) have been disrupting the natural flow regime and thus impacting habitat [43].

As pressure grows on the water resources of the world, international cooperation between countries is essential to manage the shared rivers, aquifers, and lakes. The World Bank Group has been supporting initiatives to enhance cooperation in water to ensure water security needed for sustainable development and job creation in the countries concerned. Management of these shared water basins are becoming more complex but more urgent due to climate change and increasing water demand [44].

## 1.8 Water—The Global Challenge and Everybody's Business

From the discussions so far, we can conclude that

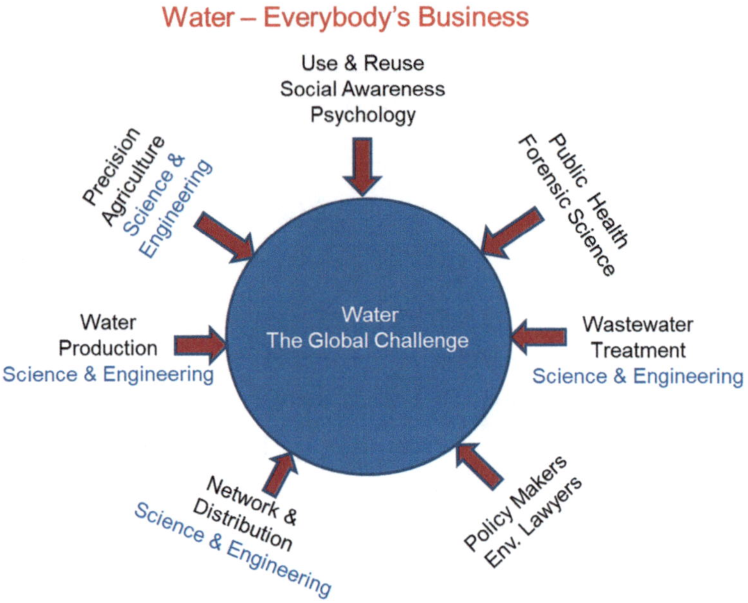


**Fig. 1.10** Water- everybody's business

- Water affects everybody (personal, social, political). Water insecurity cannot be understood from the perspective of one discipline.
- Apart from the technological, scientific and engineering dimensions, there is an essential social dimension to water insecurity.
- Although some of the technological problems being faced regarding water security and water management could easily be resolved in a matter of years, social and political issues regarding water management will take much longer time to resolve (may take several decades).

Figure 1.10 highlights the importance of involving everyone in the society to resolve the water issues of the world.

## 1.9 Conclusions

Due to climate change, population growth, and geopolitical instability, water, energy, and food resources are critical to communities' livelihoods and their health. As highlighted in this article, although water is linked to all global challenges (Food, Energy, Health), sustainable supply of water, food, and energy together with education are absolutely essential to create a smart, resilient and a cohesive society with zero pollution in air, water and soil. However, war, pandemics, and any other disasters (natural or manmade) and EDI (equality, diversity & inclusion) non-compliant behavior of

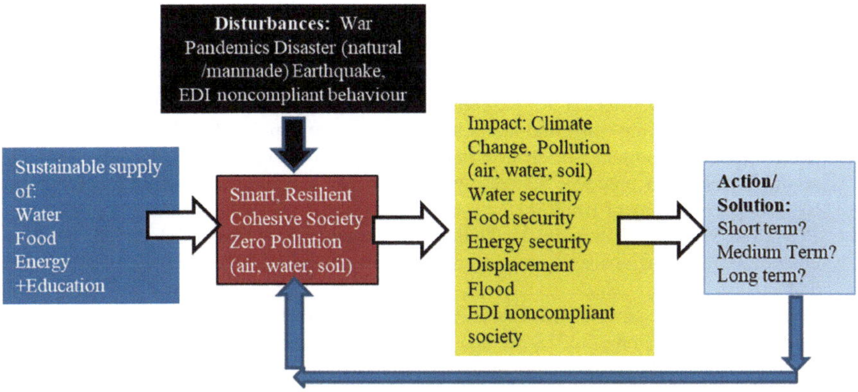


**Fig. 1.11** Smart resilient society water-energy-food-environment (+education) nexus [45]

the society results in climate change and environmental pollution leading to food insecurity, energy insecurity, displacement (refugees), flood and greater EDI non-compliant society (Fig. 1.11). Any short-, medium- and long-term actions/solutions require involving everyone in the society.

## References

1. Mekonnen MM, Gerbens-Leenes PW, Hoekstra AY (2015) The consumptive water footprint of electricity and heat: a global assessment. Environ Sci Water Res Technol 1:28
2. Gerbens-Leenes PW, Vaca-Jiménez SD, Holmatov B, Vanham D (2024) Spatially distributed freshwater demand for electricity in Africa. Environ Sci Water Res Technol 10:1795
3. Gerbens-Leenesa W, Hoekstra AY, van der Meer TH (2009) The water footprint of bioenergy. PNAS 106(25):10219–10223
4. de Fraiture C, Giordano M, Yongsong L (2007) Biofuels: implications for agricultural water use, Colombo, Sri Lanka. International Water Management Institute
5. Buabeng-Baidoo E, Mafukidze N, Tiwari S, Kumar A, Srinivasan B, Majozi T, Srinivasan R (2018) Exploring water reuse opportunities in a large-scale milk processing plant through process integration. In Mujtaba IM, Majozi T, Amosa M (eds) Water management: social and technological perspectives. CRC Press, USA, pp 457–474
6. Mujtaba IM, Srinivasan R, Elbashir N (2017) The water-food-energy nexus: processes, technologies and challenges. CRC Press, USA
7. Al-Zu'bi M, Weston S, Shahid S, Brouziyne Y, Mujtaba I, Al Naber M (2024) Conceptualizing the intersection of the water-energy-food-environment (WEFE) nexus, human security, and inclusive society: insights from Middle East and North Africa (MENA), Sub-Saharan Africa, and Asia-Pacific regions. Colombo, Sri Lanka. International Water Management Institute (IWMI)
8. Mohamed ES, Belal AA, Abd-Elmabod SK, El-Shirbeny MA, Gad A, Zahran MB (2021) Smart farming for improving agricultural management. Egyp J Rem Sens Space Sci 24:971–981
9. Walter A, Finger R, Huber R, Buchmann N (2017) Smart farming is key to developing sustainable agriculture. PNAS 114(24):6148–6150
10. Abdelgeliel M (2024) Automation system in water energy food nexus. In: NATO advanced research workshop, 9–12 July, Rome

11. Rhue SJ, Torrico G, Amuzie C, Collins SM, Lemaitre A, Workman CL, Rosinger AY, Pearson AL, Piperata BA, Wutich A, Brewis A, Stoler J (2023) The effects of household water insecurity on child health and well-being. WIREs Water 10:e1666
12. Currie J, Zivin JG, Meckel K, Neidell M, Schlenker W (2013) Something in the water: contaminated drinking water and infant health. Can J Econ 46(3):791–810
13. Sefat F (2024) Importance of clean water in managing bacterial vagionosis (BV). Presented at the Theory into practice: addressing water-energy-health-environment (WEHE) Nexus workshop. University of Bradford
14. Peebles K, Velloza J, Balkus JE, McClelland RS, Barnabas RV (2019) High global burden and costs of bacterial vaginosis: a systematic review and meta-analysis. Sex Transm Dis 46(5):304–311
15. Roggenburg M, Warsinger DM, Evans HB, Castillo L (2021) Combatting water scarcity and economic distress along the US-Mexico border using renewable powered desalination. Appl Energy 291:116765
16. Al-Obaidi MA, Kara-Zaitri C, Mujtaba IM (2020) Wastewater treatment by reverse osmosis process. CRC Press, USA
17. Tizaoui C, Odejimi OO, Abdelaziz A (2017) Occurrence, effects, and treatment of endocrine-disrupting chemicals in water. In: Mujtaba IM, Srinivasan R, Elbashir NO (2017) The water-food-energy nexus: processes, technologies, and challenges. CRC Press, USA, 157–182
18. Schwarzenbach RP, Escher BI, Fenner K, Hofstetter TB, Johnson CA, Gunten U, Wehrli B (2006) Review: the challenge of micropollutants in aquatic systems. Science 313:1072–1077
19. Bolong N, Ismail AF, Salim MR, Matsuura T (2009) A review of the effects of emerging contaminants in wastewater and options for their removal. Desalination 239:229–246
20. Bueno MJ, Martínez GMJ, Herrera S, Hernando MD, Agüera A, Fernandez-Alba AR (2012) Occurrence and persistence of organic emerging contaminants and priority pollutants in five sewage treatment plants of Spain: two years pilot survey monitoring. Environ Pollut 164:267–273
21. Fijalkowski K, Rorat A, Grobelak A, Kacprzak MJ (2017) The presence of contaminations in sewage sludge—the current situation. J Environ Manage 203:1126–1136
22. Yang J, Monnot M, Sun Y, Asia L, Wong-Wah-Chung P, Doumenq P, Moulin P (2023) Microplastics in different water samples (seawater, freshwater, and wastewater): removal efficiency of membrane treatment processes. Water Res 232:119673
23. Bisung E, Elliott SJ (2017) A scoping review of psychosocial impacts of water and sanitation. J Water Health 15(1):17–30
24. Sultana R, Nahar N, Luby SP, Swarna ST, Gurley ES, Tamason CC, Khan S, Rimi NA, Kabir H, Saifullah MK et al (2022) Measuring water quantity used for personal and domestic hygiene and determinants of water use in a low-income urban community. Int J Environ Res Public Health 19:15656
25. Jones ER, van Vliet MTH, Qadir M, Bierkens MFP (2021) Country-level and gridded estimates of wastewater production, collection, treatment and reuse. Earth Syst Sci Data 13:237–254
26. Qu J, Wang H, Wang K, Yu G, Ke B, Yu H, Ren H, Zheng X, Li J, Li W, Gao S, Gong H (2019) Municipal wastewater treatment in China: development history and future perspectives. Front Environ Sci Eng 13(6):88
27. Nyamayedenga N (2018) Regeneration-recycling of industrial wastewater to minimise fresh-water usage with water cascade analysis. In: Mujtaba IM, Majozi T, Amosa M (eds) Water management: social and technological perspectives. CRC Press, USA, pp 427–442
28. Bhavsar A, Srinivasan B, Srinivasan R (2023) Freshwater minimization in multistage multiproduct batch processes through intentional waiting. Chem Eng Res Des 196:204–213
29. Miller GW (2006) Integrated concepts in water reuse: managing global water needs. Desalination 187:65–75
30. Chen W, Bai Y, Zhang W, Lyu S, Jiao W (2015) Perceptions of different stakeholders on reclaimed water reuse: the case of Beijing, China. Sustainability 7:9696–9710
31. Ravishankar C, Nautiyal S, Seshaiah M (2018) Social acceptance for reclaimed water use: a case study in Bengaluru. Recycling 3:4

32. Baawain MS, Al-Mamun A, Omidvarborna H, Al-Sabti A, Choudri BS (2020) Public perceptions of reusing treated wastewater for urban and industrial applications: challenges and opportunities. Environ Dev Sustain 22:1859–1871
33. Mujtaba IM (2024) Water-energy-food-environment-health nexus: opportunities for PSE community, plenary talk at the Asia PSE 2024 Conference, 5–8 August, Malaysia
34. Gosling SN, Arnell NW (2016) A global assessment of the impact of climate change on water scarcity. Clim Change 2016(134):371–385
35. Stringer LC, Mirzabaev A, Benjaminsen TA, Harris RMB, Jafari M, Lissner TK, Stevens N, der Pahlen CT (2021) Climate change impacts on water security in global drylands. One Earth 4:851–864
36. Vanderhoof MK, Christensen JR, Alexander LC, Lane CR, Golden HE (2024) Climate change will impact surface water extents and dynamics across the central United States. Earth's Future 12:e2023EF004106
37. Mujtaba IM, Majozi T, Amosa M (2018) Water management: social & technological perspectives. CRC Press, USA
38. Mujtaba IM, Sowgath MT (2022) Desalination technology: design and operation. Elsevier, Netherlands
39. Myers SL (2012) U.S. intelligence report warns of global water tensions. The New York Times (2012)
40. Shahi A, Vachkova M (2018) Water security and the rise of sectarian conflict in Yemen. In: Mujtaba IM, Majozi T, Amosa M (eds) Water management: social and technological perspectives, CRC Press, USA, pp 3–9
41. Demerew K (2021) From Red Sea to the Nile: water, power, and politics in Northeast Africa. Third World Quart 42(12):2883–2901
42. Shumilova O, Tockner K, Sukhodolov A, Khilchevskyi V, De Meester L, Stepanenko S, Trokhymenko G, Hernández-Agüero JA, Gleick P (2023) Impact of the Russia-Ukraine armed conflict on water resources and water infrastructure. Nature Sustain 6:578–586
43. Islam ARMT, Talukdar S, Akhter S, Eibek KU, Rahman MM, Pal S, Naikoo MW, Rahman A, Mosavi A (2022) Assessing the impact of the farakka barrage on hydrological alteration in the Padma River with future insight. Sustainability 14:5233
44. Jha SK (2023) Water knows no borders: transboundary cooperation is key to water security and avoiding conflict. Water Blog. https://blogs.worldbank.org/en/water/water-knows-no-borders-transboundary-cooperation-key-water-security-and-avoiding-conflict. Accessed 3 Mar 2025)
45. Mujtaba IM (2024) Implementing the WEFE nexus: the crucial role of education, pp 380–381. https://www.openaccessgovernment.org/publication/open-access-government-october-2024/183238/

# Chapter 2
# Toward a Single Interactive and Integrative Planning and Resource Allocation Platform for Food-Energy-Water Nexus

**Ahmed Hamza H. Ali**

**Abstract** Integrative planning for interconnecting water, food, energy, and ecosystems is a significant worldwide security challenge amplified by different parameters, including future population growth, geopolitical risks, and climate change. These challenges are tightly interconnected, which requires a systems approach to address them. Each country has national strategies for developing its water, food, and energy sectors, which must be integrated with the ecosystem. However, the tool for the technical coordination of these resources is based on the national security level of each sector, and the economic and social development needs to exist. This concept aims to pave the way toward the need for a single interactive decision-making platform to support evidence-based inter-sectoral planning among water, energy, and food resource systems integrated with the ecosystem and guide future investments and risk mitigation strategies across these domains. Such a platform must be co-created with close stakeholder engagement throughout the multi-discipline of expertise, bringing together different disciplines to evaluate trade-offs associated with technological and policy interventions on the interconnected food-energy-water resource systems integrated with the ecosystem. This can be achieved through an internationally sufficiently founded project focusing on four interrelated objectives to improve food-energy-water-systems' sustainability, resilience, equitability, and integration with the ecosystem.

**Keywords** Water · Food · Energy · Ecosystems Nexus · Interactive and integrative planning · Resource allocation platform · Interactive decision · Making platform

A. H. H. Ali (✉)
Department of Mechanical Power Engineering, Faculty of Engineering, Assiut University, Assiut, Egypt
e-mail: ah-hamza@aun.edu.eg



A. Angelini et al. (eds.), *Security Enhancement for Climate Change Impacting Urban Resources - SECCURe*, NATO Science for Peace and Security Series C: Environmental Security, https://doi.org/10.1007/978-94-024-2345-7_2

## 2.1 Introduction

Many developing nations face several water, energy, and food security challenges. Various factors, including population growth, the threat to water resources, and the effects of climate change, compound these challenges. The following points summarize some of the world's SDG challenges (https://www.un.org/en/exhibits/page/sdgs-17-goals-transform-world).

- **Undernourishment**: 900 million people
- 108 countries with over 5% of people deprived
- 28 countries with over 20% of people deprived
- **No access to electricity**: 1.1 billion people
- 50 countries with over 33% of people deprived
- **No modern fuel, cooking, or heating**: 3 billion people
- **No safe water**: 900 million people
- 36 countries with over 20% of people deprived
- **No adequate sanitation**: 2.6 billion people
- 66 countries with over 20% of people deprived
- **Risk of climate change, pollution, health, and environmental hazards**
- Risks are higher where development needs are greater

For food, the development challenges are sustainable agriculture and food supply, efficient use of water/fertilizer, adaptation to climate change, urbanization, and food versus fuel. However, environmental concerns include incursion into natural ecosystems (biodiversity), greenhouse gas emissions (e.g., deforestation), and land degradation from unsustainable agricultural practices. For water, the development challenges include population growth, economic growth, urbanization, ensuring sufficient water to feed a growing population, providing safe drinking water and sanitation for all, promoting more efficient use of water resources, and protecting groundwater resources. The environmental concerns are climate change (droughts/floods/sea level rise), pollution, and eutrophication. The challenges for energy development are access to sustainable energy for all, support for industrial development, energy for the needs of the twenty-first century, and efficient use of land and water. Environmental concerns include greenhouse gas emissions, mining (mountaintop removal, other), environmental emissions ($NO_x$, $SO_x$, lead, particulates), indoor air pollution, solid and liquid wastes, and radioactive waste.

### *2.1.1 Water, Energy, and Food Systems Interconnections*

Water, energy, and food systems are tightly interconnected and referred to as the Nexus, as shown in Fig. 2.1, and how they are managed directly impacts the environment and ecosystems in which they exist, as well as economic growth and human development. The water for food is interlinked and integrated, as shown in Fig. 2.2.

Production of food accounts for the largest component of the human water footprint; about 70% of freshwater withdrawals worldwide are for agriculture and can be as high as 90% in some low-income countries, and it is lower in most higher income countries. While the water for energy is interlinked and integrated, as shown in Fig. 2.2; for example, approximately 20% of freshwater withdrawals are for industrial purposes such as cooling thermal processes in the power and manufacturing sectors, primary fuel extraction, fuel refining and conversion, emissions control, and biofuel/biomass production. The energy for water is interlinked and integrated, as shown in Fig. 2.2. The water supply chain is energy-intensive utilization, as around 7% of global electricity is used to supply surface water. For example, up to 2400 kWh of energy is needed per million liters for transportation, depending on distance and change in elevation. Meanwhile, groundwater varies in depth (e.g., 140 kWh per million liters at 35 m depth and 530 kWh per million liters at 120 m depth). For water treatment, the energy needs are about 26 kWh per million liters for high-quality groundwater, 300–1400 kWh per million liters for brackish groundwater desalination, and 3600–4500 kWh per million liters for seawater desalination. For water distribution, the average energy needs are about 290 kWh per million liters but vary widely depending on distance and change in elevation. The food for energy is interlinked and integrated, as shown in Fig. 2.2, as a growing share of food is dedicated to supplying energy as the bioenergy accounts for a little more than 10% (60 EJ) of the world's total primary energy supply on average 11% of coarse grains, 11% of oil seeds and 21% of sugarcane were used for biofuel production from 2008 to 2010 495 TWh of electricity was produced from biomass in 2014 or 2% of world electricity generation. The energy for food is interlinked and integrated, as shown in Fig. 2.2. For example, energy is required to power agricultural machinery for field preparation, crop harvesting, drying, and processing, as direct energy use accounts for roughly 2.1% of the total final energy demand. Energy is required to produce fertilizers, pesticides, and other agricultural inputs. The indirect energy use in agriculture is about 1.2% of the total final energy demand [1]. In addition, energy is needed to process, transport, store, package, and market food and food products. For example, the agricultural sector in Egypt consumes the most significant proportion of water resources at 80%. It also contributes to food security, providing 40–50% of the nation's self-sufficiency. However, the government's plan to expand agricultural land by 20% to increase food security is limited by the availability of water, land, and energy resources [2, 3].

Many land reclamation projects in Egypt rely on irrigation with fossil groundwater, and the Nile, the country's only renewable water source, is threatened by the yearly filling of the Grand Ethiopian Renaissance Dam.

To address these challenges, the government is considering a range of technical interventions, such as using renewable energy for irrigation systems and desalination. However, these solutions will also impose additional energy demands, and their potential benefits must be carefully coordinated across sectors to avoid unintended competition and ensure sustainable economic growth. Additionally, plans to increase water supply through desalination and wastewater reuse will also require significant amounts of energy. Therefore, the interdependent issues of water, food, and energy

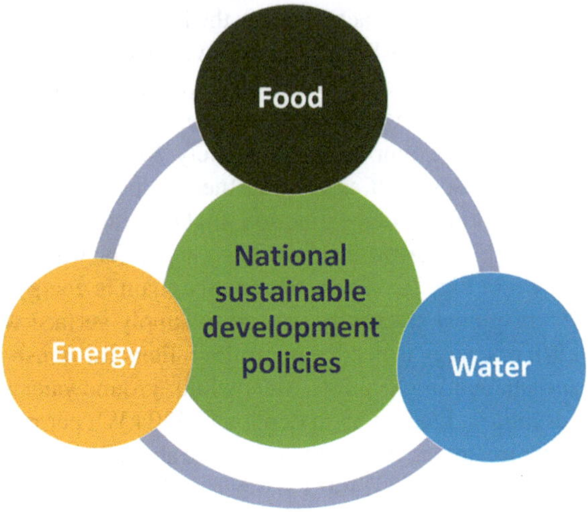


**Fig. 2.1** The food-energy-water resources are highly interlinked and integrated with the national sustainable policies

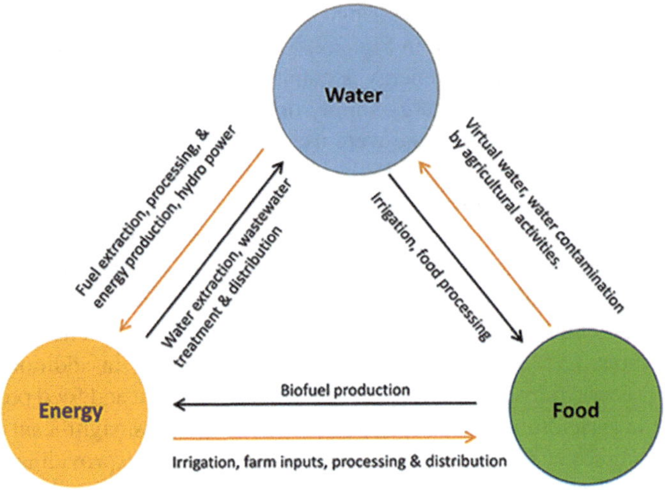


**Fig. 2.2** The food-energy-water resources interlinked [10]

security are closely linked and directly impact the nation's social, economic, and environmental sustainability and the well-being of its citizens.

Egypt's Strategies for Water 2037 [4–8], Food 2030 [4, 5, 7–9], and Energy 2035 [4, 5, 7] identified the national action plan that addresses four priority sectors, including Water, Energy, and Agriculture. The strategic goal aims to translate the green economy's (GE) conceptual framework from a broad development vision into operationally focused plans and projects. Additionally, it looks at how priority sectors are connected and work together to form "Sustainable Integrated Communities."

## 2.2 Integrated Planning

There are few professional platforms for strategic planning for water and other energy, and many for food. However, despite growing evidence of the tight interconnectedness between water, energy, and food systems and the interdependence of their challenges at different scales, coordination exists for planning, financing, and governing these sectors [8], but without the existence of a single platform for interactive and integrative planning and resource allocation platform for food-energy-water Nexus. Plans are often developed within institutional silos with little recognition of interdependence. Given the growing pressures on resource systems, there is a need to create new innovative frameworks and coordination mechanisms that properly evaluate the synergies and trade-offs associated with different technical and policy interventions being considered across sectors. This requires developing a shared vision and successful metrics to ensure coherence between cross-sectoral policies. It also involves the development of innovative shared financing and governance models that can provide the implementation of solutions that address the tight interconnectedness between sectors.

### *2.2.1 Research Gaps*

Water, energy, and food systems are complex, tightly interconnected, and impacted by different pressures, including a growing population, climate change, and urbanization. This tight interconnectedness can make it difficult to understand and address the impacts of actions taken in one sector on others. Involving all relevant stakeholders in the decision-making process can be challenging, mainly when there are conflicting interests or a lack of trust between sectors. Additionally, limited data and information may be available to inform decision-making and coordination between sectors, which can hinder effective cross-sectoral collaboration. Ample evidence in the literature refers to the sub-optimal, potentially competing strategies that may result from planning within institutional silos. They hypothesize that cross-sectoral coordination can be catalyzed and improved through a combination of systems trade-off analytics, with continuous stakeholder engagement and capacity-building activities, to guide evidence-based, coordinated decision-making leading to improved societal outcomes. The proposed project will allow the evaluation of the level of coherence and potential competition between national water, energy, and food strategies. It will also provide a platform for cross-sectoral stakeholders to evaluate different scenarios and interventions being considered across the various sectors.

Therefore, a single interactive decision-making platform that supports evidence-based intersectoral planning among water, energy, and food resource systems is needed. This platform will guide future investments and risk mitigation strategies across these domains and consider the trade-offs and synergies associated with different technical, policy, or behavioral interventions. Such a platform must be co-created with close stakeholder engagement throughout the multi-discipline of expertise, bringing together various disciplines to evaluate trade-offs associated with technological and policy interventions on the interconnected food-energy-water resource systems integrated with the ecosystem. This can be achieved through an internationally sufficiently founded project focusing on four interrelated objectives to improve food-energy-water systems' sustainability, resilience, equitability, and integration with the ecosystem.

## 2.3 Conclusion

Integrative planning for interconnecting water, food, energy, and ecosystems is a significant worldwide security challenge amplified by different parameters, including future population growth, geopolitical risks, and climate change.

- This concept aims to pave the way toward the need for a single interactive decision-making platform to support evidence-based inter-sectoral planning among water, energy, and food resource systems integrated with the ecosystem and guide future investments and risk mitigation strategies across these domains.
- Such a platform must be co-created with close stakeholder engagement throughout the multi-discipline of expertise, bringing together different disciplines to evaluate trade-offs associated with technological and policy interventions on the interconnected food-energy-water resource systems integrated with the ecosystem.
- This can be achieved through an internationally sufficiently founded project focusing on four interrelated objectives to improve food-energy-water-systems' sustainability, resilience, equitability, and integration with the ecosystem.

## References


1. Towards climate-smart agriculture in Egypt: Scaling up sustainable practices for enhancing agrifood system resilience and adaptive capacity (2022). [Online]. Available: https://openknowledge.fao.org/server/api/core/bitstreams/8327142e-2479-4035-ad46-718396257db3/content. Last accessed 3 Mar 2025
2. https://d3n8a8pro7vhmx.cloudfront.net/foe/legacy_url/1868/Water-energy-2009CERA.pdf?1471405077. Last accessed 3 Mar 2025
3. Sustainable Development Strategy: Egypt's Vision 2030 | Planipolis. https://planipolis.iiep.unesco.org/en/2016/sustainable-development-strategy-egypt%E2%80%99s-vision-2030-6301,https://mped.gov.eg/Highlights?id=1&lang=en&Egypt-Vision-2030-Launch#:~:text=The%20updated%20vision%20emphasizes%20addressing,strategies%20for%20various%20government%20agencies. Last accessed 3 Mar 2025

4. Egypt's 2021 Voluntary National Review|Arab Development Portal. https://www.arabdevelopmentportal.com/publication/egypts-2021-voluntary-national-review https://sustainabledevelopment.un.org/content/documents/279512021_VNR_Report_Egypt.pdf. Last accessed 3 Mar 2025
5. Egypt's Strategy for Managing Water Resources—AmCham Egypt. https://www.amcham.org.eg/cyberlink/events/view/1133/egypts-strategy-for-managing-water-resources. Last accessed 3 Mar 2025
6. Report on National Efforts to Ensure Access to Water on the Occasion of World Water Day on March 22, 2022. https://sschr.gov.eg/media/ysalpy5h/english742022.pdf. Last accessed 3 Mar 2025
7. National Action Plan for Sustainable Consumption and Production (SCP) in Egypt|2015. https://www.oneplanetnetwork.org/sites/default/files/national_action_plan_for_sustainable_consumption_and_production_scp_in_egypt_2015.pdf. Last accessed 3 Mar 2025
8. Updating the Sustainable Agricultural Development Strategy 2030 and Preparing a Medium-Term Plan of Action—TCP/EGY/3701. https://www.fao.org/documents/card/en/c/CB6249EN/. Last accessed 14 Jan 2023
9. Daher B, Hannibal B, Portney KE, Mohtar RH (2019) Toward creating an environment of cooperation between water, energy, and food stakeholders in San Antonio. Sci Total Environ 651:2913–2926. https://doi.org/10.1016/J.SCITOTENV.2018.09.395
10. Lawford RG (2019) A design for a data and information service to address the knowledge needs of the water-energy-food (W-E-F) nexus and strategies to facilitate its implementation. Front Environ Sci 7:56. https://doi.org/10.3389/fenvs.2019.00056

# Chapter 3
# Enhancing Security and Sustainability in Latin America: Circular Economy Approaches for Water, Energy and Agricultural Residue Utilization

**Jalelys Liceth Leones-Cerpa, Jimena Sánchez-Ojeda, Karina A. Ojeda, and Eduardo Luis Sánchez-Tuirán**

**Abstract** Latin America is one of the most biodiverse regions on Earth. Its location and climatic characteristics provide favorable conditions for the development of agriculture, whose crops generate waste and pollution but contribute mainly to Gross Domestic Profit (GDP). Latin America is a region that meets its energy requirements using a diversity of sources. These sources are both non-renewable and renewable. As time passes and fossil fuels deplete, the share of renewable energy sources grows. In this chapter, a brief overview of the energy-water-social situation in this region is presented. Latin America has high water availability; however, pollution generates great concern because the effluents are of anthropogenic origin, and the availability of drinking water services is low. The social panorama of Latin America from the education sector indicates a decrease in government support at different educational levels. At the same time, job creation is not increasing, and the gap between countries is increasing. The health sector does not offer good service provision, and there is rising inequality in achieving Universal Health Coverage (UHC) while basic sanitation is not available to more than half of the population, which can be linked to the high pollution faced by this region. Finally, sustainable opportunities in Latin America to meet energy needs, improve access to water resources, and strengthen agricultural

J. L. Leones-Cerpa · K. A. Ojeda · E. L. Sánchez-Tuirán (✉)
Process Design and Biomass Utilization Research Group (IDAB), College of Engineering, Universidad de Cartagena, Cartagena, Colombia
e-mail: esanchezt2@unicartagena.edu.co

J. L. Leones-Cerpa
e-mail: jleonesc@unicartagena.edu.co

K. A. Ojeda
e-mail: kojedad@unicartagena.edu.co

J. Sánchez-Ojeda
College of Engineering, Universidad de los Andes, Bogotá, Colombia
e-mail: j.sanchezo2@uniandes.edu.co



A. Angelini et al. (eds.), *Security Enhancement for Climate Change Impacting Urban Resources - SECCURe*, NATO Science for Peace and Security Series C: Environmental Security, https://doi.org/10.1007/978-94-024-2345-7_3

activities are discussed to enhance the quality of life for vulnerable populations in the region.

**Keywords** Latin America · Circular economy · Sustainable · Drinking water · Bioenergy

## 3.1 Introduction

Latin America and the Caribbean is composed of over 40 countries. However, some authors classify only 19 as Latin America, according to the language, covering an area of approximately 2754 million square kilometers. This region is distinguished by its biodiversity, natural resources, variety of thermal floors, and geographical position, which favors agriculture. As a result, Latin America is a global leader in agricultural production, with the agricultural sector, aquaculture, and forestry accounting for 6.5% of its Gross Domestic Product (GDP) in 2023, compared to an average of 4.1% in the same categories worldwide [1]. Historically, the countries in this region have been prominent in producing and exporting raw materials such as sugarcane, coffee, timber, and fruits.

Furthermore, products such as coffee, sugarcane, and tropical fruits account for nearly 50% of global production, with Latin America and the Caribbean playing a central role. This gives the region a significant capacity for the export of these raw materials, with export rates ranging from 3.47 to 14.09% of the global production based on its thermal floors, which allows for the cultivation of a wide range of products within a single country [1].

Nature provides goods and services that are essential for human well-being and our economies; from goods that humanity can harvest and extract (i.e., water, food, timber, biofuel, medicines) to ecosystem services that underpin life on Earth as we know it (i.e. nutrient cycling, production of atmospheric oxygen, flood control, hydrological cycle, and climate regulation) [2]. Research suggests that 55% of the world's total GDP, amounting to approximately US$ 58 trillion, depends moderately or highly on nature and its ecosystem services.

Biodiversity and ecosystems are declining globally, requiring an urgent need to reverse natural resource loss and avoid a 6th mass extinction [2]. There has been a global average decline of 69% in species and populations since 1970. Latin America and the Caribbean are in an extremely alarming position, having a 94% drop [3]. The biodiversity crisis is mainly caused by the regular production methods and excessive consumption standards, exceeding nature's capacity to regenerate properly.

For instance, food production is responsible for about 60–70% of the loss of different species on land and around 50% in freshwater [4]. On the other hand, wood extraction is the primary contributor to forest biodiversity degradation, accounting for 5–10% of global species loss. Grey solutions in water management have led to a 20% decrease in species in rivers and 15% in floodplains.

This unsustainable pattern occurs because current economic strategies prioritize higher production rather than the sustainability of the natural resources, leading to severe depletion of natural assets' reserves on which our well-being and economies depend on [4]. Between 1992 and 2014, the GDP/person doubled, while the natural capital stock/person was reduced by almost 40%.

It is for these reasons that sustainability has become a topic of relevance globally within the last few years, aiming for the preservation of biodiversity, resource management, and waste reduction.

Latin America and the Caribbean face a big need to implement sustainability practices; not only for the protection of its natural resources, but for the positive outcome for its population. For instance, in the year 2016, only 56.6% of the rural population had access to energy, compared to 99.2% of the urban population in the same year [5]. In 2022, only 53.05% of the rural population had access to safely managed drinking water services compared to 80.22% of the urban population [1]. This highlights the current inequality that exists regarding basic needs such as energy and drinking water. Through the following chapter, there is going to be an analysis for the opportunities to implement circular economy and sustainability practices in Latin America and the Caribbean, in order to enhance security and preservation of its agriculture, energy and water sources.

## 3.2 Agriculture Sector

Latin America is a region with environmental conditions that favor agricultural development. Some crops thrive in the predominantly warm and tropical climate, while others thrive in temperate seasonal variations. According to Flachsbarth et al., the countries of these regions have the greatest agricultural land and water availability per capita globally, with 15% of the world's land area, 29% of global precipitation, and 33% of globally available renewable water resources [6]. Hence, there is a vast diversity in crops and livestock. Latin American countries are responsible for an important part of products that are used around the world. These products come from the main crops in the region such as maize, soybean, sugarcane, coffee, wheat, fruits, and others. Table 3.1 shows the contribution of these crops in the environment, economic and social aspects.

The production and use of these agricultural products generate residues during all the stages of the process. Several technologies have been developed to take advantage of these residues to overcome the issues of contamination and to improve the overall economy of the processes. Amongst the most frequent alternatives are combustion, bioconversion via fermentation, biorefining process [19], and composting [20]. All these treatments aim to valorize waste biomass through physicochemical transformations that have to be economically feasible and environmentally conscious.

This important volume of production of agricultural crops and its international commerce has an impact on the economy of Latin American countries and it can be seen, for instance, in the Gross Domestic Profit (GDP) which represents from 5

**Table 3.1** Main crops in Latin American and Caribbean countries

| Type of crops | Country/countries | Economic, social, environmental or cultural impacts | References |
|---|---|---|---|
| Coffee (*Coffea*) | Brazil; Mexico, Nicaragua, Guatemala, El Salvador, Costa Rica, Honduras, Guatemala, Jamaica Colombia, | Integral role in the region's economy, and high carbon footprint | [7, 8] |
| Cocoa (*Theobroma cacao*) | Mexico, Peru, Bolivia, Ecuador, Colombia, Venezuela, Trinidad and Tobago, Jamaica, Brazil | Social and economic progress in the countries, soil erosion, and compaction | [9, 10] |
| Sugarcane (*Saccharum officinarum* L.) | Brazil, and Colombia | Support for regional economies, concentration of families with cultural tradition in sugarcane planting | [11] |
| Rice (*Oryza* spp.) | Brazil, Columbia, Ecuador, Peru, Argentina, Uruguay, Guyana, and Paraguay | Economic growth, Pollution of water sources, and soil erosion | [12, 13] |
| Potato (*Solanum tuberosum* L.) | Peru, Chile, Colombia, Argentina, Bolivia, Ecuador, Venezuela | Economic growth through exports, and soil erosion | [14, 15] |
| Tropical fruits (avocado, mango, banana, and papaya) | Mexico, Colombia, Peru, Chile, Brazil | Gastronomic diversity, and its influence on economic growth | [16–18] |

to 18% in 20 countries, and an even larger share when broader contributions across food systems are considered [1].

Due to the variety of climates, extensions, social and economic reasons, the GDP in the Latin American region varies from tenths of $US billion and up to thousands of $US billion. Figure 3.1 shows the GDP for the strongest and weakest economies in Latin America:

The impact of agriculture on the economy of each country depends on several factors such as export revenues, employment generation, linkage with other industries, and so forth. In the case of export revenues, agriculture has a great impact on the GDP of some countries. For instance, in Paraguay, it accounts for approximately 20% of GDP, in Nicaragua and Argentina is around 15%, and around 6% in Brazil [4].

While many countries in the region share similar climatic conditions, the rich biodiversity allows for a wide range of crops and varieties to thrive. As a result, each country develops unique strategies to utilize its resources to meet local needs. This variation is influenced not only by environmental factors but also by social and cultural considerations that shape the decisions of policymakers.

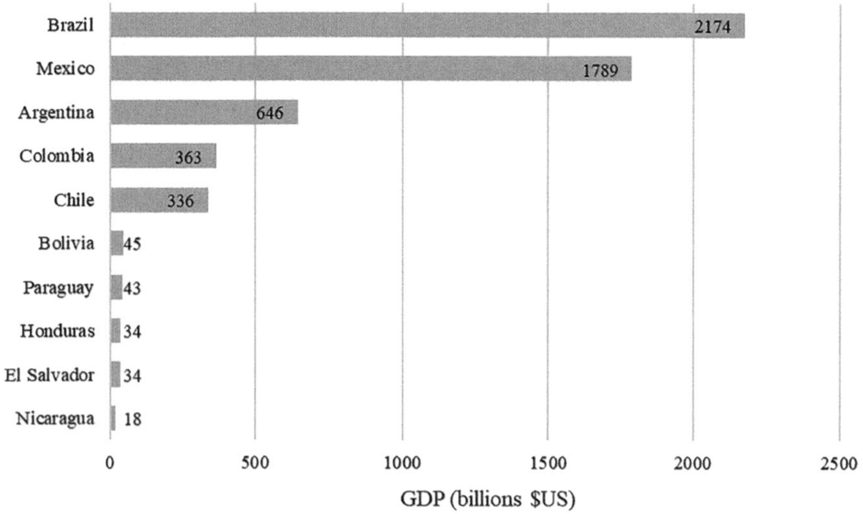


**Fig. 3.1** Gross domestic product of the top-5 and bottom-5 countries in Latin America in 2023 (*Source* based on data from Database World Bank [5])

## 3.3 Energy Overview

As mentioned earlier, Latin America is a diverse region rich in natural resources. Countries like Venezuela, Brazil, Colombia, Argentina, and Mexico are well-known for their significant oil and natural gas production. Additionally, countries such as Brazil, Venezuela, Mexico, Colombia, Argentina, and Paraguay have harnessed their water resources to generate hydroelectric power, while Brazil, Mexico, Chile, and Argentina benefit from favorable conditions for solar and wind energy generation [21].

Energy from biomass has been surely the main source of energy for non-interconnected zones where access to electric energy is a main concern (FEMP) [22]. More recently biofuels such as bioethanol and biodiesel from first and second generations have gained more popularity to overcome the possible depletion of fossil fuels in the mid-term [23].

Other forms of energy, such as geothermal and tidal have been studied but there are not great advances in that field because of the high initial cost risk, technical, regulatory, and legal challenges, and competition from other energy sources [24]. Access to electric energy in Latin America has improved significantly over the past decades. The percentage of the population in this region with access to electricity is presented in Fig. 3.2.

This notorious increase means that the region is on the right path and yet there is still much effort needed to bridge the gap between rural and urban areas, especially in non-interconnected zones. Several countries like Brazil have implemented programs such as «*Luz para Todos*» to provide electricity to over 10 million people living

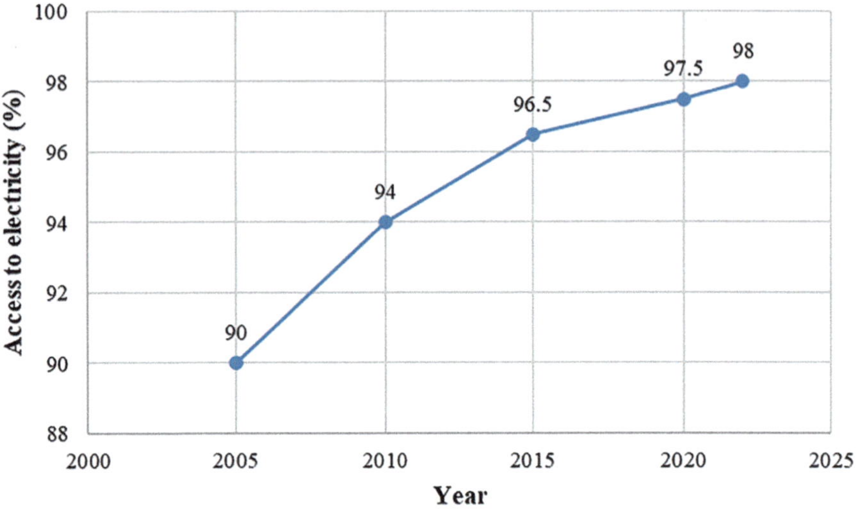


**Fig. 3.2** Access to electricity in Latin America (2005–2022) (*Source* Based on data from Database World Bank [26])

in rural areas by 2008. In Peru, there are projects supported by the World Bank to increase access to efficient and sustainable electricity in rural areas [25].

The above highlights that Latin America has made significant efforts to achieve energy self-sustainability and diversify its energy matrix by adopting strategies focused on alternative renewable energy sources. These efforts have allowed the region to make significant strides in expanding its energy infrastructure over the past decade, improving coverage, and diversifying energy sources. As a result, Latin America now boasts some of the highest levels of electrical coverage in the world [26]. However, many communities still lack access to electricity at night, which prevents them from utilizing essential services like food refrigeration [27].

## 3.4 Water Outline

Water is a natural resource essential for the development of life on Earth. In recent years, it has become necessary to meet increasingly higher standards that guarantee the population access to water in a safe, clean, and dignified manner [28]. Latin America and the Caribbean have more than 30% of the world's water availability, making this territory one of the areas with the greatest water availability in terms of inhabitants number [29]. Water availability in this territory represents a contribution to the life not only of human beings but also of flora and fauna species that enrich biodiversity and cultural heritage [30]. Although Latin America has a wealth of water, there is inequality in the availability of drinking water between some countries due to their geographic location, infrastructure in the distribution networks, or the lack of purification systems [31].

Latin America, a territory with great biodiversity and a varied mountain structure, has two main sources of drinking water: surface and underground. Surface water sources are mostly rivers, lakes, lagoons, and reservoirs, representing approximately 80% of water sources. Their easy availability provides resources and interactions for the development of living beings and ecosystems [30]. Underground water, on the other hand, represents about 30% of the water used in Latin America, becoming an important source for the development of human activities that suffer the effects of pollution and overexploitation [32].

Among the most important sources of surface water in Latin America is the Amazon basin, where four of the world's largest rivers converge and which discharges nearly 18% of the world's freshwater into its estuary and the Atlantic Ocean, making it one of the most complex and structured arteries on Earth [33]. Lakes and reservoirs also represent sources of fresh water for Latin America, among which we can find Glacial Lakes and Piedmont Lakes located mainly in mountain ranges and mountainous areas [34]. On the other hand, reservoirs are mostly used for the generation of electric energy and are considered economic engines of the regions where they are located.

In Latin America, access to drinking water has become a problem that requires more attention every day. According to information from the World Bank [35], there is inequality in safely managed drinking water services between Latin American countries. Figure 3.3 shows the average percentages of the population in Latin America (LA) that used safely managed drinking water services from 2012 to 2022.

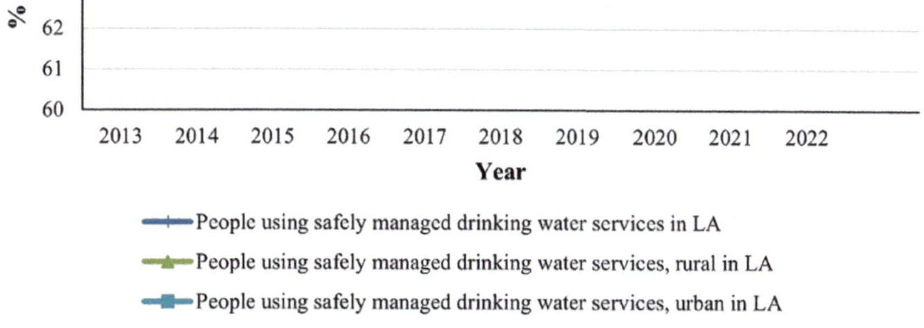


**Fig. 3.3** Average percentages of the population in Latin America (LA) using safety-managed drinking water service between 2013 and 2022 in urban, rural, and national items (*Source* Based on data from Database World Bank [35])

Figure 3.3 differentiates the trend of the rural population from the urban population, indicating that the average percentages between Latin American countries follow a similar percentage trend between 63 and 66%, which means an average level in the use of this service, the possible lack of infrastructure to supply it and, therefore, the gap between countries with larger economies, and the divergence between their rural and urban areas.

Pollution is another factor that influences the distribution of water. Water pollution of sources is increasing in Latin America, mainly due to the various types of pollutants that enter water sources by different means, such as emerging pollutants found in different urban water sources [36]. Furthermore, water sources in Latin America are threatened by waste when it is used for multiple purposes (like mining, domestic tasks, industrial use, and agriculture) along with the lack of government policies that indicate the limits of pollutants in the different phases of the water cycle [37].

In Latin America, various methods have been used to purify water, some traditional and others innovative, including disinfection by chlorination in homes as an alternative for rural areas, although the quality of the water is not considered, which is why the doses may vary [37]. Techniques such as flocculation, coagulation, and filtration have also been used in addition to disinfection, and in the last decade, the incursion of nanotechnology. However, these technologies still face challenges for science as more resistant, eco-friendly materials continue to be developed that guarantee the effective elimination of pathogens [38].

Another factor of concern is the wastewater that is used as a point source of pollution for surface water sources. In Latin America, only about 20% of wastewater from urban areas is treated efficiently in Wastewater Treatment Plants (WWTP) where the complete absence of contaminants is not guaranteed, and the disposal of the contaminants they retain is uncertain [39].

For several decades, aquatic ecosystems have been affected by pollution, which is caused by the accelerated growth of technology and the indiscriminate use of natural resources and the environment [40]. Several sources of water pollution must be identified to avoid further damage and make decisions to avoid future damage [41].

Water pollution sources can be classified as point-based and non-point-based [42], among which are wastewater discharges [43], oil spills, plastics [44], discharge and decomposition of animal waste [45], diffuse source [46], submarine discharge of groundwater [47], combustion of biomass, wood, fossil fuels and coal [48]; industrial and urban effluents [49], terrestrial pollutants [50], surface runoff from rain and agricultural activities [51].

Some pollutants come from natural cycles, but most come from anthropogenic sources and are characterized by not being biodegradable or recyclable [52]. Among the most common pollutants in aquatic environments are Emerging Pollutants (EP), whose particularity is that they are difficult to monitor and their capacity to generate environmental impacts is relatively high [53]. Among them are microplastics (MP), which are portions of plastic whose diameter is less than 5 mm, considered harmful environmental pollutants that can be found in different ecosystems and the least expected places [54].

Pharmaceutical products derived from hydrophilic drugs are also released into ecosystems as metabolites [55], now collectively referred to as pharmaceuticals and personal care products [56]. On the other hand, Persistent Organic Pollutants (POPs) are highly toxic organic chemical compounds that do not degrade easily, including pesticides and organochlorine pesticides [57].

Heavy metals are metallic elements of high density, toxicity, and danger. Among those most found in wastewater are copper (Cr), mercury (Hg), lead (Pb), chromium (Cr), cadmium (Cd), nickel (Ni), and zinc (Zn) [31]. The main nutrients that affect water bodies when they are found in excess are nitrates and phosphates because they generate the rapid growth of aquatic plants that lead to eutrophication [13]. Petroleum hydrocarbons are among the most dangerous environmental pollutants, especially polycyclic aromatic hydrocarbons (PAH) due to the complexity of their structure [58].

Anthropogenic pollution sources have brought with them the presence of pollutants in water bodies that affect the aquatic ecosystem to a greater or lesser extent. In recent years, various investigations have revealed the presence of different types of contaminants in water sources in several Latin American countries. Table 3.2 presents some investigations that identified the contaminants that generate the greatest concern due to their capacity to affect the environment and bioaccumulate.

Table 3.2 shows that the most important sources of pollution in the rest of the world also affect water quality in Latin America. In general, in a ranking of these pollutants, those that generate the greatest concern in Latin America are heavy metals, mainly because they are toxic and difficult to manage. In addition, they can pose risks to human health by entering the body through different exposure routes [63].

For all the above, it is necessary to accurately and urgently identify the contaminants in water sources that impact the population and its health, mainly due to the high exposure resulting from poor management in the disposal of solid, liquid, and gas waste. The environment has suffered aggressively over the years from the effects

**Table 3.2** Main contaminants found in water sources in Latin American countries

| Water source | Location | Contaminant | References |
|---|---|---|---|
| Paraná river | Argentina | Physicochemical: phosphates, nitrates, nitrites and sulfates | [59] |
| Surface water and wastewater | Costa Rica, and Latin America | Pharmaceutical products, personal care products | [36, 60] |
| Surface water | Latin America | Microplastics: fibers (approx. 62%) | [61, 62] |
| Surface water | Latin America | Heavy metals (arsenic) | [63, 64] |
| Groundwater (aquifers) | Latin America | Persistent organic pollutants (POPs): pesticides | [65] |
| Surface water and wastewaters | Latin America | Hydrocarbons | [66] |
| Surface water and wastewaters | Latin America and the world | Inorganic nutrients: fertilizers and nitrate | [67] |

of pollution; in most cases, water sources that have animal and plant species act as a focus for the proliferation of diseases and conditions in the general population [36]. In the economic sphere, water availability provides stability to markets in urban areas and fundamentally in rural areas, as it encourages the production and sale of food and supplies.

The need to seek improvements in the availability of drinking water for the entire population of Latin America lies in four items: the migration of the population from urban to rural areas, the influence on people's health, mitigating the influence on quality due to the pandemic (COVID-19) and compliance with the 2030 Agenda [68].

## 3.5 Social Panorama in Latin America and the Caribbean

### *3.5.1 The State of Education*

Education in Latin America has been marked by social, economic, and cultural aspects. The gaps in access to education between countries are marked and are increasingly widening. Currently, in Latin America, there are four stages of schooling: pre-primary education (preschool), primary, secondary, and higher education (tertiary) [69]. Education coverage in Latin America has focused on increasing access to primary education, which has been reflected in an increase of 8.1% [69]. Between 2000 and 2019, in Latin America, attendance opportunities at all educational stages increased to reach more than 90% coverage [70].

One of the indicators that quantifies political support for education in a country is the average government expenditure on education as a percentage of GDP. Figure 3.4 shows the Government expenditure on education as a percentage of GDP (%) in Latin America considering the different education levels.

Figure 3.4 shows that the largest government contribution to the Gross Domestic Product (GDP) in Latin America has been, over the years, in primary education. Since 2021, the percentage allocated to education at three of the four levels has decreased, while the contribution to pre-primary education has remained generally constant. The above is a discouraging panorama considering the exponential growth of the population and the few opportunities that future generations could have to access quality education. Besides, in Latin America, there has been an exploration in various fields such as Learning Analytics (LA) to improve learning and the quality of education in student environments in Latin American countries by "doing" [71].

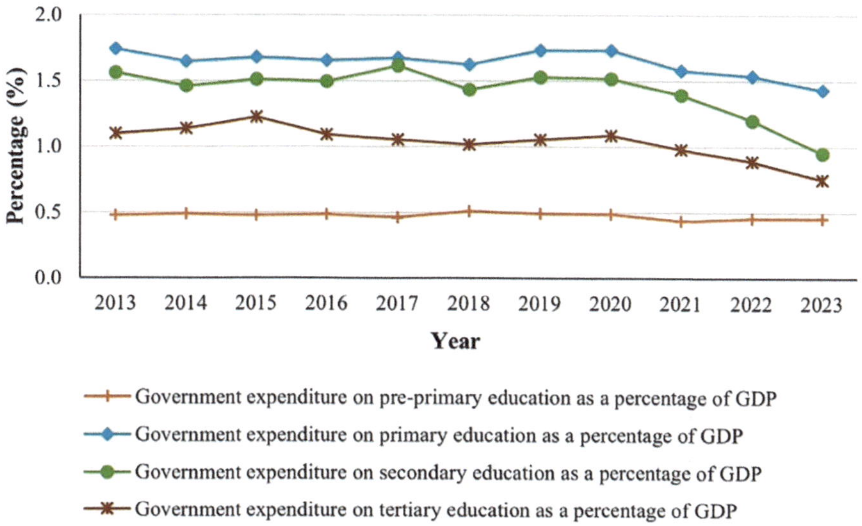


**Fig. 3.4** Average government expenditures on different levels of education as a percentage of GDP (%) in Latin America between 2013 and 2023 (*Source* Based on data from UNESCO Institute of Statistics (UIS) Database [70])

### *3.5.2 Poverty and Employment*

Poverty is a concept strongly linked to terms such as growth and inequality. With the economic variety of some countries, the difference in income and sources of financing increases inequality throughout the world. Developing countries have the greatest inequality in income, opportunities, access to health and education, employment, and lifestyles, which relates all these characteristics to Latin America [72].

Latin America and the Caribbean present very different levels of poverty, although the middle class has increased in Latin America and the Caribbean, poverty reduction showed little or almost no increase (30%) by 2023 [73].

According to the World Bank, the average unemployment rate in Latin America and the Caribbean was between 7 and 10% for the period 2013–2023 [35]. Poverty goes hand in hand with job creation, which is why the World Bank has developed the Employment Quality Index (EQI), which relates benefits, income, satisfaction, and security to quantify the generation of guaranteed jobs [74].

Figure 3.5 shows the Employment Quality Index (EQI) for some Latin American countries in urban and rural areas and the national average in 2023, whose quantification range is between 0 and 1, where favorability is presented when approaching 1.

Among the Latin American countries listed in Fig. 3.5, Costa Rica has the highest national EQI for 2023, while El Salvador has the lowest. The situation of these two countries can be explained as a stage of growth for the first and a stabilization for the second. Another possible hypothesis is the stabilization in job creation for both

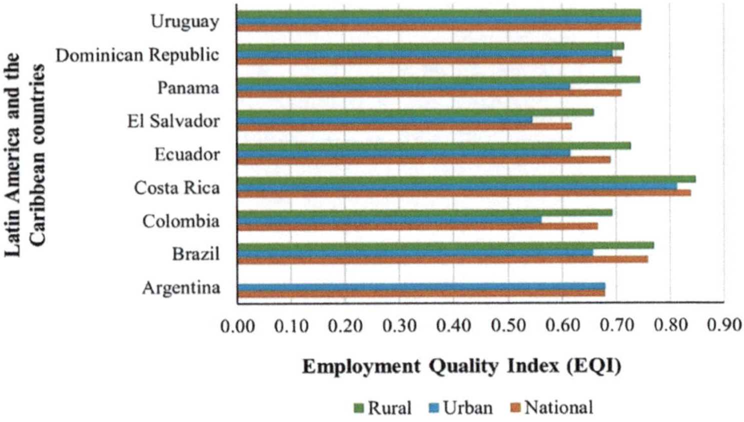


**Fig. 3.5** Employment Quality Index (EQI) of some Latin American countries in 2023 (*Source* Based on data from Database World Bank [74])

countries, as has generally happened for Latin America and the Caribbean in the last ten (10) years [73].

### *3.5.3 Addressing Health Inequalities*

Health is another of the fundamental factors that indicate inequality between countries. In Latin America and the Caribbean, health services coverage generates a larger gap with the world-developed countries. Social inequalities mark the tendency to associate it directly with health status; Inequalities in the health field are seen in different situations, like in the proper management of procedures for physical and mental illnesses treatment, and the monitoring of child and maternal development processes [75]. Although in the 1990s, some Latin American countries implemented changes in the provision of their health services, like improvements in efficiency and care, social protection led by health as the first service is not going through good times providing comprehensive care to the population [76].

Nowadays, it is necessary to achieve the Universal Health Coverage (UHC) identified as a comprehensive way to guarantee the provision of health services. According to the World Health Organization (WHO), with the application of Universal Health Coverage (UHC), people have access to all the health services they require, prioritizing primary health care to meet needs and consolidate as an equitable health system [77]. Some Latin American countries such as Mexico, Argentina, Colombia, and Brazil have made progress to achieve Universal Health Coverage (UHC). However, failures can still be identified in the implementation of this system caused by changes in government regulations, the lack of commitment between health service providers (and distributors), and debts in the system [76].

Thus, while health services coverage in Latin America and the Caribbean is high, the population faces enormous inequality, starting with the free provision of general health services in some countries and excessive charges for the same services in others [75]. The concerns focus on the lack of monitoring and treatment of chronic diseases such as cancer, orphan diseases, metabolic diseases, and organ transplants [78].

### 3.5.4 Importance of Basic Sanitation Coverage

Basic sanitation services are the central axis to avoid sources of contamination that lead to the impact on natural resources. This component constitutes a sector for the management of solid waste, its availability, and the use and management of water for consumption and waste [79].

The coverage of basic sanitation services in Latin America and the Caribbean is very low, with less than 50% of the population having this service optimally and safely. Some people perform these functions autonomously but without adequate knowledge of how to manage water and waste systems [80].

Although Latin America does not have a high level of basic sanitation system coverage, there are some services to comply with sanitation objectives. Figure 3.6 shows the average number of people in Latin America who used at least basic sanitation services between 2012 and 2022. The information summarized in Fig. 3.6 indicates a clear increase in the average percentages of people who have access to basic sanitation services. However, the percentage increase is low (between 6 and 7%) after ten years, which indicates that, although coverage increased in parallel in urban and rural areas, there was no accelerated population growth, nor exacerbated inclinations to implement other basic sanitation systems.

In 2015, only 22% of the population of Latin America and the Caribbean used basic sanitation services that involved the correct disposal of fecal matter through pipes to avoid contaminating nearby water sources, corresponding to people living in urban areas [81].

It is essential to relate the contamination of different water sources, soil, and biota due to the lack of availability of a system that adequately treats waste produced by domestic wastewater and domestic solid waste such as batteries, which causes cross-contamination. On the other hand, industrial waste management is also a cause for concern because the traditional treatments used are inefficient and generate adverse effects.

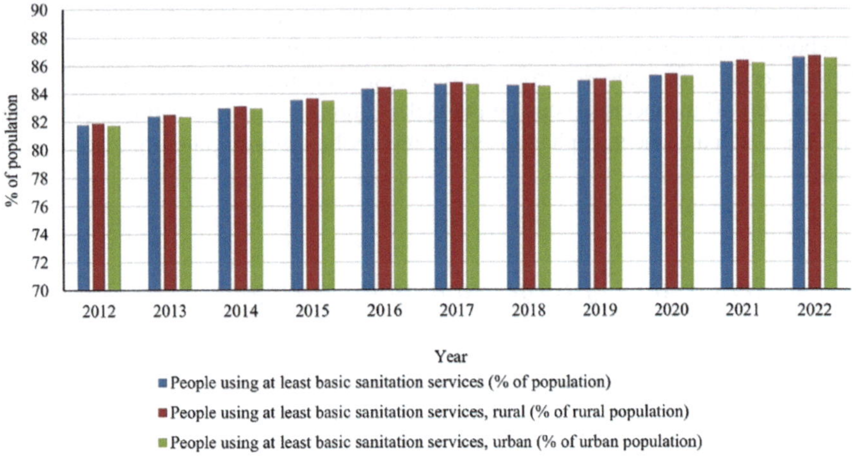


**Fig. 3.6** Average percentages of people using at least basic sanitation services in Latin America between 2012 and 2022 (*Source* Based on data from Database World Bank [35])

## 3.6 Circular Economy and Sustainability: Opportunities in Latin America

The implementation of circular economy models in Latin America faces several challenges, largely due to the region's socio-cultural context. To advance in implementing circular economy practices in the region, it is essential to assess the current situation and identify available opportunities [82]. Therefore, strategies are needed to reduce waste generation in both communities and the industrial sector, while promoting the sustainable use of resources. These efforts are essential to improving the quality of life, promoting social and economic changes, and ensuring sustainability in the region. Latin American countries have made significant progress in both individual and collective initiatives aimed at reducing inequalities in access to essential resources, while also focusing on pathways that contribute to the socioeconomic growth of the agricultural sector. However, challenges remain in various areas, which must be prioritized to ensure equitable conditions for all populations.

### *3.6.1 Enhancing Access to Water*

Latin America has made significant strides in increasing access to drinking water and water resources in general. However, many communities, particularly vulnerable populations in rural areas, still face challenges in accessing safe water. These areas also often struggle with deficiencies in infrastructure, healthcare, education, and other essential services.

In some regions, communities lack reliable access to drinking water and basic sanitation, as well as to water sources that are suitable for consumption. To address these issues, alternative strategies must be developed that are adapted to the specific needs of Latin America's vulnerable communities.

While most Latin American countries source their water from rivers, dams, and dikes, and recognize access to water as a fundamental right, the reality of water access is inconsistent. In Colombia, for example, access to water is not only a right but also a public service (as established by Sentence T-740/11 of the Constitutional Court). This ensures that all people should have access to sufficient quantities of clean water. The government is responsible for ensuring its provision. However, in some vulnerable communities, the lack of distribution and purification systems means that access to drinking water is limited or intermittent, making it difficult for people to meet their daily water needs. On average, a person needs between 1.5 and 2 L of water per day, depending on their body mass, without this, health issues may arise [83]. Therefore, it is crucial not only to achieve universal water coverage but also to ensure that its supply is continuous.

To improve water access, Latin America could benefit from systems focused on water collection, desalination, and innovative water supply network technologies that meet the needs of underserved communities.

**Rainwater Harvesting**

Rainwater harvesting systems are designed to capture and store large amounts of water from rainfall, which can then be used for domestic or agricultural purposes [84]. These systems are especially suitable for areas lacking drinking water infrastructure but with favorable climatic conditions [85].

The development of rainwater harvesting systems can offer a valuable solution in areas distant from traditional water supply sources. While many Latin American countries are rich in water resources, including surface and groundwater, this is not true for all regions. Some areas may lack direct water sources but experience high rainfall due to their geographical and climatic conditions. In such cases, rainwater collection systems can help address water access challenges.

The Food and Agriculture Organization of the United Nations (FAO) [86] has identified six major techniques for rainwater collection systems: 1. Microcollection 2. Macrocollection 3. Diversion of water through intakes 4. Water harvesting of housing roofs and other impermeable structures 5. Collection of groundwater and phreatic water 6. Collection of atmospheric water (Mists) (very little used). By analyzing historical rainfall data from the past five years, average rainfall patterns in the target areas can be defined to identify the most suitable technology for rainwater harvesting. This information can help determine whether there is sufficient rainfall to support the implementation of a rainwater harvesting system. The analysis should also include estimates of water supply, the size of the area, the runoff coefficient, and the monthly water demand, considering the number of people who will benefit and the number of days per month. Additionally, these studies must assess the quality of the available water and evaluate the risk of diseases associated with water that does not meet the required physical, chemical, and microbiological standards for human consumption.

Rainwater harvesting systems can be effectively implemented in both arid and semi-arid regions, as well as in sub-humid areas, where competition for water resources is rising. Rainwater harvesting systems help reduce the impact of rainfall fluctuations and water scarcity on crop production [87], especially in Latin America, where only 10% of agricultural land is irrigated. These systems are particularly valuable in the large agricultural, livestock, and forestry areas of the region's arid and semi-arid zones.

**Desalinization**

Seawater purification by evaporation methods has been widely used, however it requires high energy consumption [88]. Nevertheless, recent advancements in desalination technologies have made these processes more efficient, reducing both energy use and overall costs. The choice of desalination system depends on the type of separation process it employs, which can be either distillation, based on evaporation techniques, or physical–chemical processes involving chemical treatments or membrane permeability [89].

The most commonly used desalination technologies include distillation methods such as Multi-Stage Flash (MSF) Distillation [90], Multiple-Effect Distillation (MED) [91], and Solar Distillation [92]. Additionally, Reverse Osmosis, which uses semipermeable membranes, is the most widely adopted method, accounting for nearly three-quarters of the global production of desalinated water [93].

While desalination technologies have been successfully implemented, there are ongoing concerns about their environmental impacts. These concerns primarily involve the construction and operation of desalination plants, which can affect coastal areas and the waste generated by the process. Wastewater discharges are often released directly into the sea, potentially harming marine life. Additionally, the energy consumption of desalination technologies can lead to greenhouse gas emissions [93]. Despite these challenges, desalination offers a viable, non-conventional solution to meet the growing demand for freshwater, improving access to this essential resource while helping to reduce pressure on conventional water sources, which are often insufficient.

In Latin America, desalination technologies have primarily been implemented on a small scale, though there is significant potential for larger projects, particularly in countries like Chile and Mexico. In Chile, large desalination plants have been established to supply water for both domestic and industrial use in the Atacama Desert. In 2023, the Chilean National Congress passed Law 21,639 to promote the development of water infrastructure and desalination projects [94]. The law seeks to ensure the long-term sustainability of water availability for human consumption, sanitation, and irrigation. In Mexico, the National Water Law, which previously required concessions for extracting inland marine waters and territorial sea waters for desalination purposes, was modified in 2022. The Senate approved a reform to the law, regulating seawater desalination. Under the new provisions, desalination projects for agricultural and domestic use are exempt from concession requirements. The reform was aimed at streamlining the process and ensuring more accessible water resources for these purposes [95].

In Colombia, desalination systems are primarily implemented in coastal areas facing freshwater shortages, with a focus on small communities. In 2022, a desalination plant using ultrafiltration and reverse osmosis technology was installed in the San Andrés, Providencia, and Santa Catalina archipelago. This plant can supply 137 L/s of water [96]. Additionally, new desalination plant construction projects have recently begun in the upper Guajira region. In Brazil and Argentina, desalination plants have become a promising solution to supply drinking water to semi-arid regions and communities facing water shortages [97]. This highlights that Latin America is increasingly turning to desalination technologies as a viable alternative to meet the growing demand for water in various communities, many of which are vulnerable. However, challenges remain in the development of new materials, process optimization, and reducing energy consumption. Addressing these issues is crucial for ensuring desalination can be a sustainable and effective solution for the region.

### *3.6.2 Drinking Water Networks*

The implementation of water distribution systems connected to treatment plants or direct sources is a significant challenge in Latin America. Despite efforts to deploy treatment and purification technologies, many vulnerable communities face limited or nonexistent distribution networks, especially in remote areas or new suburban settlements. In some regions, forced displacement due to social issues, such as migration from other countries or regions, exacerbates the problem. Migrants often settle in suburban areas of large cities or communities lacking full access to public services, including water. Additionally, as mentioned in previous sections, certain regions lack direct access to water sources. In these areas, the design and implementation of drinking water distribution systems, using optimization approaches that ensure sustainability, are crucial. Integrating rainwater harvesting and desalination technologies with the optimized design of distribution networks offers promising solutions for vulnerable communities across Latin America [98].

## 3.7 Technologies for Improved Access to Energy

The climatic, geographical, and agricultural characteristics of Latin America create several opportunities to diversify the region's energy matrix through renewable alternatives. It is important to note that many rural areas in certain Latin American countries remain part of "non-interconnected zones," where there is no connection to the national electricity distribution system. Instead, these areas rely on small-scale energy generation, some of which still depend on fossil fuels. For example, the Comprehensive Sustainability, Operation, and Market Report (2021) [99] from the Carbon Neutral Alliance, led by the Ministry of Mines of Colombia, highlights that the majority of the daily energy supply to the National Interconnected System

(SIN) comes from renewable sources (83.4%), primarily hydraulic energy, with only 16.6% coming from fossil fuels. However, the situation in non-interconnected zones (ZNI)—which make up 53% of Colombia's territory and include vast agricultural areas—is different. These zones primarily rely on diesel as their main energy source, accounting for 84% of their energy needs [100]. Therefore, alternative energy sources could be an option for this region.

**Solar Energy**

Solar radiation is a widely available and highly attractive resource in Latin America, offering significant potential for infrastructure development aimed at a low-carbon energy transition. Solar photovoltaic power plants convert solar radiation into electricity, making them an ideal solution in many regions. In several Latin American countries, there is little variation in seasons, which means that solar radiation is consistently favorable throughout much of the year; therefore, these systems provide a valuable alternative for areas without access to the energy distribution network and can serve as a complementary strategy to expand energy coverage.

Additionally, in developing countries with high solar radiation, small- and medium-sized industries represent a promising market for solar thermal systems. These industries often rely heavily on fossil fuels, and solar thermal systems can help reduce this dependence while lowering operating costs [101].

Countries such as Bolivia, northern Argentina, northern Chile, and southern Peru—located in the Andes region of Latin America—boast some of the largest solar energy resources in the world. These areas have significant photovoltaic energy generation potential due to consistently clear skies, clean air, low temperatures, and high altitudes, which result in a thinner atmosphere compared to lower-altitude regions [102]. Additionally, other Latin American countries like Colombia, Mexico, El Salvador, Cuba, Costa Rica, and Nicaragua have regions where solar radiation levels exceed the global average, creating favorable conditions for solar energy development.

Other forms of energy, such as wind, tidal, and geothermal, have been explored in the region as alternative strategies for achieving energy sustainability. Mexico leads wind energy efforts in Latin America, along with Brazil and Chile. Mexico has also explored geothermal energy and conducted studies on thermal energy. Other countries in the region are similarly exploring these energy sources, reflecting a growing interest and commitment to diversifying energy options to promote sustainability.

**Bioenergy**

Biomass can be used to generate energy in various forms, including direct combustion, gasification, anaerobic digestion, and the production of bioethanol, biodiesel, or biohydrogen. In Latin America, the region's agricultural potential produces a substantial amount of agricultural waste throughout the year, with some crops generating one to two times more waste than the main products (Table 3.3). Due to the lack of technology in some regions, these residual biomasses are often not used for energy generation or the production of processed products. Instead, they are primarily disposed of as animal feed, used as natural fertilizer, or simply discarded, resulting

**Table 3.3** Agricultural waste factors (*Source* Based on data from The Mining and Energy Planning Unit—UPME [103])

| Crop | Waste Product | Waste factor (t waste/t main product) |
|---|---|---|
| Palm oil | Empty fruit bunch (EFB) | 1.06 |
| | Mesocarp fiber (MF) | 0.63 |
| | Palm kernel shell (PKS) | 0.22 |
| Sugarcane | Leave and others | 3.26 |
| | Bagasse | 2.68 |
| Corn | Agricultural waste | 0.93 |
| | Cob | 0.27 |
| | Bract | 0.21 |
| Rice | Husk | 0.20 |
| | Straw | 2.35 |

in a loss of value within the production chain. Also, burn biomass is a common practice in some communities contributing to greenhouse gases (GHG). Additionally, residual agricultural biomass left in the fields can contribute to pest problems and other environmental issues. However, this biomass holds significant potential for generating renewable energy, presenting an opportunity to diversify energy sources in these areas.

However, using agricultural waste for energy generation still presents several challenges. These include the collection and storage of biomass for large-scale use, transportation costs to remote agricultural areas or regions with difficult access, a lack of on-site technology, and the need for flexible technologies that enable the sustainable use of biomass. It is also crucial to ensure that these strategies do not negatively impact the biodiversity of the regions. Therefore, the development of a multi-feedstock approach offers flexibility and ensures a steady supply of raw materials, even if weather or other factors disrupt the harvest of a specific crop. This approach also allows for selecting the most suitable feedstock (or combination) for the desired product, potentially improving conversion efficiency. Additionally, using residual biomass as feedstock reduces the environmental impact of landfills and supports a more sustainable, circular bioeconomy.

### *3.7.1 Enhancing Sustainable Agricultural Practices*

The agricultural sector is an important component of Latin America's economy, playing a significant role in the GDP of many countries (Fig. 3.7). However, this sector also highlights a contrast in inequality between rural and urban areas. In some regions, the average income of a rural worker is one-third lower than that of an urban resident [104], leading to social and economic imbalances.

**Fig. 3.7** Agricultural sector's contribution to GDP in Latin American countries (*Source* Based on data from World Bank, 2021 [35])

The rural sector, primarily driven by agricultural activities, needs greater access to technology that supports sustainable practices. The improper use of fertilizers, poor management of crop and post-harvest residues, and other factors highlight the need for alternatives that promote good agricultural practices. These practices should focus on reducing water and air pollution, preventing future food safety issues, and mitigating biodiversity loss, water scarcity, and the increase of greenhouse gas emissions [105]. Implementing strategies focused on rural areas can help reduce social gaps between communities, enhance regional economies, and produce more competitive agricultural products with lower environmental impacts.

One important consideration is the implementation of water and carbon footprint assessments for crops, which help identify associated environmental impacts and develop strategies tailored to the type of crops and specific characteristics of the region. The Water and Carbon Footprint methodologies are valuable tools for strengthening supply chains, reducing emissions, and promoting the efficient use of water resources in production processes [106]. These footprints—carbon (CF) and water (WF)—serve as key indicators of environmental impacts, allowing for the quantification of GHG emissions and water usage in agricultural systems. They also encourage actions to improve cultivation, harvesting, and waste management practices in the region.

The Carbon Footprint (CF) measures the total greenhouse gas emissions produced by an individual, production system, or organization, expressed in $CO_2$ equivalent units [107]. The Water Footprint (WF) is an indicator of both the consumption and contamination of freshwater, encompassing both direct and indirect dimensions. It was first introduced in 2002 by Dr. Arjen Hoekstra and has since been promoted by the Water Footprint Network (WFN) [108]. The analysis includes the quantification of blue, green, and gray water footprints. It measures the consumption of blue water resources (freshwater, from either surface or groundwater) throughout the production chain of a product. It also accounts for green water resources (rainwater that is absorbed by crops, rather than becoming runoff) and the volume of freshwater needed to assimilate a contaminant load, based on natural concentrations and environmental water quality standards. In recent years, the use of these analytical methodologies applied to cultivation systems has been growing in Latin America, aiming to develop measures and actions that reduce GHG emissions in production systems and promote the efficient use of water resources [106, 107, 109, 110]. These analyses have highlighted the need for technologies such as systematized irrigation in crops and implementing renewable energy for agricultural machinery, among other solutions, which can reduce water consumption during droughts and improve crop productivity with less environmental impacts.

Finally, Latin America is making a global contribution by adopting sustainability and circular economy strategies. While gaps remain between countries and regions in achieving sustainable communities, the individual and collective efforts of Latin American nations according to the context of each country have resulted in policies that can positively impact food security, access to energy and water, and overall quality of life. These approaches are especially beneficial in areas with higher concentrations of vulnerable communities.

## References

1. World Bank. DataBank (2023) World development indicators. https://databank.worldbank.org/source/world-development-indicators. Accessed 18 Jan 2025
2. IPBES (2019) Global assessment report on biodiversity and ecosystem services of the Intergovernmental Science-Policy Platform on Biodiversity and Ecosystem Services. IPBES secretarial, Bonn
3. WWF (2022) Living planet report 2022—building a naturepositive society, Gland, Switzerland
4. Inter-American Development Bank (IDB). Biodiversity in Latin America and the Caribbean. https://www.iadb.org/en/news/biodiversity-latin-america-and-caribbean. Accessed 18 Jan 2025
5. World Bank. DataBank (2016) Sustainable energy for all. https://databank.worldbank.org/source/sustainable-energy-for-all. Accessed 18 Jan 2025
6. Flachsbarth I, Willaarts B, Xie H, Pitois G, Mueller ND, Ringler C, Garrido A (2015) The role of Latin America's land and water resources for global food security: environmental trade-offs of future food production pathways. PLoS ONE 10:1–24
7. Volsi B, Telles TS, Caldarelli CE, da Camara MRG (2019) The dynamics of coffee production in Brazil. PLoS ONE 14:1–15
8. Pham Y, Reardon-Smith K, Mushtaq S, Cockfield G (2019) The impact of climate change and variability on coffee production: a systematic review. Clim Change 156:609–630
9. Díaz-Valderrama JR, Leiva-Espinoza ST, Catherine Aime M (2020) The history of cacao and its diseases in the Americas. Phytopathology 110:1604–1619

10. Gateau-Rey L, Tanner EVJ, Rapidel B, Marelli JP, Royaert S (2018) Climate change could threaten cocoa production: effects of 2015–16 El Niño-related drought on cocoa agroforests in Bahia, Brazil. PLoS ONE 13:1–17
11. Cherubin MR, Karlen DL, Franco ALC, Tormena CA, Cerri CEP, Davies CA, Cerri CC (2016) Soil physical quality response to sugarcane expansion in Brazil. Geoderma 267:156–168
12. Chauhan BS, Jabran K, Mahajan G (2017) Rice production worldwide. Rice Production Worldwide, pp 1–563
13. Pittelkow CM, Zorrilla G, Terra J, Riccetto S, Macedo I, Bonilla C, Roel A (2016) Sustainability of rice intensification in Uruguay from 1993 to 2013. Glob Food Sec 9:10–18
14. Haan S, Rodriguez F (2016) Potato origin and production. Adv Potato Chem Technol 1–32
15. Scott GJ, Kleinwechter U (2017) Future scenarios for potato demand, supply and trade in South America to 2030. Potato Res 60:23–45
16. Taramuel-Taramuell JP, Montoya-Restrepo IA, Barrios D (2024) Challenges in the avocado production chain in Latin America: a descriptive analysis. Agronomia Colombiana 42:12
17. Lieb VM, Schuster LK, Kronmüller A, Schmarr HG, Carle R, Steingass CB (2019) Fatty acids, triacylglycerols, and thermal behaviour of various mango (*Mangifera indica* L.) kernel fats. Food Res Int 116:527–537
18. Storey WB (2020) Papaya (*Carica papaya*). In: CRC handbook of flowering, pp 147–157
19. Artola A, Font X, Moral-Vico J, Sánchez A (2024) The role of solid-state fermentation to transform existing waste treatment plants based on composting and anaerobic digestion into modern organic waste-based biorefineries, in the framework of circular bioeconomy. Front Chem Eng 6:1–11
20. Noor RS, Shah AN, Tahir MB, Umair M, Nawaz M, Ali A, Ercisli S, Abdelsalam NR, Ali HM, Yang SH, Ullah S, Assiri MA (2024) Recent trends and advances in additive-mediated composting technology for agricultural waste resources: a comprehensive review. ACS Omega 9:8632–8653
21. International Energy Agency (2023) Latin America Energy Outlook. https://iea.blob.core.windows.net/assets/1055131a-8dc4-488b-9e9e-7eb4f72bf7ad/LatinAmericaEnergyOutlook.pdf. Accessed 20 Jan 2025
22. Whole Building Design Guide—WBDG. Biomass for Electricity Generation. Whole Building Design Guide—WBDG 1. https://www.wbdg.org/resources/biomass-electricity-generation. Accessed 20 Jan 2025
23. El-Araby R (2024) Biofuel production: exploring renewable energy solutions for a greener future. Biotechnol Biofuels Bioprod 17:32
24. Polagye B, Van Cleve B, Copping A, Kirkendall K (2010) Environmental effects of tidal energy development: proceedings of a scientific workshop. In: Tidal energy workshop, Seattle, Washington
25. World Bank. Promoting Rural Electrification in Peru. World, https://www.worldbank.org/en/results/2019/05/13/promoting-rural-electrification-in-peru. Accessed 20 Jan 2025
26. Relevant data from the energy sector of Latin America and the Caribbean validated by international experts. Energy hub. Latin America and the Caribbean 1. https://hubenergia.org/. Accessed 20 Jan 2025
27. Inter-American Development Bank [IDB]. Five things you should know about the energy sector in Latin America and the Caribbean, https://blogs.iadb.org/energia/es/cinco-cosas-que-debes-saber-sobre-energia-en-america-latina-y-el-caribe/#:~:text=El58%25delageneración,biomasas%2Cy1%25geotérmica. Accessed 20 Jan 2025
28. Martínez-Espiñeira R, Pérez-Urdiales M (2025) Water affordability challenges in Latin America and the Caribbean: accounting for coping costs due to reliance on multiple, non-exclusive sources. World Bank Group 186:18
29. Rodríguez C, García B, Pinto C, Sánchez R, Serrano J, Leiva E (2022) Water context in Latin America and the Caribbean: distribution, regulations and prospects for water reuse and reclamation. Water 14:26
30. Herrera FS, Rubiano DG, Garrido FB, Hernández PC, Polo SG, Romero J, Gomez-Charris Y (2020) Evaluation of water quality state through regulations and physicochemical indicators

for the administration of water resources in the integrated management District of Salto del Tequendama. Proc Comput Sci 177:300–307
31. Mekonnen MM, Pahlow M, Aldaya MM, Zarate E, Hoekstra AY (2015) Sustainability, efficiency and equitability of water consumption and pollution in latin America and the Caribbean. Sustainability 7:2086–2112
32. Mendieta-Mendoza A, Hanson RT, Renteria-Villalobos M (2021) Potential adverse impacts on vulnerability and availability of groundwater from climate-change and land use. J Hydrol 594:125978
33. Latrubesse EM, Arima EY, Dunne T, Park E, Baker VR, D'Horta FM, Wight C, Wittmann F, Zuanon J, Baker PA, Ribas CC, Norgaard RB, Filizola N, Ansar A, Flyvbjerg B, Stevaux JC (2017) Damming the rivers of the Amazon basin. Nature 546:363–369
34. Alfonso MB, Scordo F, Seitz C, Mavo Manstretta GM, Ronda AC, Arias AH, Tomba JP, Silva LI, Perillo GME, Piccolo MC (2020) First evidence of microplastics in nine lakes across Patagonia (South America). Sci Total Environ 733:139385
35. World Bank. Database. World Development Indicators (2022) World Bank Group 2. https://databank.worldbank.org/. Accessed 20 Jan 2025
36. Peña-Guzmán C, Ulloa-Sánchez S, Mora K, Helena-Bustos R, Lopez-Barrera E, Alvarez J, Rodriguez-Pinzón M (2019) Emerging pollutants in the urban water cycle in Latin America: a review of the current literature. J Environ Manage 237:408–423
37. Torremorell A, Hegoburu C, Brandimarte AL, Rodrigues EHC, Pompêo M, da Silva SC, Moschini-Carlos V, Caputo L, Fierro P, Mojica JI, Matta ÁLP, Donato JC, Jiménez-Pardo P, Molinero J, Ríos-Touma B, Goyenola G, Iglesias C, López-Rodríguez A, Meerhoff M, Navarro E (2021) Current and future threats for ecological quality management of South American freshwater ecosystems. Inland Waters 11:125–140
38. Pooi CK, Ng HY (2018) Review of low-cost point-of-use water treatment systems for developing communities. NPJ Clean Water 1:8
39. Benavides L, Avellán T, Caucci S, Hahn A, Kirschke S, Müller A (2019) Assessing sustainability of wastewater management systems in a multi-scalar, transdisciplinary manner in Latin America. Water 11:51
40. Isiuku BO, Enyoh CE (2020) Pollution and health risks assessment of nitrate and phosphate concentrations in water bodies in South Eastern, Nigeria. Environ Adv 2:100018
41. Wu W, Ren J, Zhou X, Wang J, Guo M (2020) Identification of source information for sudden water pollution incidents in rivers and lakes based on variable-fidelity surrogate-DREAM optimization. Environ Model Softw 133:104811
42. Rahman K, Barua S, Imran HM (2021) Assessment of water quality and apportionment of pollution sources of an urban lake using multivariate statistical analysis. Clean Eng Technol 5:100309
43. Han J, Xin Z, Han F, Xu B, Wang L, Zhang C, Zheng Y (2021) Source contribution analysis of nutrient pollution in a P-rich watershed: Implications for integrated water quality management. Environ Pollut 279
44. Häder DP, Banaszak AT, Villafañe VE, Narvarte MA, González RA, Helbling EW (2020) Anthropogenic pollution of aquatic ecosystems: emerging problems with global implications. Sci Total Environ 713:1–8
45. Zhou L, Li LZ (2021) Huang JK (2021) The river chief system and agricultural non-point source water pollution control in China. J Integr Agric 20:1382–1395
46. Huang C, Zhao D, Fan X, Liu C, Zhao G (2022) Landscape dynamics facilitated non-point source pollution control and regional water security of the Three Gorges Reservoir area, China. Environ Impact Assess Rev 92:106696
47. Luo M, Zhang Y, Li H, Hu W, Xiao K, Yu S, Zheng C, Wang X (2022) Pollution assessment and sources of dissolved heavy metals in coastal water of a highly urbanized coastal area: the role of groundwater discharge. Sci Total Environ 807:151070
48. Potapowicz J, Szopińska M, Szumińska D, Bialik RJ, Polkowska Ż (2022) Sources and composition of chemical pollution in Maritime Antarctica (King George Island), part 1: sediment and water analysis for PAH sources evaluation in the vicinity of Arctowski station. Chemosphere 288

49. Dey M, Akter A, Islam S, Chandra Dey S, Choudhury TR, Fatema KJ, Begum BA (2021) Assessment of contamination level, pollution risk and source apportionment of heavy metals in the Halda River water, Bangladesh. Heliyon 7:e08625
50. Alam R, Ahmed Z, Seefat SM, Nahin KTK (2021) Assessment of surface water quality around a landfill using multivariate statistical method, Sylhet, Bangladesh. Environ Nanotechnol Monit Manage 15:100422
51. Tokatl C, Varol M (2021) Variations, health risks, pollution status and possible sources of dissolved toxic metal(loid)s in stagnant water bodies located in an intensive agricultural region of Turkey. Environ Res 201:111571
52. Parra-Luna M, Martín-Pozo L, Hidalgo F, Zafra-Gómez A (2020) Common sea urchin (*Paracentrotus lividus*) and sea cucumber of the genus Holothuria as bioindicators of pollution in the study of chemical contaminants in aquatic media. A revision. Ecol Indicat 113:106185
53. Ouda M, Kadadou D, Swaidan B, Al-Othman A, Al-Asheh S, Banat F, Hasan SW (2021) Emerging contaminants in the water bodies of the Middle East and North Africa (MENA): a critical review. Sci Total Environ 754:142177
54. Mammo FK, Amoah ID, Gani KM, Pillay L, Ratha SK, Bux F, Kumari S (2020) Microplastics in the environment: interactions with microbes and chemical contaminants. Sci Total Environ 743:140518
55. Akerman-Sanchez G, Rojas-Jimenez K (2021) Fungi for the bioremediation of pharmaceutical-derived pollutants: a bioengineering approach to water treatment. Environ Adv 4:100071
56. Dey S, Bano F, Malik A (2019) Pharmaceuticals and personal care product (PPCP) contamination-a global discharge inventory (chap 1). In: Pharmaceuticals and personal care products: waste management and treatment technology. Elsevier Inc
57. Vasseghian Y, Hosseinzadeh S, Khataee A, Dragoi EN (2021) The concentration of persistent organic pollutants in water resources: a global systematic review, meta-analysis and probabilistic risk assessment. Sci Total Environ 796:149000
58. Saravanan A, Kumar PS, Hemavathy RV, Jeevanantham S, Harikumar P, Priyanka G, Devakirubai DRA (2021) A comprehensive review on sources, analysis and toxicity of environmental pollutants and its removal methods from water environment. Sci Total Environ 812:15
59. Peluso J, Pérez Coll CS, Cristos D, Rojas DE, Aronzon CM (2021) Comprehensive assessment of water quality through different approaches: physicochemical and ecotoxicological parameters. Sci Total Environ 800:9
60. Ramírez-Morales D, Masís-Mora M, Montiel-Mora JR, Méndez-Rivera M, Gutiérrez-Quirós JA, Brenes-Alfaro L, Rodríguez-Rodríguez CE (2024) Pharmaceuticals, hazard and ecotoxicity in surface and wastewater in a tropical dairy production area in Latin America. Chemosphere 346:25
61. Kutralam-Muniasamy G, Pérez-Guevara F, Elizalde-Martínez I, Shruti VC (2020) Review of current trends, advances and analytical challenges for microplastics contamination in Latin America. Environ Pollut 267:15
62. Orona-Návar C, García-Morales R, Loge FJ, Mahlknecht J, Aguilar-Hernández I, Ornelas-Soto N (2022) Microplastics in Latin America and the Caribbean: a review on current status and perspectives. J Environ Manage 309:15
63. Bundschuh J, Schneider J, Alam MA, Niazi NK, Herath I, Parvez F, Tomaszewska B, Guilherme LRG, Maity JP, López DL, Cirelli AF, Pérez-Carrera A, Morales-Simfors N, Alarcón-Herrera MT, Baisch P, Mohan D, Mukherjee A (2021) Seven potential sources of arsenic pollution in Latin America and their environmental and health impacts. Sci Total Environ 780:29
64. Zamora-Ledezma C, Negrete-Bolagay D, Figueroa F, Zamora-Ledezma E, Ni M, Alexis F, Guerrero VH (2021) Heavy metal water pollution: a fresh look about hazards, novel and conventional remediation methods. Environ Technol Innov 22:26
65. Grondona SI, Lima ML, Enrique Massone H, Miglioranza KSB (2023) Pesticides in aquifers from Latin America and the Caribbean. Sci Total Environ 901:27

66. Tursi A, Beneduci A, Chidichimo F, De Vietro N, Chidichimo G (2018) Remediation of hydrocarbons polluted water by hydrophobic functionalized cellulose. Chemosphere 201:530–539
67. Bijay-Singh, Craswell E (2021) Fertilizers and nitrate pollution of surface and ground water: an increasingly pervasive global problem. SN Appl Sci 3:1–24
68. Munoz-Jofre J, Hinojosa S, Mascle-Allemand AL, Temprano J (2023) A selectivity index for public-private partnership projects in the urban water and sanitation sector in Latin America and the caribbean. J Environ Manage 335:6
69. United Nations Educational Scientific and Cultural Organization (UNESCO). Education in Latin America and the Caribbean at a crossroads: regional monitoring report SDG4—Education 2030; summary (spanish). https://www.unicef.org/lac/media/37791/file/EducationinLatinAmericaandtheCaribbeanatacrossroads.pdf. Accessed 20 Jan 2025
70. United Nations Educational Scientific and Cultural Organization (UNESCO) (2024) UNESCO Institute of Statistics (UIS). https://uis.unesco.org/bdds. Accessed 20 Jan 2025
71. Salas-Pilco SZ, Yang Y (2020) Learning analytics initiatives in Latin America: implications for educational researchers, practitioners and decision makers. Br J Edu Technol 51:875–891
72. Adeleye BN, Gershon O, Ogundipe A, Owolabi O, Ogunrinola I, Adediran O (2020) Comparative investigation of the growth-poverty-inequality trilemma in Sub-Saharan Africa and Latin American and Caribbean Countries. Heliyon 6:11
73. World Bank. Poverty in Latin America: 10 facts we need to know in 2024 (Spanish), https://blogs.worldbank.org/es/opendata/poverty-in-latin-america--10-facts-you-need-to-know-for-2024. Accessed 20 Jan 2025
74. World Bank. LAC Equity Lab: Labor Markets—Employment Quality Index (IQE) (Spanish). https://www.bancomundial.org/es/topic/poverty/lac-equity-lab1/labor-markets/indice-de-la-calidad-del-empleo. Accessed 20 Jan 2025
75. Abramo L, Cecchini S, Ullmann H (2020) Addressing health inequalities in Latin America: the role of social protection. Ciencia e Saude Coletiva 25:1587–1598
76. Gilardino RE, Valanzasca P, Rifkin SB (2022) Has Latin America achieved universal health coverage yet? Lessons from four countries. Arch Public Health 80:1–11
77. WHO (2024) Universal health coverage. https://www.who.int/health-topics/universal-health-coverage#tab=tab_1. Accessed 20 Jan 2025
78. Barrios CH, Werutsky G, Mohar A, Ferrigno AS, Müller BG, Bychkovsky BL, Castro ECJ, Uribe CJ, Villarreal-Garza C, Soto-Perez-de-Celis E, Gutiérrez-Delgado F, Kim JS, Ismael J, Delgado L, Santini LA, Teich N, Chavez PC, Liedke PER, Exman P, Cazap E (2021) Cancer control in Latin America and the Caribbean: recent advances and opportunities to move forward. Lancet Oncol 22:e474–e487
79. Zhou X, Li Z, Zheng T, Yan Y, Li P, Odey EA, Mang HP, Uddin SMN (2018) Review of global sanitation development. Environ Int 120:246–261
80. Cruvinel VRN, Marques CP, Cardoso V, Novaes MRCG, Araújo WN, Angulo-Tuesta A, Escalda PMF, Galato D, Brito P, Da Silva EN (2019) Health conditions and occupational risks in a novel group: waste pickers in the largest open garbage dump in Latin America. BMC Public Health 19:1–15
81. Mahlknecht J, González-Bravo R, Loge FJ (2020) Water-energy-food security: a Nexus perspective of the current situation in Latin America and the Caribbean. Energy 194:1–17
82. De Miguel C, Martínez K, Pereira M, Kohout M (2021) Circular economy in Latin America and the Caribbean: opportunity for a transformative recovery (Spanish). Comisión Económica para América Latina y el Caribe (CEPAL), United Nations, Santiago, Chile
83. WHO (2015) World Health Statistic. World Health Organization. https://www.who.int/docs/default-source/gho-documents/world-health-statistic-reports/world-health-statistics-2015.pdf. Accessed 20 Jan 2025
84. Campisano A, Modica C (2014) Selecting time scale resolution to evaluate water saving and retention potential of rainwater harvesting tanks. Proc Eng 70:218–227
85. Solano C, Gonzaga F, Espinoza F, Espinoza J (2017) Rainwater collection system for domestic use, Jambelí Island, cantón Santa Rosa (Spanish). Cumbres 3:151–159

86. FAO (2013) Rainwater harvesting and storage. Technical options for family farming in Latin America and the Caribbean (Spanish). Oficina Regional de la FAO para América Latina y el Caribe, Santiago, Chile
87. Ricardo Calzadilla MP, Mendez Fernández M, Bonet Pérez C, Sierra Castellanos LO, Cutié V (2015) Evaluating the efficiency of rainwater harvesting in farm houses (Spanish). Revista Ingeniería Agrícola 5:3–9
88. Contreras M, Reventós M (2012) Obtaining drinking water from seawater by freezing and melting (Spanish). Universitat Politècnica de Catalunya
89. Mentis D, Karalis G, Zervos A, Howells M, Taliotis C, Bazilian M, Rogner H (2016) Desalination using renewable energy sources on the arid islands of South Aegean Sea. Energy 94:262–272
90. Ibrić N, Ahmetović E, Nemet A, Kravanja Z, Grossmann IE (2022) Synthesis of heat-integrated water networks using a modified heat exchanger network superstructure. Energies 15:23
91. Ramilo LB, Gomez de Soler SM, Coppari NR (2013) Process technologies for water desalination. In: 29th annual meeting of the argentine association of nuclear technology (AATN 2002), vol 13
92. Valero A, Uche J, Sierra L (2001) Desalination as an alternative to NHP (Spanish). CIRCE
93. Nava Escudero C (2021) Water desalination and energy justice.The 2030 agenda for sustainable development. Revista Mexicana De Derecho Constitucional 1:243–272
94. Ministry of Public Works of Chile. Law 21639. Modifies regulations for the development of water infrastructure and desalination projects, in order to allocate water to fulfill the function of subsistence and irrigation (Spanish). In: Library of the National Congress of Chile, vol 3. https://www.bcn.cl/leychile/navegar?idNorma=1199507. Accessed 20 Jan 2025
95. Ministry of Environment and Natural Resources. National Water Law and its Regulations (Spanish), https://www.gob.mx/cms/uploads/attachment/file/105137/Ley_Aguas_Nacionales.pdf. Accessed 20 Jan 2025
96. Findeter. Territorial development bank. Desalination plant, a reality to improve the lives of San Andres residents (Spanish). https://www.findeter.gov.co/blog/casos-de-exito/planta-desalinizadora-una-realidad-para-mejorar-la-vida-de-los-sanandresanos. Accessed 20 Jan 2025
97. Grueso-Dominguez MC, Castro-Jiménez CC, Correa-Ochoa MA, Saldarriaga-Molina JC (2019) State of the Art: desalination using membrane technologies as an alternative for the problem of fresh water shortage (Spanish). Revista Ingenierías Universidad de Medellín 18:69–89
98. González-Bravo R, Nápoles-Rivera F, Ponce-Ortega JM, El-Halwagi MM (2015) Involving integrated seawater desalination-power plants in the optimal design of water distribution networks. Resour Conserv Recycl 104:181–193
99. Sistema Interconectado Nacional (SIN). Comprehensive sustainability, operation and market report 2021 (Spanish). XM Sumando energías 1. https://informeanual.xm.com.co/informe/pages/home.html. Accessed 20 Jan 2025
100. Institute for Planning and Promotion of Energy Solutions for Non-Interconnected Areas—IPSE. Energy characterization of the ZNI (Spanish). IPSE 1. https://ipse.gov.co/cnm/caracterizacion-de-las-zni/. Accessed 20 Jan 2025
101. Farjana SH, Huda N, Mahmud MAP, Saidur R (2018) Solar process heat in industrial systems—a global review. Renew Sustain Energy Rev 82:2270–2286
102. World Bank. Global Photovoltaic Power Potential by Country. World Bank Group 1. https://documents.worldbank.org/en/publication/documents. Accessed 20 Jan 2025
103. Mining-Energy Planning Unit, Institute of Hydrology Meteorology and Environmental Studies (Spanish), Atlas of the energy potential of residual biomass in Colombia (Spanish). https://www1.upme.gov.co/siame/Paginas/atlas-del-potencial-energetico-de-la-biomasa.aspx. Accessed 20 Jan 2025
104. United Nations Development Programme-UND. Rural Colombia: Reasons for hope. https://www.undp.org/sites/g/files/zskgke326/files/migration/co/undp-co. Accessed 20 Jan 2025

105. Food and Agricultue Organization—FAO. Estimating greenhouse gas emissions from agriculture (Spanish). https://www.fao.org/climatechange/42726-0916f811ec1c96f58db3555f0361732f7.pdf. Accessed 20 Jan 2025
106. Andrade H, Campo O, Segura M (2015) Carbon footprint of the rice (*Oryza sativa*) production system in the municipality of Campoalegre, Huila, Colombia (Spanish). Ciencia & Tecnología Agropecuaria 15:25–31
107. Arunrat N, Sereenonchai S, Wang C (2021) Carbon footprint and predicting the impact of climate change on carbon sequestration ecosystem services of organic rice farming and conventional rice farming: a case study in Phichit province, Thailand. J Environ Manag 289:112458
108. Hoekstra AY, Chapagain AK, Aldaya MM, Mekonnen MM (2011) The water footprint assessment manual. Setting the global standard. Social Environ Account J 31 (Library of Congress Cataloging)
109. Segura M, Andrade H (2012) Carbon footprints in the coffee (*Coffea arabica* l.) productive chains with different certification standards in Costa Rica (Spanish). Luna Azul 60–77
110. Iriarte A, Yáñez P, Villalobos P, Huenchuleo C, Rebolledo-Leiva R (2021) Carbon footprint of southern hemisphere fruit exported to Europe: the case of Chilean apple to the UK. J Clean Prod 293:9

# Chapter 4
# Impacts of Climate Change on Water Resources and Land Degradation in MENA Region with Focus on Egypt

**Nader Noureldeen Mohamed**

**Abstract** Global warming will directly affect water resources, land degradation, forestry, and food production. Energy, which represents 33% of the total agriculture cost, industry and Delta ecosystem, will also be affected indirectly. Developing countries are more vulnerable to climate change. MENA countries are some of the most vulnerable regions to climate change, though they have the least water resources and high dryness and drought waves. MENA's temperatures after climate change are expected to increase by 2–2.5 °C by 2050, and precipitation is expected to decrease by 10.5% and may reach up to 30–40%, which will increase desertification processes. Egypt will be affected by the rise in sea level by the end of this Century. Sea level rises in the northern Delta will affect 4 million people who have to migrate and an area of almost 1800 $km^2$ of fertile lands will be lost. Egypt will also be impacted indirectly by dryness as a result of high evaporation potential and low precipitation in the Nile basin countries, especially in Ethiopia, which shares 85% of the total Nile water. Climate change may affect Egypt's agricultural production according to the rise in temperatures, which will affect the yields of various crops, causing a decrease in yield crop by 30% for essential crops such as wheat, rice, maize, barley, sunflowers, and soybeans. Higher temperatures mean higher water consumption by the plants to get the same regular crop yield, in addition to increased evaporation from irrigation canals, water contamination, and both water and soil salinity.

**Keywords** Climate change · MENA · Egypt · Agriculture · Water resources and land degradation

---

N. N. Mohamed (✉)
Faculty of Agriculture, Cairo University, Giza, Egypt
e-mail: nader.mohamed@agr.cu.edu.eg



A. Angelini et al. (eds.), *Security Enhancement for Climate Change Impacting Urban Resources - SECCURe*, NATO Science for Peace and Security Series C: Environmental Security, https://doi.org/10.1007/978-94-024-2345-7_4

## 4.1 Introduction

Since the Nobel Peace Prize 2007 awarded jointly to the Intergovernmental Panel on Climate Change IPCC and Albert Arnold Gore Jr, climate change has become the truth. The IPCC in 2007 stated that "...global warming and climate change is an expectation that may be considered as a level of certainty..." [1]. Nowadays, climate change is a certain true fact that should be coped and adapted with. Climate change has direct effects on all agriculture environments such as temperature, precipitation, wind and soil salinity. Climate change also affects agriculture indirectly through its impacts on energy, industry (fertilizers and pesticide manufacturers), marketing, financial services, human settlement, health, coastal zone and seawater intrusion [2]. On the other hand, the World Development Report (2010) [3] draws on an analysis of the Intergovernmental Panel on Climate Change [1] to calculate that agriculture directly accounts for 14% of global GHG emissions in $CO_2$ equivalents. Moreover, the indirect impact accounts for another additional 17% of emissions when land use and conversion for crops and pasture are included in the calculations. Agriculture's share in global GDP by 4% means that agriculture is a highly intensive greenhouse gas producer [1, 4]. Earth's average temperature is expected to rise even if the amount of greenhouse gases decreases. But the rise would be less if greenhouse gas amounts remain the same or decrease: climate models predict that Earth's average temperature will keep rising over the next 100 years even though, in some coming years, the Earth's average temperature is steady or falls. Still, the overall trend is expected to increase [5].

Climate vulnerability is often used as an integrative measure of the threats to a system and it can act as a proxy for monitoring aspects of adaptation [6, 7]. Vulnerability has been defined as "The propensity or predisposition to be adversely affected [8]. Vulnerability includes a variety of concepts, such as sensitivity or susceptibility to harm and lack of capacity to cope and adapt". O'Brien reported another definition in 2001 [9]: "The degree to which a system is susceptible to, or unable to cope with, adverse effects of climate change, including climate variability and extremes. Vulnerability is a function of the character, magnitude, and rate of climate variation to which a system is exposed, its sensitivity, and its adaptive capacity".

Developing countries are more vulnerable to climate change because they rely on agriculture for their GDP. These countries suffer from a lack of infrastructure that can respond well to increased adaptation and the lack of capital investment that is needed for adaptations. Stern report [10] projected that a 2 °C increase in average temperatures would reduce world GDP roughly by 1%. Moreover, the 2009 World Development Report [11] focused on developing countries, and expected that, without adaptation and mitigation, climate change will cause a decrease in annual GDP by 4% in Africa and the Middle East.

Rainfed agriculture represents about 83% of global agriculture and 60% of total production. Rainfed agriculture represents 95% of farmed land in sub-Saharan Africa, 90% in Latin America, 75%, 65% and 60% in the Near East, North Africa, East Asia and South Asia respectively. The water productivity (the amount of crops produced per unit of water) is low in rainfed farming systems due to the impacts of global warming and mismanagement. Land and water are often degraded, and crops may be lost due to frequent droughts or floods. In parts of sub-Saharan Africa and South Asia, productivity is particularly low, which results in food insecurity and poverty for rural communities [12, 13].

In this chapter, after a brief overview of the general concepts related to the climate change, water scarcity and agricultural connections, a more specific description of criticisms will be first reported on the MENA Region, with particular attention on the Egyptian case.

## 4.2 Effect of Global Warming on the Agriculture Sector

### *4.2.1 General View*

Agricultural production directly follows the weather and soil conditions. Weather conditions can be modified by irrigation in case of rainfall deficit or by replacing the weather humidity, such as cultivating rice in Egypt under an irrigation system, not summer rain in the high humidity areas. Greenhouses agriculture is another solution to adapt to weather conditions and water shortages, which govern temperature, humidity and solar radiation inputs. Some extreme weather cannot be managed, which leads to severe damage, such as strong winds, hail, frost, and drought [14].

### *4.2.2 Climate Change, Water, Energy, and Food Security*

The "water, food, and energy" nexus should be changed in "water, energy, food, population growth, and climate change" nexus. These nexuses are the major leading factors regulating world food production, which directly impacts food security in developing countries. The impacts of climate change on agricultural production are expected to intensify over time and vary across countries and regions. By 2030, the negative impacts of climate change on crop productivity will increase severely, especially in arid areas. This would have profound implications for food security and would lead to rising food prices [15].

### *4.2.3 Climate Change and Agriculture Sector*

The relationship between climate, water, soil, and crop production is complicated and looks like a circle, whereas each item affects the other. Global warming will increase the evaporation from all water delivery bodies, causing a lack of irrigation

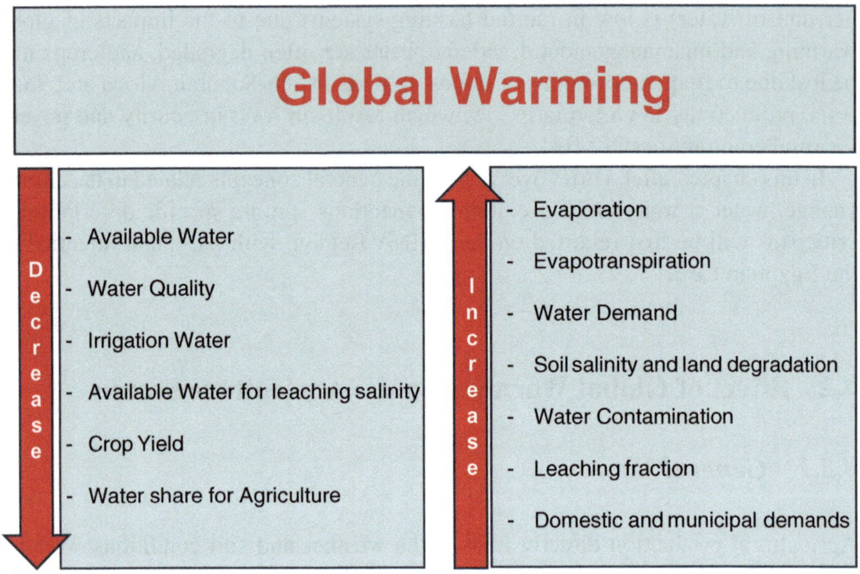


**Fig. 4.1** Global heating and water resources

water, especially in high-potential evaporation countries such as MENA regions. It will evaporate the moisture content in the soil surface. Thus, the plants will need more water and more frequent irrigation to produce the same yield crops. The soil will also need more water due to the salinity building up and salt accumulation due to the high evaporation under water shortage as a leaching fraction. Deterioration of water quality due to increased pollutant concentration under high evaporation may also exist. This will also increase land degradation, requiring more water for rehabilitation. Water scarcity in arid regions will lead to expansion in the reusing of wastewater, which contains appreciable amounts of salts, residual chemical fertilizers, and residual pesticides. This will be classified as problematic water and will increase land degradation. The following drawing diagram (Fig. 4.1) summarizes the relationship between global warming and agriculture.

### *4.2.4 Climate Change and Water Resources*

**Soluble Oxygen, water quantity and quality.** Temperature is a driver of all hydrological variables. The increase in air temperature will cause an increase in water temperatures. As water temperatures increase, water contamination will increase and the pollutants will be more concentrated. Thus, many aquatic habitats will be negatively affected. The Effects of Climate Change on Water Resources and Programs [16] can be listed as follows:

- Lowering dissolved oxygen levels in water due to the inverse relationship between dissolved oxygen and temperature.
- Increase the activity of pathogens, such as microbes, trace elements, nutrient concentration, and invasive species.
- Increase the concentrations of some pollutants such as ammonia and pentachlorophenol, due to their chemical response to warmer temperatures.
- Increase algal blooms.
- Harm of some aquatic species which are classified as temperature dependent.
- Change in the abundance of coastal and marine species and the decline in populations of some species.
- Increase rates of evapotranspiration from water bodies, resulting in shrinking of some water bodies such as rivers, irrigation canals and fresh lakes.
- Changes and shifts in the location and amount of precipitation will affect water availability and water quality.

**Sea Level Rise.** The factors driving sea level rise include:

1. Ocean water expansion due to ice melting of the Northern and Southern Pole ocean because of increases of ocean temperature.
2. The melting of mountain icecaps.
3. The melting of the Greenland Ice Sheet and the West Antarctic Ice Sheet.
4. Icecap shrinking and rising sea levels by 0.18–0.59 m.

**Precipitation Changes.** As the air temperature increases due to global warming, the evaporation rate from soil surfaces and water bodies increases, raising the humidity per cent of the atmosphere. More atmospheric moisture content means heavier downpours when it rains. While moderate increases in precipitation, there is a wider variation in the rainfall pattern, specifically during dry periods punctuated by more intense rainfall.

**Cycle of global warming in agriculture.** The impact of global warming on agriculture is expected to be that the production of the same amount of current food will need more water, according to high evapotranspiration. Moreover, human consumption of water will increase. Accordingly, the competition of the main sectors on water demands (domestic & municipal, agriculture, industry and environment) will be increased. This obliged trend will be against agriculture as the primary and most consuming sector.

**Effect of climate change on land degradation.** Agriculture relies on water first, energy second (electricity, fuel, marketing, transportation), and finally, soils. The cultivated area depends upon the amount of available and accessible water, even from rainfall or irrigation.

The IPCC Third Assessment Report [17] estimates a global increase of mean annual temperature of 0.8–2.6 °C by 2050 and 1.4–5.8 °C by 2100. The study reported a rise in yearly precipitation in high and mid-latitudes and most equatorial regions. On the contrary, a decrease in the subtropics. The report predicts flood frequency is likely to increase due to the concentration of precipitation in winter in most areas of the globe. Simultaneously, the decrease of low flows in many regions associated with higher temperatures constitutes a serious threat to the water quality. As a dynamic ecosystem, soil is the only producer of food, forage, livestock meat, fiber, pharmacist and aromatic productions. That means we should keep the soil healthy and fertile. Soil health refers to soil quality and the continued capacity of soil to function as a vital living ecosystem that sustains agricultural productivity.

Land degradation is "the lowering of the productive capacity of land" [18–20]. It thus covers the various forms of land degradation, such as adverse natural or human impacts on water resources, deforestation, and reducing the productive capacity of rangelands. From the last definitions, it is clear that land degradation is a natural or human-induced reduction of the biophysical capacity of the land, which in turn negatively affects the ability to operate at its sustainable, full productive potential and provide biodiversity services. Land degradation includes water erosion, wind erosion, soil and water salinity, nutrient depletion, physical and chemical properties degradation [18, 19]. Along with land mismanagement, climatic variability is a major driving force of land degradation. High temperatures, limited rainfall, water, and wind erosion are the main reasons for damaging interactions between land degradation and climate change.

Land degradation processes includes [21]: Landscape modification, Water and wind erosions, Soil compaction and surface sealing, Formation of salt-affected soils (salinity–alkalinity), Soil acidification, Nutrient depletion and fast burning of organic matter, Soil contamination, impacts of aridity and drought, reduce vegetation cover, reduce vegetation community functioning, Decline in biomass, Decline in biodiversity, Depletion of seed bank, Increase in weeds and grassed, Increase in invasive species, Habitat loss, Hydrological modification, Change in groundwater level/quality, Water logging, over-extraction of ground water.

## 4.3 Impact of Climate Change on MENA Region

MENA region includes 16 Arabian countries in addition to Israel and Iran (and recently Turkey), as seen in Fig. 4.2.

MENA Region is characterized by:

- One of the most vulnerable regions, though it has the least water resources and a high aridity index.
- Temperatures are expected to increase by 2–2.5 °C by 2050 [23]
- Precipitation is expected to decrease by 10.5%, but up to 30–40% in Morocco, Saudi Arabia, Yemen and UAE.
- An intensive repeat of the drought cycle will increase desertification processes.
- Runoffs are expected to be decreased by 20–30%.
- Reducing the infiltration rate and soil permeability will affect groundwater.
- Increasing urbanization (93% up to 2050) and population growth to more than 2% will put more pressure on water resources.

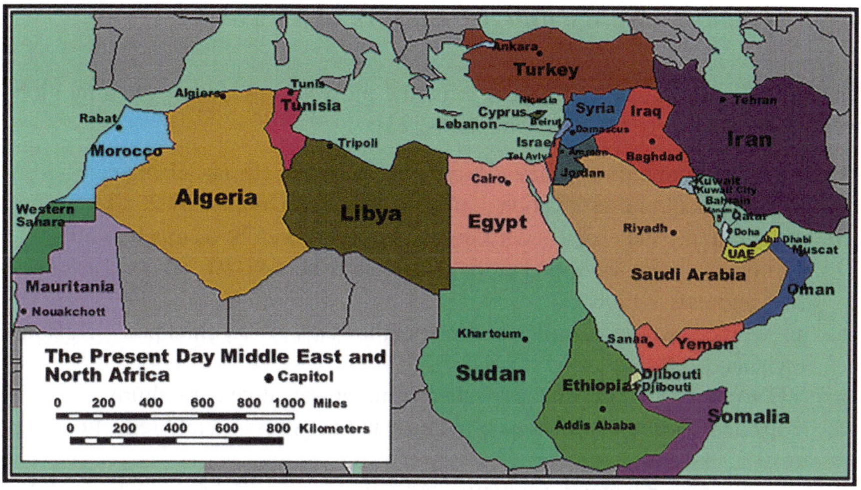


**Fig. 4.2** Representation of the MENA region [22]

- Water availability is expected to decrease by 30–50% by 2050.
- Share renewable water per capita will decrease to less than 550 $m^3$.
- Water access, efficiency and water-related disease indicators also deteriorate.
- The impacts of rising sea levels are expected to be higher than the rest of the world.
- Low-lying areas in Tunisia, Qatar, Libya, UAE, Kuwait and Nile Delta of Egypt are at risk (Fig. 4.3).

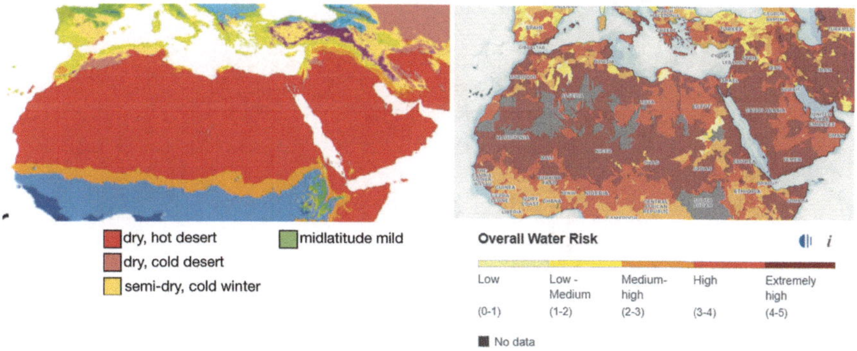


**Fig. 4.3** Köppen climate classification (extracted from [24], *left*) and water stress in the MENA Region [25]

### 4.3.1 Climate Change and Agriculture in MENA

The agriculture sector is nominated to be the most negatively impacted by global warming according to the following items [23]:

1. the agriculture sector is the most sensitive sector to climate change;
2. global agricultural production could be reduced up to 25% by 2080;
3. reduction in yield crops under rainfed agriculture can be as much as 50%;
4. the biggest losers are winter crops such as strawberry -32%, wheat -18%, rye—16% and oats—12%;
5. the winners are pearl millet + 31%, sunflower + 18%, chickpea + 15% and soybean + 14%;
6. MENA has a high dependency rate on climate-sensitive agriculture;
7. crops and livestock will face an increase in heat stress—yield reduction by up to 10%;
8. increase in crop water consumptive use will be up to 8% by 2070, with a fall in the productivity of water units;
9. high evaporation potential will increase the building up of soil salinity;
10. increase in pests under warmer weather –will increase the use of chemical pesticides and its bad effect to the soil, ground water and human;
11. the rise in sea level will threaten agriculture in low-lying deltas, particularly the Nile Delta in Egypt;
12. rangelands and forests will be more vulnerable (Fig. 4.4).

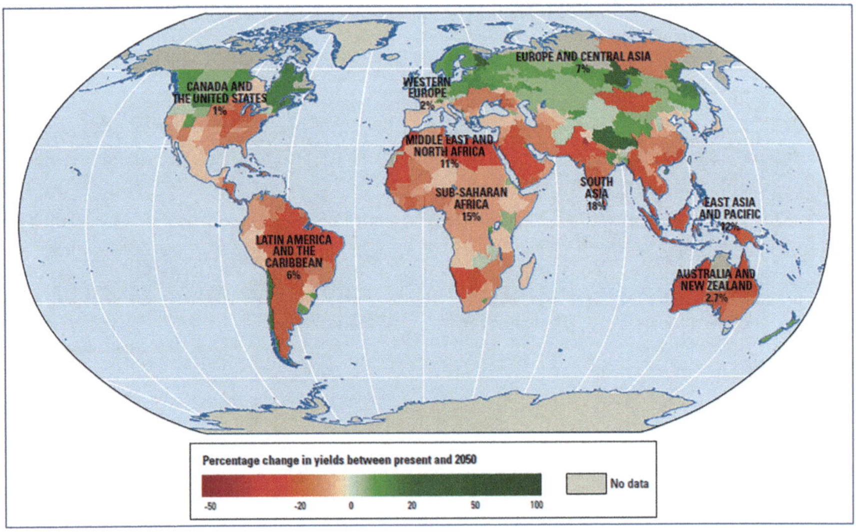


**Fig. 4.4** Climate change impacts on agricultural yield crop by 2050 [26]

### 4.3.2 Climate Change and Small Farmers

Small or family farms that have an area range between 1 and 3 hectares which represent 70–90% from the total land tenure in Africa and 75% of the in rural areas and sub-Saharan countries [27]. The community of the family farms usually lives under the poverty level and low yield crops which barely cover their demands. Thus, the adding of chemical fertilizers in these farms is usually at its minimum, and the soil suffers from nutrient depletion, decline of organic matter, physical degradation, soil compaction and high salinity.

The following impacts of climate change are expected to happen on the family farms:

- Low crop yields
- Increase plant disease and mortality of livestock, create low-quality crops and undervalue prices of their products.
- Increase food insecurity, poverty, and hunger dependency on food subsidies.
- Increase migration from rural to urban area.

Water resources in the MENA region are limited. Water scarcity in the region is already critical and it will reach severe levels by 2030. Some experts even predict that the Fertile Crescent, from Iraq, Syria to Lebanon, Jordan and Palestine, will lose its fertility before the end of the Century due to deteriorating and shortage of water supply from the major rivers. Mismanagement of soil and water will worsen the situation [28].

With continuing rising temperatures, water discharge in the Euphrates River may decrease by 30% and for the Jordan River by 80% before the end of the Century. The same trend will repeat in Egypt with water flow in the Nile River because the climate change impacts may decrease the precipitation in Ethiopia as the main upstream of the Nile River (by 85%) by 70% [29].

### 4.3.3 Water Character in the MENA Region

MENA Region is considered the world's arid region, where per capita water is predicted to be 50% by 2050 even without the effects of climate change. The region has limited options for increasing water storage since almost 90% of its fresh water resources are already stored in reservoirs. The increased water scarcity combined with more significant variability will threaten the agriculture sector, which consumes 85% of the region's water. Vulnerability is compounded by a high population growth rate and economic activity in flood-prone coastal zones and by social and political tensions that resource scarcity could heighten [3, 30].

**Egypt in General View.** Egypt belongs to the hyper-arid region, with an average precipitation of 20 mm/year [31] and evapotranspiration of 2000 mm/year, which means a hundred times more than the precipitation. The alluvial loamy land represents

3.5% of its total area and the settlement around the arable lands represents another 3%. The desert represents 93.5% of the total area. Winter precipitation is common in a Mediterranean country in the strip along the northern Mediterranean coastline, with no rains in upper Egypt. The aridity of Egypt doesn't mean it has no other sources of water, such as Nile River with a quota of 55.5 billion cubic meters a year, groundwater of 5.5 billion/year and winter rain that precipitates above the delta fertile land (4.5 million acres) reached 1.3 billion cubic meters a year. All the country and human activity rely on the Nile water by more than 90% in addition to reuse of 20 billion cubic water of wastewater, even treated or diluted by fresh water. The absolute total fresh water is about 62.3 billion $m^3$ per year. The agriculture sector consumes about 62.5 bcm/year or 80 to 85% of the total available water (fresh and reuse) and 103% of the Nile share water [32, 33]. Agriculture is one of Egypt's most critical sectors due to its effective food production and national income. It is one of the most labor-intensive industries, directly and indirectly providing employment for 55% of Egyptians labor power and utilizing 30% of the official labor force [34]. Agriculture contributes about 12% of the national GDP (in 2017; decreased to 11% in 2019) and 20% of all outside trade profit [34]. From the total cultivated area, about 3,276,000 ha (7.8 million acre) are located in the Nile valley and Delta, while 210,000 ha (500,000 feddan) are in some valleys in the western desert (oases). The alluvial soils surrounded the Nile stream and its branches with concentrates in the Delta Regions by 1.89 million ha (4.5 million acre). The Nile Valley has another 2 million acres of alluvial soil, and the newly reclaimed soil in the Nile fringes and desert has reached 2.5 million acres [35]. For crops in Egypt, a majority of farmland is cultivated annually, with only 6% of total crops planted permanently. Almost all of the nation's cropland is irrigated agriculture. Irrigated areas in Egypt total about 3,422,178 ha, of which 88.5% use surface irrigation [36]. Drip irrigation is seven percent, and five percent is sprinkler irrigation. The main source of irrigation water is surface water (85%), 11% groundwater, and the remaining six percent is a mix of surface water and groundwater [35, 36]. The next Table 4.1 shows the progress of cultivated area and crop group from 1980 to 2009.

The impacts of climate change will hit little rainfall, but it will increase the optionality of evapotranspiration, as the country primarily counts on irrigated agriculture. The effective impacts of climate change will be in the upper Nile, especially in Ethiopia, which will affect the water discharge to Egypt's. El-Sayed [40] identified water resources as one of the three most vulnerable sectors to climate change in Egypt, along with the Mediterranean coastal strips and agricultural production. Climate change will affect the whole 11 countries of the Nile River basin.

**Vulnerability of Egypt and Agriculture Sectors to Climate Change.** Several reasons respond to Egypt's vulnerability to climate change, especially in the Delta lands: high population growth, urbanization extension on agricultural land, slow economic rate, seawater intrusion to the groundwater and soil and waterlogging. Climate change may affect Egypt's agricultural production in different ways. Warmer weather will affect essential crop yields, causing a 10–30% decrease in wheat, rice, maize, barley, sunflowers, and soybeans [41]. However, some research expected that

Table 4.1 Total harvested area by crop group (M ha) from 1980 to 2009 [37–39]

| Crop group | Year 1980–84 | | 1990–91 | | 2000–01 | | 2006–07 | | 2007–08 | | 2008–09 | |
|---|---|---|---|---|---|---|---|---|---|---|---|---|
| | Area ha | % | Area ha | % | Area ha | % | Area ha | % | Area ha | % | Area ha | % |
| Cereals | 2 | 42.6 | 2.25 | 46.2 | 2.65 | 46.1 | 2.96 | 43.6 | 2.97 | 42.4 | 2.98 | 42.9 |
| Legumes | 0.14 | 2.9 | 0.15 | 3.2 | 0.17 | 3 | 0.12 | 1.8 | 0.11 | 1.6 | 0.13 | 1.9 |
| Fibers | 0.48 | 10.2 | 0.4 | 8.3 | 0.31 | 5.4 | 0.27 | 4 | 0.25 | 3.6 | 0.22 | 3.2 |
| Sugar crops | 0.11 | 2.4 | 0.13 | 3.6 | 0.19 | 3.3 | 0.21 | 3.1 | 0.23 | 3.3 | 0.25 | 3.6 |
| Oil crops | 0.08 | 1.8 | 0.03 | 1.9 | 0.12 | 2.1 | 0.03 | 0.4 | 0.02 | 0.3 | 0.02 | 0.3 |
| Fodder crops | 1.28 | 27.3 | 1.13 | 23.1 | 1.18 | 20.7 | 1.18 | 17.4 | 1.15 | 16.4 | 1.17 | 16.9 |
| Fruits | 0.17 | 3.6 | 0.23 | 4.8 | 0.48 | 8.4 | 1.37 | 20.2 | 1.4 | 20 | 1.43 | 20.6 |
| Vegetables | 0.43 | 9.2 | 0.48 | 9.9 | 0.7 | 11 | 0.63 | 9.4 | 0.87 | 12.4 | 0.74 | 10.6 |
| Totals | 4.58 | 100 | 4.81 | 100 | 5.75 | 100 | 6.77 | 100 | 6.37 | 100 | 6.94 | 100 |

the impacts on potato and cotton yields are positive, with increases ranging from 15 to 40% [41, 42]. In addition, higher temperatures will likely negatively affect crops indirectly by increasing water consumption. The ESNC [43] stated several studies expected to increase the amount of irrigation water by 5–13%. The same study predicted that climate change would cause shifts in agro-climatic zones in Egypt, with other shifts in crop suitability, pests and diseases. These impacts are still in their beginning, but Fahim [44] reports that the incidence of late blight, the most damaging disease-causing yield hazards in tomatoes, has increased in the last three decades, probably due to climate change.

Scientific data on the impact of climate change on livestock and fisheries in Egypt is still limited. A study [43] reported that blue tongue disease, common in humid cold regions in addition to the fever of the Rift Valley disease of equatorial Africa, has recently emerged in Egypt. These may be related to climate change [45], even though one is related to cold weather and the other is related to equatorial weather. Decreasing green forage crop production (Egyptian berseem) would affect the needed feed for livestock [43]. Climate change may also impact coastal resources through sea level rise (SLR), which is already noticed in the Nile Delta coast due to coastal subduction and reduced sediment loads [46]. This may lead to the loss of some productive agricultural lands in the north Delta along the Mediterranean coast, with negative impacts on fish production in the interference area between fresh water from the Nile River and saline Mediterranean water since 20% of Egypt's fish production comes from open fishing on the Mediterranean and its lakes (Egypt has five Mediterranean lakes) [34, 43–47].

**The reasons and types of land degradation in Egypt**

1. Water contamination: According to the report of the Ministry of Environment 2009 [48], the percentage of sanitary access in the rural area ranged between 10 and 12%. Most of these rural areas drain this sanitation to the agriculture drains and some irrigation canals. In addition, they discharge some of its household sanitation into relatively shallow groundwater (less than 20 m in depth). This contaminated shallow groundwater is used again, even in irrigation or drinking (by house pumping). The same report also stated that almost 102 industrial factories drain their wastewater in the drainage and irrigation canals (reduced in 2022 to 12 only).
2. Salinity from irrigation water: The average salinity concentration of Nile water is 0.5 kg/m at Giza Nile [49], and the rate of surface irrigation ranged between 6000 and 7000 $m^3$ acre/year. This amount adds 3–3.5 ton of salts to each acre a year. The reuse of agricultural drainage water in the Delta lands at a rate of 10 billion cubic meters/year, with an average salinity concentration of 1.5 kg/ $m^3$ (1500 ppm), adds about 10.5 tons of salts/acre *per* year. The Nile valley of Upper Egypt, delivers its drainage water to the Nile stream, which means that the Nile stream works as both an irrigation and drainage canal. The deep water shortage in Egypt doesn't provide any additional water over the irrigation quota to use as a leaching fraction or as a leaching requirement.
3. Water logging: delta lands represent 62% of the total Egyptian alluvial land, where the depth of the water table ranges between 1 and 1.5 m with poor drainage conditions. The capillary rise in these clayey soils under warm weather brings this poor-quality water to the soil surface, creating salinity conditions [49].
4. Sea water intrusion: All the Delta is completely intruded by the saline water of the Mediterranean and reaches the middle of the Delta 100 km long from the Mediterranean. Thus, the saline soils in the North Delta reach 47%, and in the middle Delta 37%, with 24% in the South Delta, while the average soil salinity in all Delta lands (4.5 million acres) is 37% [50] (Fig. 4.5).
5. Overuse of chemical Fertilizers which have a high salt index, especially Nitrogen fertilizers.
6. Mismanagement practice of alluvial soils due to intensive cultivation.

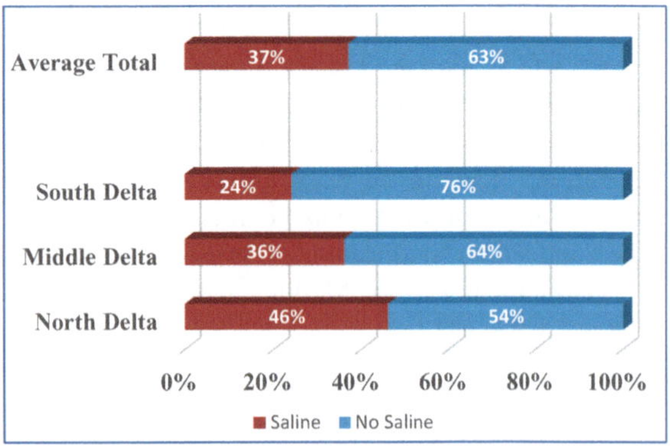


**Fig. 4.5** Salinity distribution in the Nile delta under effect of sea water [41]

7. Urbanization in the rural areas which reached 12% of the Delta lands instead of 3% in 60 years [41].
8. Nutrient depletion: especially under intensive and continuous cultivation and over-cultivation in the small farm. The small farms in Egypt that own less than one acre represent 50–70%, and from one to less than three acres ranged between 70 and 90%. The high poverty percent in these areas prevent them from adding the recommended amounts of chemical fertilizers to keep the soil healthy and fertile.
9. Groundwater exploitation: where most of the desert farms which rely on the groundwater wells suffer from the lowering of the groundwater level due to the high extraction rate of groundwater that exceeds the natural recharge capacity. This leads to the depletion and salinization of groundwater.
10. Reduction in soil organic matter or burn it according to hyper-arid weather (organic matter usually less than 1% in Egypt land), with an associated decline in soil biodiversity.
11. Degradation of soil physical properties as a result of reduction of organic matter or sodicity results in soil compaction (structure, aeration and water-holding capacity may be affected).
12. Formation of soil sealing and surface soil crust.
13. Wind erosion in the newly reclaimed desert land.
14. Buildup of toxic substances—e.g. pollution, overdose of fertilizers, especially phosphorus fertilizers, which contain appreciable amounts of cadmium.
15. The creation of salt-affected soils (saline, saline-sodic, sodic, alkaline soils), especially in the lands adjacent to the Mediterranean coast and in some other interior area in deep Egypt is very common. The main reason is the use of Nile water in irrigation, which is rich in carbonate and bicarbonate ions under the condition of low saline content in the soil.
16. Heavy clay percentage in several areas of the Delta lands, which have low infiltration rates and low soil permeability, causes less aeration and reduces conditions that affect some nutrients such as iron and manganese.

## 4.4 Egypt and Climate Change

### *4.4.1 Water Resources in Egypt*

Water use in the agriculture sector exceeds the total Nile water share by 103% (62.5 BCM/year of the Nile quota of 55.5 BCM) and the total water use is 117% of the total water resources (groundwater, rain, reuse), with a dependency ratio of 97% [50]. Irrigation water is delivered through a net of irrigation delivery systems that reaches 33 thousand kilometres long of open non-cemented irrigation canals. Surface irrigation is legally restricted in the newly re-claimed desert soils. However, overuse of water at the farm fields is still a major problem for wasting water in Egypt. According to the Central Agency for Public Mobilization and Statistics (CAPMAS)

[51] the wastewater within the irrigation delivery system from south Egypt up to Delta through 33.000 longs reaches 19 billion cubic meter/year in addition to low irrigation efficiency of surface flood irrigation not exceeding 50%, increased to 74% by reuse.

### *4.4.2 Climate Change Impacts on Water Resources Inside Egypt*

Water shortage in Egypt will increase due to the impact of climate change. These impacts will affect food, energy, urban, and environmental and biodiversity systems. Some other drivers will increase water demand and deepening the water gap under the effects of climate change which are [33]:

1. High population growth rate
2. Expecting increasing incomes of people and country (such as emerging economies in south Asia) which affects by:
    a. Change in food habits toward meat, poultry, fish, sugar, and fruits.
    b. Overall increase of demand in products and services.
    c. Higher consumption
    d. Increase of wastes
3. Migration to cities and urbanization will converge upon the water demand, which rises exponentially while the supply becomes more uncertain.

The World Development Report [26] and UNDP [46] expect a decline in the GDP in all Arab and Middle East regions (including Egypt) by 6% by the year 2050 and even apply the modern technology of rising water use efficiency in all sectors according to impacts of climate change and water scarcity, as seen in Fig. 4.6. With better water resource management, some regions stand to see growth accelerate as much as 6%. This effect will be worse in the rain-fed agriculture regions, which are occupied mostly by poor people and family farms due to the difficulty of achieving better water management for this type of cultivation.

### *4.4.3 Climate Change Impacts on the Nile Basin and Nile River Flows*

The Nile Basin is highly vulnerable to the impacts of climate change for several reasons. Poor basin communities have limited ability against adverse impacts of climate variability. The total Nile water discharge is limited and out of the top 10 world rivers (84 BCM/year compared with 5550 BCM for Amazon River). These discharges are highly sensitive to any little rainfall, but the Nile Basin consists of a

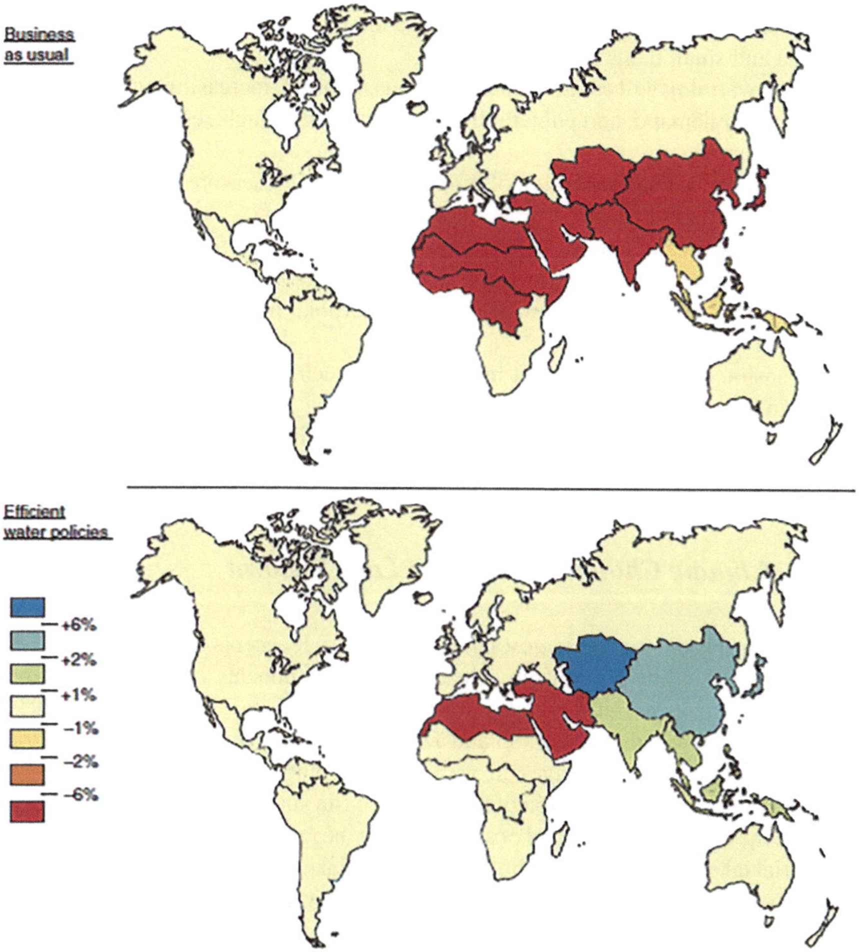


**Fig. 4.6** Impacts of water scarcity on GDP in year 2050 under current policy (up) and increasing of water uses policy (down) [52]

number of distinct sub-basins that each respond quite differently to possible climatic variations [53].

### *4.4.4 Possible Impacts of a Warming Trend in the Nile Basin*

Climate change will primarily affect the average temperature and precipitation, which drives the seasonal flows of water, lakes, rivers, wetlands, and groundwater aquifers. Global warming will have several impacts, which include [53] such as:

- Higher expected evaporation will increase losses from dam reservoirs (almost 10 mega and small dams in the Nile basin).
- Global warming will create higher evapotranspiration, increasing water consumption, water demand, and vulnerability to aridity and drought, especially in rainfed agriculture.
- More frequency and repetition of drought and long dry season spells will increase drought impacts.
- More frequent and intense severe rainstorms will increase flood risk and damage.
- Higher water temperature in the Equatorial Lakes strengthens thermal stratification, increases algal bloom, accelerates microbial mineralization, and reduces oxygen dissolution.
- Expansion of the range of heat-borne diseases such as malaria to higher altitudes due to the warm temperatures.
- Sea-level rise that could threaten fertile soils in the Nile Delta along the Mediterranean and the cities along the Indian Ocean coast in Kenya and Tanzania.

### *4.4.5 Climate Change Impacts on Lake Victoria*

Lake Victoria is the second largest lake in the world. It accepts their importance due to its role as a White Nile origin. The principal components of the Lake Victoria water balance are rainfall and evaporation from lake water. For the period 1956–78, the balance was contributed 85 and 78% of its water. This makes Lake Victoria highly sensitive to any changes in warming and climate change, as recorded by the large inter-annual fluctuation of historic net basin supply. Global warming would increase the evaporation rate and possibly impact the rain pattern on the plateau of the equatorial lakes. The impact of climate change on lake levels will affect Victoria Nile flows (the name of the upper white Nile river). However, climate change will change Lake Victoria outflow, impacting hydropower production on the Victoria two dams (own and Kiira dams) [54]. It will also affect the extent and seasonal fluctuation of the Sudd swamps in South Sudan. However, the swamp and its wetlands will considerably dampen the effects of possible changes in the Victoria Nile, or the migration of the Intertropical convergence zone, which critically affect the weather vulnerability in the Nile region.

The Eastern sub-basin and the other eastern Equatorial Lakes Plateau sub-basin of the Nile River, including Ethiopia, Eritrea, Kenya, Uganda and Tanzania, are highly vulnerable to climate change in the agriculture sector that significantly contributes to the sub-region's economies. About 80% of the population in East Africa relies primarily on agriculture, contributing to 40% of the sub-region's GDP. Climate change will significantly affect the agricultural sector and ultimately reduce yields of subsistence crops, cash crops and the livestock sector, especially the rainfed ones. Areas almost face water stress conditions and are predicted to become drier, which may cause disputes and conflicts [54].

### 4.4.6 Climate Change Impacts on Blue Nile

The predicted impacts on climate and potential evapotranspiration and streamflow of the main Abay (Blue Nile) and Gilgel Abay rivers (means Junior or Abay baby) of the eastern Nile sub-basins are evaluated using downscaled (medium-high) and (medium-low) emission scenarios [55]. The study showed a consistent increasing trend for both minimum and maximum temperature for all time horizons (2030s, 2060s and 2080s) with a higher rate of increase towards the end of the Century [55]. Results also show that the rate of change in maximum temperature is higher than the rate of change in minimum temperature. This trend will be responsible for alterations of the hydrological cycle by increasing the evapotranspiration by 2% in the 2030s, 4.7% in the 2060s and 7.8% by 2100. The increase in evapotranspiration suggests increasing crop water consumption. Moreover, larger storage structures are needed to offset the decline in rainfall during the major rainfall season. The results indicated that this change in the water balance component could severely affect the stream flow of both Abay rivers (main and Junior). The average annual streamflow could decrease in future time horizons compared to the current period. On a seasonal basis, the wet season flow will decrease for the coming Century, reducing by 19% and 17% for the main Nile River and Gilgel Abay (which represent 59- 64% of the total main Nile water), respectively, at 2100. The dry season flow may increase by 64% and 65% for main Abay and Gilgel Abay, respectively. The monthly and seasonal version of stream flow is relatively higher than the annual flow variation [56].

### 4.4.7 Climate Change and Nile Flow to Egypt

According to Hany Mostafa et al. [57], the construction of 10 dams on the Eastern Nile subbasin would reduce the average annual inflow and energy production at Lake Nasser of High Aswan Dam by 18%, 4% and 4%, respectively in addition to increase the evaporation from the lake too. However, climate change can cause complete depletion and dryness of Lake Nasser. The average annual inflow received at Aswan High Dam could be reduced due to the impacts of climate change are estimated to be 24%, 35% and 36% for the near future (2011–2040), intermediate future (2041–2070) and far future (2071–2100), respectively. Moreover, the average annual evaporation reductions due to climate change will be increased from 4 to 29%, 62% and 63% for the near, intermediate and far future, respectively.

Increasing the water loss by evaporation will affect the Nile water quality by increasing the concentration of Nile water salinity. These impacts may affect the input of drinking water stations, industrial pump station irrigation, navigation and energy generation.

A study by Strategic Foresight Group [58] confirmed that the total flow of the Nile River through the last 150 years was 42 BCM at its minimum (1913–14), and 150 BCM as maximum (1878–79), with a mean annual flow from 1900 to 2004 of 85.31

BCM/year. Climate change models predicted an increase or decrease of 15–20% in precipitation on a yearly basis. Climate change in the Nile Basin, which will mainly affect the agriculture sector, can lead to a decline in food production by up to 50% in Ethiopia, Kenya, Tanzania, Uganda and Sudan, which applied rainfed agriculture by more than 85%. The temperature across the Nile Basin is expected to increase by 1.5–2.1 °C by 2050, which means that almost the entire Nile region may belong to arid to semiarid areas instead of subhumid within the next 30–40 years. This could significantly reduce agricultural land production [58].

The study concluded that the impacts of climate change will affect water flow and quality and weaken food security in the region. Food insecurity is expected to increase in the next 30–40 years due to a decline in food production by 30–50%. This will be due to the expected to be caused by rising temperature, climate change, drought and reduction with shortening in the rainy season to a narrow and intense period. This can also lead to a drop in food production by 30 to 50%. During the same period, the population of the Nile Basin will be doubled [26], which will deep the food insecurity.

## 4.5 Conclusions and Recommendations

Egypt suffers from a deep gap in its water resources reaching 45 billion cubic meter/year. This water shortage pushes Egypt to import 65% of its essential needed food. Climate change will hit Egypt hard, the Nile Delta which represents 62% of the Egyptian cultivated lands, and 72% of agriculture production, becomes under threat from the Mediterranean sea level rise which may affect 1800 $km^2$ of its fertile lands by being submerged, causing a migration of 4 million people. Global warming and high temperatures may reduce internal water resources by 6% especially from evaporation from its long open irrigation canal which extends forto 33,000 km long. Moreover, food production may decrease by 11–20% due to high temperature and water shortage. On the other hand, the decline of rainfall in both Blue Nile, which is driven by Ethiopia and Victoria Lake, the origin of White Nile, may decrease the total Nile flow by 20%. An additional reduction in Nile water flow is expected due to the effect of the mega new dam in Ethiopia (Great Ethiopian Renaissance Dam) with its large water storage capacity of 74.5 BCM from a small Blue Nile with a total flow as low as 49 BCM/year. This mega-dam is projected to cause a decrease in the Egyptian water share quota by 12 BCM/year. Accordingly, Egypt needs to adopt smart and modern agriculture to cope with climate change and deep water shortage and produce more food with less water.

## References


1. Solomon S, Qin D, Manning M, Chen Z, Marquis M, Averyt KB, Tignor M, Miller HL (2007) Climate change 2007: the physical science basis. In: Contribution of working group I to the fourth assessment report of the intergovernmental panel on climate change (IPCC). Cambridge University Press, Cambridge, United Kingdom and New York, NY
2. Edame GE, Fonta WM (2007) Climate change, food security and agricultural productivity in Africa: issues and policy directions
3. Jones PD, Mann ME (2004) Climate over past millennia. Rev Geo-phys 42:RG2002
4. Kundzewicz ZW, José Mata L, Arnell N, Asanuma J, Betts R, Co-hen S, Becker A, Bruce J, Mata L, Arnell N, Döll P, Kabat P, Jimé-nez B, Miller K, Oki T, Sen Z, Shiklomanov I, Parry M, Canziani O, Palutikof J, van der Linden P, Hanson C (2007) Freshwater resources and their management. In: Kundzewicz ZW, Mata LJ, Arnell NW, Döll P, Kabat P, Jiménez B, Miller KA, Oki T, Sen Z, Shiklomanov IA (eds) Climate Change 2007: impacts, adaptation and vulnerability. Contribution of working group II to the fourth assessment report of the intergovernmental panel on climate change. Cambridge University Press, Cambridge, pp 173–210
5. NASA. https://www.nasa.gov/audience/forstudents/5-8/features/nasa-knows/what-is-climate-change-58.html
6. Agassiz A (2001) Climate change 2001: impacts, adaptation, and vulnerability
7. Kelly PM, Adger WN (2000) Theory and practice in assessing vulnerability to climate change and facilitating adaptation. Clim Change 47(4):325–352
8. Field CB, Barros VR, Dokken DJ, Mach KJ, Mastrandrea MD, Bilir TE, Chatterjee M, Yuka KLE, Estrada O, Genova RC, Girma B, Kissel ES, Levy AN, Maccracken S, Mastrandrea PR, White LL (2014) Climate change 2014 impacts, adaptation, and vulnerability part A: global and sectoral aspects working group II contribution to the fifth assessment report of the intergovernmental panel on climate change (IPCC)
9. O'Brien CM, Fox CJ, Planque B, Casey J (2000) Climate variability and North Sea cod. Nature 404:142
10. Stern NH (2008) The economics of climate change: the stern review. Cambridge University Press, Cambridge; New York
11. World development Report (2008) Development and climate change, a strategic framework for the World Bank Group
12. Energy Transport and Water Department Water Anchor (ETWWA) (2010) Improving water management in rainfed agriculture: issues and options in water constrained production systems. http://siteresources.worldbank.org/INTWAT/Resources/ESWWaterManagementRainfed_final.pdf
13. Wani SP, Rockström J, Oweis TY (2009) Rainfed agriculture: unlocking the potential. CABI. http://www.iwmi.cgiar.org/Publications/CABI_Publications/CA_CABI_Series/Rainfed_Agriculture/Protected/Rainfed_Agriculture_Unlocking_the_Potential.pdf
14. Parry ML, Canziani OF, Palutikof JP, van der Linden PJ, Hanson CE (2007) Summary for policymakers. Climate change 2007: impacts, adaptation and vulnerability. In: Contribution of working group II to the fourth assessment report of the intergovernmental panel on climate change, IPCC. Cambridge University Press, Cambridge
15. Hayduk D (2013) The state of food and agriculture: climate change, agriculture and food security. Food and Agriculture Organization of the United Nations Rome, FAO
16. Watershed Academy Web and US Environmental Protection Agency (2013) The effect of climate change on water resources and programs. https://cfpub.epa.gov/watertrain/pdf/modules/Climate_Change_Module.pdf
17. Mackay A (2008) Climate change 2007: impacts, adaptation and vulnerability. Contribution of working group II to the fourth assessment report of the intergovernmental panel on climate change. J Environ Qual 37:2407–2407
18. FAO (1979) A provisional methodology for soil degradation assessment. FAO, Rome, 73 pp

19. UNEP (1992) Desertification, land degradation [definitions]. Desertif Control Bull 21
20. Darkoh MBK (1998) The nature, causes and consequences of desertification in the drylands of Africa. Land Degrad Dev 9:1–20
21. Orr AL, Cowie VM, Castillo Sanchez P, Chasek ND, Crossman A, Erlewein G, Louwagie M, Maron GI, Metternicht S, Minelli AE, Tengberg S, Walter S (2017) United Nations convention to combat desertification (UNCCD), scientific conceptual framework for land degradation neutrality, a report of the science-policy interface, Bonn, Germany
22. https://www.pinterest.co.uk/pin/717901996820657072/
23. Barghouti S (2009) Climate change in the MENA region: effects on water and agriculture. addressing climate change in the Middle East and North Africa (MENA) regions launch conference, 13–14 May 2009, Rome, Italy
24. Beck H, Zimmermann N, McVicar T, Vergopolan N, Berg A, Wood EF (2018) Present and future Köppen-Geiger climate classification maps at 1-km resolution. Sci Data 5:180214
25. World Resources Institute (WRI) (2019) Aqueduct tools, WRI Aqueduct Web. https://www.wri.org/aqueduct. Accessed Mar 2025
26. World Bank (2010) World development report 2010: development and climate change. © World Bank. http://hdl.handle.net/10986/4387
27. Moyo S (2016) Family farming in sub-Saharan Africa: its contribution to agriculture, food security and rural development. International Policy Centre for Inclusive Growth (IPC-IG) Family farming in sub-Saharan Africa: its contribution to agriculture, food security and rural development. Working Paper No. 150
28. Göll E (2017) Future challenges of climate change in the MENA region. In: Project of Middle East and North Africa Regional Architecture: mapping geopolitical shifts, regional order and domestic transformations (MENARA)
29. Desta H (2015) Overview on degradation of land due to over grazing in Ethiopia. http://www.mi.imati.cnr.it/sara/biodiversita/Overview%20on%20Degradation%20of%20Land%20Due%20to%20Overgrazing%20in%20Ethiopia.pdf
30. Müller C, Bondeau A, Popp A, Waha K, Fader M (2009) Development and climate change climate change impacts on agricultural yields
31. African water atlas. United Nations Environment Programme (2010)
32. Tilahun H, Schmidt E (2012) FAO-Nile synthesis report. In: Information products for Nile Basin water resources management. FAO-Nile Basin Project GCP/INT/945/ITA 2004 to 2009. Food and Agricultural Organization (FAO) of the United Nations, Rome. Spatial analysis of livestock production pattern in Ethiopia, Ethiopia strategy support program 2, ESSP2, working paper 44. http://www.fao.org/nr/water/faonile/products/Docs/Reports/Synthesis.pdf
33. FAO (2011) Climate change water and food security; FAO water report 36
34. Sustainable Agricultural Development Strategy (SADS) Towards 2030 (2009) Agricultural research & development council. Arab Republic of Egypt, Ministry of Agriculture & Land Reclamation
35. Mohamed NN (2017) Use of groundwater in Nile Alluvial soils and their fringes. In: Negm AM (ed) Groundwater in the Nile Delta. Springer International Publishing AG (Hdb Env Chem)
36. FAO (2012) Coping with water scarcity An action framework for agriculture and food security. FAO water reports, vol 38
37. Egyptian Ministry of Agriculture and Land Reclamation (EMALR) (2019) Bulletin of agricultural statistics. Egyptian Ministry of Agriculture and Land Reclamation
38. FAO Egypt Country Profile (2011)
39. Mohamed NN (2017) Land degradation in the Nile Delta. Hdb Env Chem 55: 235–264 [Springer International Publishing AG 2016, Published online: 4 November 2016. Negm AM (ed.) The Nile Delta (2017)].
40. El Sayed LM (2012) Determining an optimum cropping pattern for Egypt Determining an optimum cropping pattern for Egypt. The American University in Cairo
41. Attaher SM, Medany MA, Abou-Hadid AF (2009) Possible adaptation measures of agriculture sector in the Nile Delta to climate change impacts. Adv Sci Res 3:123–126

42. El Ramady HR, El-Marsafawy SM, Lewis LN (2013) Sustainable agriculture and climate Changes in Egypt. In: Lichtfouse E (ed) Sustainable agriculture reviews, vol 12. Springer Netherlands, Dordrech
43. Egypt Second National Communication (ESNC) under the UNFCCC (2010) Environmental Affairs Agency, Cairo, Egypt. http://unfccc.int/resource/docs/
44. Fahim MA, Hassanein MK, Abou-Hadid AF, Kadah MS (2011) The impacts of climate change on the widespread and epidemics of some tomato diseases during the last decade in Egypt. Acta Hort 914:317–320
45. Taqi MO (2012) Milk productivity and physiological parameters of Egyptian buffalo under climatic conditions of Giza and Qena governorates. Master Thesis. Department of Animal Production, Faculty of Agriculture, Cairo University, Egypt
46. United Nations Development Programme UNDP (2013) Egypt Office. Potential impacts of climate change on the Egyptian economy
47. Sušnik J, Vamvakeridou-Lyroudia LS, Baumert N, Kloos J, Renaud FG, La Jeunesse I, Mabrouk B, Savić DA, Kapelan Z, Ludwig R, Fischer G, Roson R, Zografos C (2015) Interdisciplinary assessment of sea-level rise and climate change impacts on the lower Nile Delta, Egypt. Sci Total Environ 503–504:279–288
48. Ministry of state for environmental affairs, Egypt (2010) Egypt State of the Environment Report
49. Mohamed NN (2017) Management of salt-affected soils in the Nile Delta. In: Negm AM (ed) Groundwater in the Nile Delta. Springer International Publishing AG (Hdb Env Chem)
50. Lowder SK, Skoet J, Singh S (2014) What do we really know about the number and distribution of farms and family farms worldwide? ESA working paper no. 14-02. Food and Agriculture Organization of the United Nations
51. Central Agency for Public Mobilization and Statistics (CAPMAS) (2020) Egypt water resources
52. World Bank Group (2016) High and dry: climate change, water, and the economy. © World Bank. http://hdl.handle.net/10986/23665
53. Nile Basin Initiative (NBI) (2014) The State of the River Nile Basin 2012
54. (2014) Adapting to climate change through land and water management in Eastern Africa: results of pilot projects in Ethiopia, Kenya and Tanzania. Food and Agriculture Organization of the United Nations, FAO
55. Wale Worqlul A, Taddele YD, Ayana EK, Jeong J, Adem AA, Gerik T (2018) Impact of climate change on streamflow hydrology in headwater catchments of the upper Blue Nile Basin, Ethiopia. Water 10:120
56. Mohamed NN. CLIMATE2020—the worldwide online climate conference category 3-education, awareness & adaptation impacts of climate change on water resources and land degradation on Nile Basin
57. Mostafa H, Saleh H, El-Sheikh MA, Kheireldin KA (2016) Assessing the impacts of climate changes on the Eastern Nile flow at aswan climate change and its impacts on water resources. J Am Sci 12(1)
58. Strategic Foresight Group (2013) Blue Pease for the Nile. India. http://www.strategicforesight.com/publication_pdf/11374Nile%20concise.pdf

# Part II
# Modern Concepts of Water Depuration and Use Towards a Secure Resource

# Chapter 5
# Urban Runoffs as a Source of Emerging Micropollutants in Aquatic Environments

**Mira Petrovic, M. Isabel Cadena-Aizaga, and Sergi Sabater**

**Abstract** Freshwater ecosystems are continuously impacted by contaminants from urban runoff, a process that becomes more pronounced during rainfall events. This runoff originates from various urban surfaces, including roads, pavements, residential, industrial, and green areas, carrying a diverse array of micropollutants that pose a significant threat to surface water quality and aquatic ecosystem health. Climate change has further exacerbated these challenges by increasing the frequency and intensity of extreme weather events in urban areas. These conditions enhance the mobilization of pollutants through rainfall wash-off, leading to contamination levels that exceed normal conditions. Additionally, prolonged dry periods allow pollutants to accumulate on urban surfaces, which are then flushed into urban streams and water bodies during subsequent rainfall, further impacting biodiversity and water quality. To address these issues, we propose a comprehensive approach to assess the presence of micropollutants in urban runoff during extreme weather events. Our methodology employs liquid chromatography coupled with high-resolution mass spectrometry (LC-HRMS), integrating both target and suspect screening methods to identify and quantify a broad range of micropollutants. Understanding the composition of micropollutants in urban runoff is crucial for developing effective mitigation strategies. These findings will contribute to improved water management and the protection of freshwater ecosystems in the face of evolving climate conditions.

**Keywords** Urban runoffs · Micropollutants · Target and suspect screening · HRMS

M. Petrovic (✉) · M. I. Cadena-Aizaga · S. Sabater
ICRA-Catalan Institue Fro Water Research, Girona, Spain
e-mail: mpetrovic@icra.cat

S. Sabater
e-mail: ssabater@icra.cat

M. Petrovic
Catalan Institution for Research and Advanced Studies (ICREA), Barcelona, Spain

S. Sabater
Institut d'Ecologia Aquàtica (IEA), Universitat de Girona, Girona, Spain

 
A. Angelini et al. (eds.), *Security Enhancement for Climate Change Impacting Urban Resources - SECCURe*, NATO Science for Peace and Security Series C: Environmental Security, https://doi.org/10.1007/978-94-024-2345-7_5

**Table 5.1** Typical micropollutants in urban runoffs

| Urban runoffs | Sources | Specific groups |
|---|---|---|
| Organic micropollutants | Gardening; elution of paints and additives from building materials and surfaces; elution from the pavement surface; Elution of chemicals used at roofs and surfaces to prevent mold and plant growth; droppings from pets | Pesticides, PFAS, alkylphenols, phthalates, veterinary drugs |
| | Deposition of exhaust gases and particles; mechanical wear of pavement surface; wear of tires and brakes | Hydrocarbons, PAHs, BTEX benzotriazoles |
| Inorganic micropollutants | Washing of corroded metal surfaces, roof washing; catalytic converters; wear of tires and brakes | Metals |
| Microplastics | Elution from the pavement surface; release from artificial turfs or recycled rubber mulch (e.g., in playgrounds); wear of tires and brakes | |

## 5.1 Urban Runoffs as Diffuse Source of Pollutants

Urban runoff is an important diffuse source of pollution, as rainwater washes pollutants from urban surfaces and deposits them from the atmosphere. This results in a complex mixture of pollutants, including trace metals, salts, suspended solids, plastics, and a variety of organic micropollutants [1, 2]. These pollutants are transported in pulses during rainfall events, making their management challenging.

Urban land use strongly influences pollution, with high-traffic areas being major contributors of micropollutants. For instance, smooth asphalt pavements enhance water flow, facilitating the transport of traffic-related contaminants [3]. While traditional monitoring has focused on suspended solids, nutrients, and metals, micropollutants have gained increasing attention due to their potential toxicity, persistence, and bioaccumulative properties. These include polycyclic aromatic hydrocarbons (PAHs), plasticizers, flame retardants, corrosion inhibitors, pesticides, veterinary drugs, surfactants, and tire-related compounds [4]. Some micropollutants, such as perfluorinated substances (PFAS), are of particular concern due to their environmental persistence and long-term ecological risks (Table 5.1).

## 5.2 Advanced Chemical Methodology for Micropollutants Analysis

Urban runoff has traditionally been studied through the analysis of parameters like suspended solids, nutrients, and metals. However, emerging micropollutants have recently gained attention due to their potential toxicity even at low concentrations (ng $L^{-1}$ to μg $L^{-1}$).

The characterization of micropollutants has improved significantly with application of high-resolution mass spectrometry (HRMS), which enables the identification of both known and unknown pollutants. Wide-scope target, suspect, and non-target screening methods using HRMS offer promising approaches for identifying a broad range of compounds, including previously unrecognized contaminants, without the need for reference standards [5]. These methods provide a more comprehensive analysis of urban runoff, capturing complex and variable pollutant profiles beyond those covered by traditional target screening.

Among these approaches, wide-scope target analysis stands out for its ability to characterize complex matrices through extensive compound libraries, sometimes including over 2000 pollutants. Suspect screening further bridges the gap between targeted and non-targeted analyses by prioritizing potential contaminants based on specific filtering strategies. Coupled with HRMS, this method generates robust datasets, where well-defined workflows help reduce false positives by applying criteria such as exact mass, isotope patterns, signal-to-noise ratios, retention times, peak shapes, and blank removal. Additionally, large compound libraries help identify both conventional and previously unrecognized micropollutants.

Overall, these advanced techniques enable a more thorough understanding of urban runoff pollution, supporting better environmental monitoring and management.

### *5.2.1 Study Sites*

This study applies HRMS-based wide-scope target and suspect screening to assess urban runoff from three Spanish cities: Girona, San Sebastián, and Barcelona. A total of 48 samples were collected from urban, industrial, and residential areas to evaluate pollutant profiles and potential tracers linked to land use. Additionally, pollutants loads were monitored across different rainfall events to assess fluctuations over time. These insights will contribute to a more comprehensive understanding of urban runoff pollution, informing better water management and mitigation strategies.

### *5.2.2 Wide Scope Target and Suspect Screening Methodology*

Samples and blank extracts were analyzed using an ultra-high performance liquid chromatography (UHPLC) system coupled to a high-resolution Orbitrap Exploris 120 (Thermo Fisher Scientific, USA) with an electrospray ionization source. Chromatographic conditions were adapted from Guitierrez et al. [6].

Samples were injected in ESI + and ESI − modes, each with data-dependent acquisition (DDA) and all-ion fragmentation (AIF). AIF data was collected in full scan mode (m/z 70–1000) at a resolution of 120,000 with low collision energy (4 eV). DDA fragmentation targeted the three most intense ions at a resolution of 60,000 with the same collision energy.

For wide-scope target screening, AIF files were processed with TraceFinder v.5.2 (Thermo Fisher Scientific, USA) using an in-house database including common household and industrial chemicals. Compounds were identified based on mass error (± 5 ppm), retention time (RT) window (± 0.3 min), isotopic pattern (> 70%), and peak area (> 105). Both positive and negative modes covered 609 and 238 compounds, respectively. Identified compounds were confirmed manually and quantified using matrix-matched calibration curves when reference standards were available.

For suspect screening, DDA files were processed with Compound Discoverer v.3.3 (Thermo Fisher Scientific, USA) using the "Environmental Unknown ID" workflow. Modified parameters included mass tolerance (± 5 ppm), RT tolerance (± 0.2 min), minimum peak intensity (105), and signal-to-noise threshold (1.5). Suspects were confirmed by matching retention time (± 0.3 min), mzCloud database (> 70%), and in-house spectral library (> 75%).

Quality control and quality assurance parameters validated included linearity in matrix-matched calibration, limits of quantification (LOQs) and detection (LODs), method efficiency through recoveries at two concentration levels, and precision as relative standard deviation (RSD). Mass accuracy was ensured by Orbitrap calibration and monitoring of internal standards (ISs) with calibration standards injected periodically to verify stability.

## 5.3 Comprehensive Characterization of Micropollutants in Urban Runoffs

Urban runoff from Barcelona, San Sebastián, and Girona was analyzed using wide-scope target and suspect screening, identifying 255 compounds. Of these, 92 were quantified, and 163 were tentatively identified based on retention time, mass accuracy, and fragmentation patterns, confirmed through spectral databases.

92 quantified compounds included industrial chemicals, pharmaceuticals, pesticides, antibiotics, tyre-related compounds, and personal care products (PCPs). However, tyre-related compounds were the most abundant across all three studied locations, with notable concentrations of compounds such as benzothiazoles that are key indicators of traffic pollution.

Significant differences in pollutant concentrations were observed across cities and land uses. Samples collected after longer dry periods showed higher micropollutants loads, aligning with reports of increased pollutant accumulation preceding rain events. Highway runoff exhibited the highest concentrations of tyre-related compounds, reflecting traffic-related pollution. Temporal patterns revealed peak concentrations within the first 20–30 min of rainfall, stressing the importance of early runoff in transporting high pollutant loads.

15 compounds were detected in over 80% of samples, with some like 6PPD-quinone, 1-cyclohexyl-3 phenylurea, N,N′-Dicyclohexylurea, 1,3-diphenylguanidine, caffeine, 2-benzothiazolesulfonic acid, Bisphenol S, Methylbenzotriazole detected in over 95% of the samples. Moreover, tyre-wear related 1,3-diphenylguanidine and 2-benzothiazolesulfonic acid were consistently found at

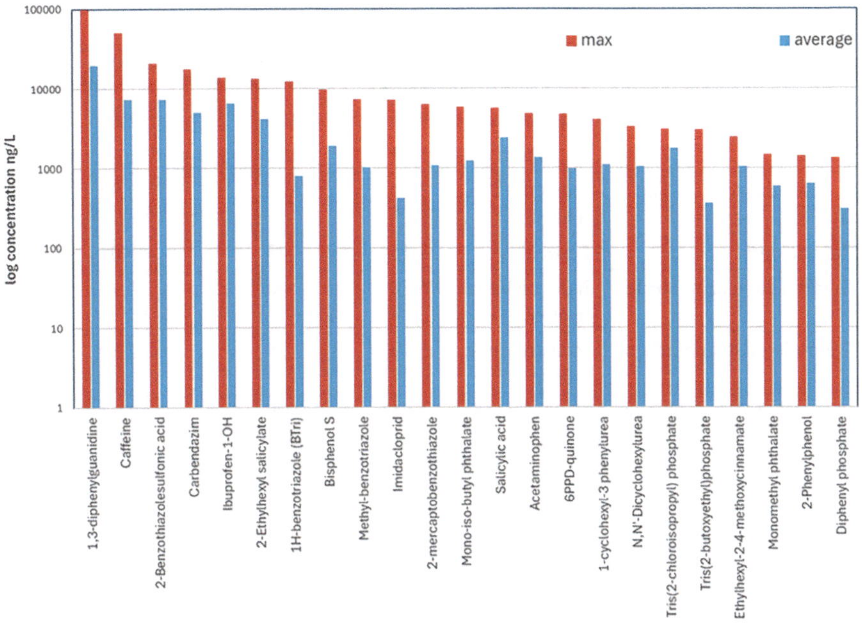


**Fig. 5.1** Overview of compounds detected at maximum and average concentrations

high concentrations of 98.5 and 38.7 μg/L. Of other non-traffic related compounds, the stimulant caffeine and the pharmaceutical metabolite O-desmethylvenlafaxine were also frequently detected, suggesting significant human impact on urban runoff quality. Figure 5.1 shows compounds found in highest average and maximum concentrations.

### *5.3.1 Ecotoxicological Importance of Micropollutants in Urban Runoff*

To evaluate the potential impact on the receiving aquatic environment, the Predicted No-Effect Concentrations (PNEC) of the most relevant compounds; those found at high detection frequencies (> 60%) and high concentrations, were reviewed (Table 5.2). PNEC values are taken from the NORMAN dataset and correspond to the most sensitive species in fresh water. While some PNEC are experimentally determined, much of the freshwater data collected corresponds to predicted PNEC (P-PNEC) values derived from Quantitative Structure–Activity Relationship (QSAR) models.

As shown in Table 5.2, the maximum detected concentrations often exceed these PNEC values. In some cases, even several orders of magnitude. Notably, the

**Table 5.2** Compounds detected at concentrations higher than PNEC values

| Compound name | Chemical group | Max conc. (μg/L) | Detection frequency (%) | PNEC (μg/L) in fresh water |
|---|---|---|---|---|
| 1,3-diphenylguanidine | Tyre related | 98.5 | 100 | 1.05 |
| Caffeine | Stimulant | 50.1 | 100 | 1.2 |
| 2-benzothiazolesulfonic acid | Industrial/tyre related | 38.7 | 90 | 9.97 |
| Imidacloprid | Pesticide | 7.15 | 42 | 0.013 |
| 2-mercaptobenzothiazole | Industrial/tyre related | 6.25 | 40 | 0.76 |
| Mono-iso-butyl phthalate | Industrial/tyre related | 5.79 | 25 | 2.39 |
| 1-cyclohexyl-3 phenylurea | Tyre related | 4.05 | 98 | 3.2 |
| 6PPD | Tyre related | 0.93 | 79 | 0.04 |
| Diuron | Pesticide | 0.52 | 69 | 0.07 |
| Terbutryn | Pesticide | 0.35 | 71 | 0.07 |

industrial tyre-related 1,3-diphenylguanidine, 2-benzothiazolesulfonic acid, methylbenzotriazole, 1-cyclohexyl-3-phenylurea, stimulant caffeine, and the pesticides carbendazim, diuron, terbutryn, and imidacloprid all surpassed their experimental PNEC values.

Important finding is detection of 6PPD quinone in 100% of the samples with detected highest concentration of 4.75 ± 0.67 μg/L). 6PPD-quinone is a highly toxic chemical byproduct that forms when 6PPD—a widely used antioxidant and anti-ozonant in vehicle tyres—reacts with ozone in the environment. 6PPD itself helps prevent tyres from breaking down and cracking, but as tyres wear down, tiny rubber particles get released onto roads. These particles interact with ozone in the air, forming 6PPD-quinone, which then washes into waterways through stormwater runoff.

What's alarming about 6PPD-quinone is its extreme toxicity to aquatic life. It's been linked to acute mortality in fish species, most notably juvenile coho salmon, even at very low concentrations [7]. Notably, these toxic effects have been observed at remarkably low concentrations, with lethal thresholds reported at 0.095 μg/L. Other species like brook trout and rainbow trout are also highly sensitive to this compound. This makes 6PPD-quinone a significant environmental concern, particularly in urban areas with heavy traffic and road runoff.

These findings highlight that the concentration of certain pollutants in urban runoff exceeds levels considered safe for aquatic ecosystems, as defined by PNEC values, thereby threatening the ecological sustainability of the receiving aquatic environments.

## 5.4 Conclusions

Urban runoff represents a significant and growing challenge to freshwater ecosystems, particularly as extreme weather events become more frequent due to climate change. Our study highlights the complexity of contaminants in urban runoff, where a wide array of micropollutants—ranging from industrial chemicals to pharmaceuticals—are mobilized by rainfall events, often at concentrations exceeding safety thresholds for aquatic organisms. The results demonstrate that compounds like those from tyre wear, such as 1,3-diphenylguanidine and 2-benzothiazolesulfonic acid, as well as pesticides and pharmaceuticals, are consistently present at high concentrations, posing substantial risks to water quality and biodiversity.

The comprehensive characterization of these micropollutants, utilizing advanced HRMS wide-scope target and suspect screening, has provided valuable insights into the pollutants most likely to impact urban water bodies. These findings reinforce the importance of addressing urban runoff as a source of water contamination, especially in light of the increasing occurrence of rainfall wash-off during intense storm events.

Our study further emphasizes the need for effective monitoring strategies and mitigation approaches that consider the full spectrum of pollutants, including those that are often overlooked. As these contaminants regularly exceed the PNEC values for fresh water, it is critical to establish more robust environmental policies and implement targeted actions to protect aquatic ecosystems. Reducing the mobilization of pollutants from urban surfaces and improving stormwater management practices will be key to safeguarding water quality in the face of ongoing climate change.

Ultimately, the results of this study contribute to a deeper understanding of urban runoff pollution and highlight the urgency of developing effective strategies to minimize its impact on freshwater ecosystems and public health.

## References

1. Müller A, Österlund H, Marsalek J, Viklander M (2020) The pollution conveyed by urban runoff: a review of sources. Sci Total Environ 709:136125
2. Peter KT, Lundin JI, Wu C, Feist BE, Tian Z, Cameron JR, Scholz NL, Kolodziej EP (2022) Characterizing the chemical profile of biological decline in stormwater-impacted urban watersheds. Environ Sci Technol 56:3159–3169
3. Mutzner L, Furrer V, Castebrunet H, Dittmer U, Fuchs S, Gernjak W, Gromaire MC, Matzinger A, Mikkelsen PS, Selbig WR, Vezzaro L (2022) A decade of monitoring micropollutants in urban wet-weather flows: what did we learn? Water Res 118968
4. Bodus B, O'Malley K, Charitha Gunawardana G, McDonald W (2024) Review of emerging contaminants in green stormwater infrastructure: Antibiotic resistance genes, microplastics, tire wear particles, PFAS, and temperature. Sci Total Environ 906:167195
5. Kang D, Yun D, Cho KH, Baek SS, Jeon J (2024) Profiling emerging micropollutants in urban stormwater runoff using suspect and non-target screening via high-resolution mass spectrometry. Chemosphere 352:141402

6. Gutiérrez-Martín D, Gil-Solsona R, Saaltink MW, Rodellas V, López-Serna R, Folch A, Carrera J, Gago-Ferrero P (2023) Chemicals of emerging concern in coastal aquifers: assessment along the land-ocean interface. J Hazard Mater 448:130876
7. Chen X, He T, Yang X, Gan Y, Qing X, Wang J, Huan Y (2023) Analysis, environmental occurrence, fate and potential toxicity of tire wear compounds 6PPD and 6PPD-quinone. J Hazard Mater 452:131245

# Chapter 6
# Automation System in Water Energy Food Nexus Applications

**Mostafa Abdelgeliel**

**Abstract** Water and food security, ongoing energy crises, and the associated environmental impact constitute global challenges that necessitate integrative and smart solutions to address Water-Energy and Food (WEF) nexus. The optimal feasible nexus can be achieved by employing advanced automation technology based on industry 4, which is defined as Cyber Physical System (CPS) based automation, specifically in agriculture. Employing industry 4 in agriculture is defined as agriculture 4. Agriculture 4 technologies are very important to mitigate many of WEF challenges. CPS based automation employing Internet of Things (IoT), cloud based services, Artificial Intelligent (AI), and Digital Twin (DT) are introduced in this chapter. Moreover, the chapter presents four implementations of advanced automation within agricultural systems: (i) smart irrigation management, (ii) smart fertilization system, (iii) solar-driven water pumping systems, and (iv) smart livestock farming operations. The illustrated examples demonstrate the considerable benefits of digitalization and advanced automation in overcoming some of WEF challenges, optimizing resource allocation, and promoting sustainable systems.

**Keywords** WEF nexus · Automation · Iot · Digital Twin · AI · Smart Irrigation

## 6.1 Introduction

The most current essential national/international challenges are scarcity of water and food, highly energy demand and environmental Impact. Therefore, it is essential to address these challenges in a global view to find an optimal and feasible solution for Water-Energy and Food (WEF) nexus. There are different interaction among Water, Energy and Food as illustrated in Fig. 6.1: (1) Water for Energy (W–E), (2) Energy for Water (E–W), (3) Energy for Food (E–F), (4) Food for Energy (F–E), and (5) Water for Food (W–F) [1]. Each interaction has different applications,

M. Abdelgeliel (✉)
Arab Academy for Science, Technology and Maritime Transport, Alexandria, Egypt
e-mail: mostafa.geliel@aast.edu



A. Angelini et al. (eds.), *Security Enhancement for Climate Change Impacting Urban Resources - SECCURe*, NATO Science for Peace and Security Series C: Environmental Security, https://doi.org/10.1007/978-94-024-2345-7_6

for example, water treatment/desalination and waste water processing need energy. Hence these applications are examples of Energy-Water (E–W) nexus. The other examples of nexus are stated in Fig. 6.1. Employing a sustainable energy resource in water desalination of underground/sea water, to be used in drinking and agriculture purposes, is one of the typical WEF nexus. Moreover, producing green hydrogen, as an energy carrier that can be stored or utilized directly to produce ammonia fertilizer, is as added benefit of WEF nexus.

The agriculture revolution, in the last decades, is one of the stem of WEF nexus. The interaction between industrial revolutions and agricultural revolutions is illustrated in the time line shown in Fig. 6.2. It is clear that the green revolution is started at 1950 in the interval between Industry 2 and industry 3, which characterized by employing machinery in agriculture [2]. After industry 3 revolution, agriculture 3 is initiated by applying industry 3 in agriculture and characterized by using digital technology, such as yield monitoring and scheduled irrigation. Consequently, Agriculture 4.0 represents the application of Industry 4.0 concepts within the agricultural sector. Agriculture 4 employs autonomous farming practices enabled by the synergistic use of the Internet of Things (IoT), Artificial Intelligence (AI), and Big Data analytics. These technologies enhance crop monitoring capabilities and reinforce various smart farm applications [2]

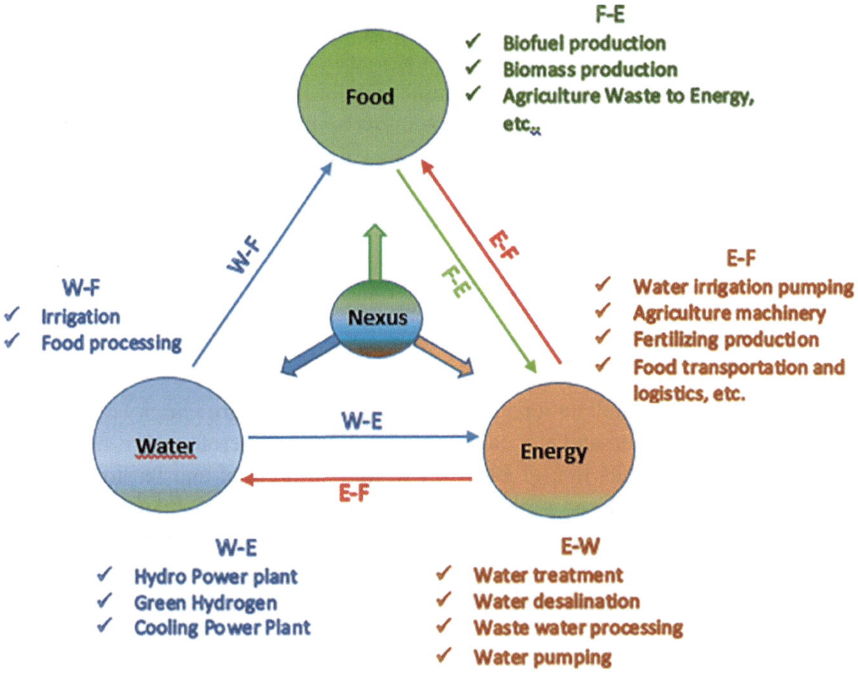


**Fig. 6.1** Interaction among water-energy-food

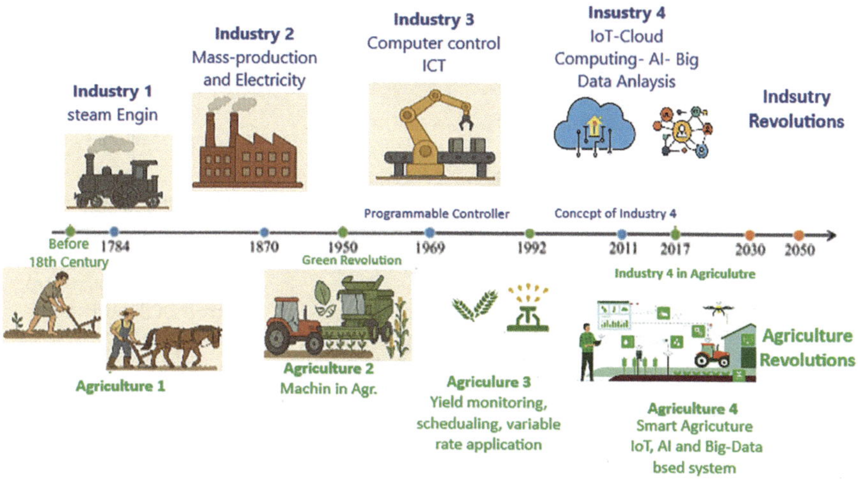


**Fig. 6.2** Timelines of agriculture revolution and industrial revolution

The optimal feasible nexus can be achieved by employing advanced automation technology in sustainable energy generation for water production and Agriculture 4. Many problems related to energy, water and the environment can be mitigated by agriculture 4 technologies. However, the dynamic interactions among food, energy and water makes solutions of WEF nexus challenges are often complex. This chapter highlights the advanced automation employing IoT and AI to obtain an optimal feasible WEF nexus. Advanced automation systems incorporate a range of functionalities, which include but are not limited to smart grid technologies for energy management, optimization techniques for efficient operation and resource allocation, condition monitoring for predictive maintenance, and fault-tolerant architectures for enhanced system reliability.

## 6.2 Automation System Evolution

The typical automation system hieratical, based on international standard for the integration of enterprise and control systems (ISA 95), is illustrated in Fig. 6.3, which has been defined by the official name ANSI/ISA-95". The Automation hierarchy is divided into 7 layers that defined as:

Level 1: Primary technology, which represents the main process components such as motor, valves, pipes, etc…

Level 2: Field level, which consisted of sensor and actuators. Sensors collect data from the physical environment. These could include temperature sensors, motion detectors, or cameras. Actuators perform actions based on instructions, like opening a valve, moving a robotic arm, or controlling a vehicle

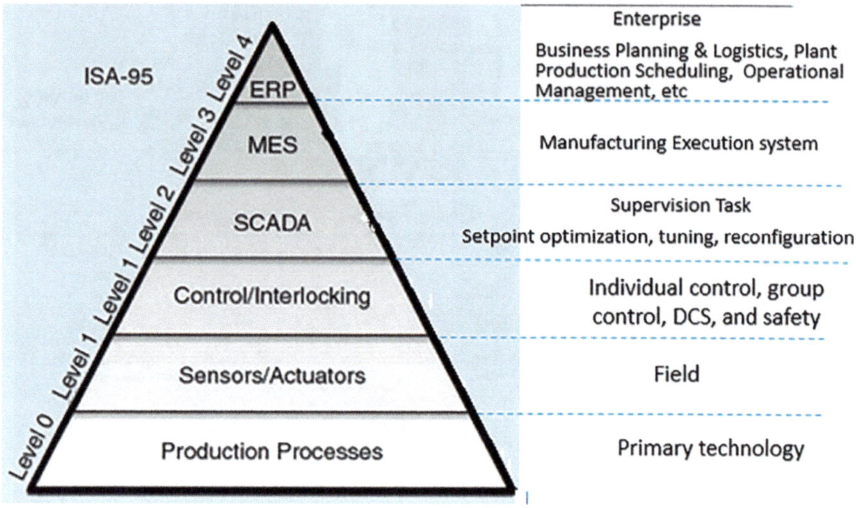


**Fig. 6.3** Typical ISA95 automation layer

Level 3: Individual control, which consists of controllers that perform the required control strategies. It represents the processing and decision-making that analyzes the sensor data and sends appropriate commands.
Level 4: Group control, which represents the masters of group of individual controllers to operate the group of individual controllers in a harmony as a cooperative team. The master represents the team coach and the individual controllers represent the players.
Level 5: The supervision level or SCADA, which performs the supervision tasks such monitoring, data analysis, storage, tuning, etc.
Level 6: Manufacturing and execution level, which represents the individual mangers interaction
Level 7: Enterprise level, which represent the master of the system as the general manager of the plant.

The response time, of each level in automation hierarchy, changes from level to level as illustrated in Fig. 6.4. It is clear that as the complexity increases (higher level) the reaction time increase too. The field devices should react very fast (*ms*) in real- time response to capture any possible transient change and execute the required action in the suitable time. In addition, level 1 (individual control) should have the capability to react very fast with the field device to take the most appropriate actions. Therefore, the levels below SCADA have critical time because they react directly with process periodically (time-based tasks), while the other levels react indirectly with the process and are mainly event based tasks.

There are different control tasks that can be implemented in control levels (Level1 to Level 4) based on the control hierarchy illustrated in Fig. 6.5 [3]. The safety control is designed for emergency and safety condition of the process to avoid accident or

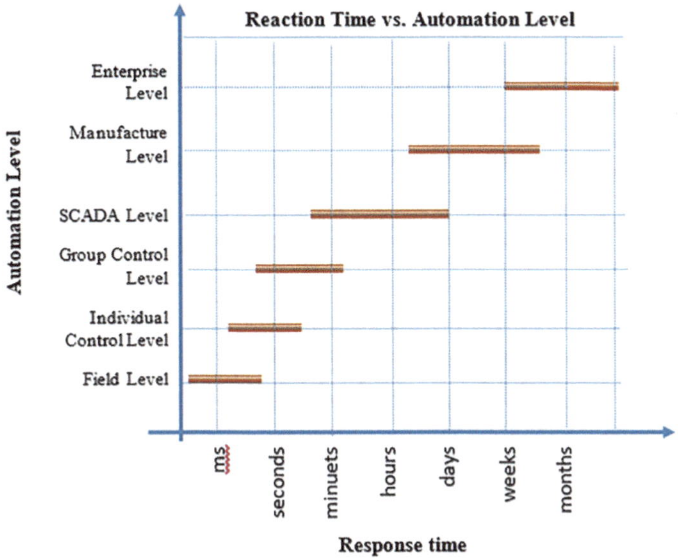


**Fig. 6.4** Response time of automation level

hazard conditions. It should react very fast and independent of process stage. While, basic regulatory control implements some basic control tasks in particular feedback control loops, such as temperature control loop, and some sequential tasks. The main functions of feedback control loop is to compare the feedback signal from the sensor of the controlled variable (such as temperature) with the desired signal (set value) and calculate the manipulated variable (actuator signal), which is transformed into the required action through actuator, based on the suitable controller structure, such as Proportional-Integral-derivative (PID) controller, as illustrated in Fig. 6.6.

In real application, there are many feedback control loops that have strong interaction, for example water supply control system. In this application, there are at least two loops: water flow and water pressure loops, it is known that as the flow increase the pressure decrease and vice versa. Hence, loops interactions require the implementation of advanced regulatory control. Another real example is industrial boiler operation, there are different loops such as steam temperature, pressure and flow loops and water level loop. The flow loop, for example, has direct effect on water level loop, so it is necessary to adjust water level loop based on the flow consumption.

Based on the process situation, it is necessary to adapt the set points (desired signals) and process parameters in an optimal manner, to maintain the system availability with accepted performance that can be implemented in SCADA level. Finally, it is necessary to add extra feature such as scheduling of operation and asset management that can be implemented in manufacturing and enterprise level [3].

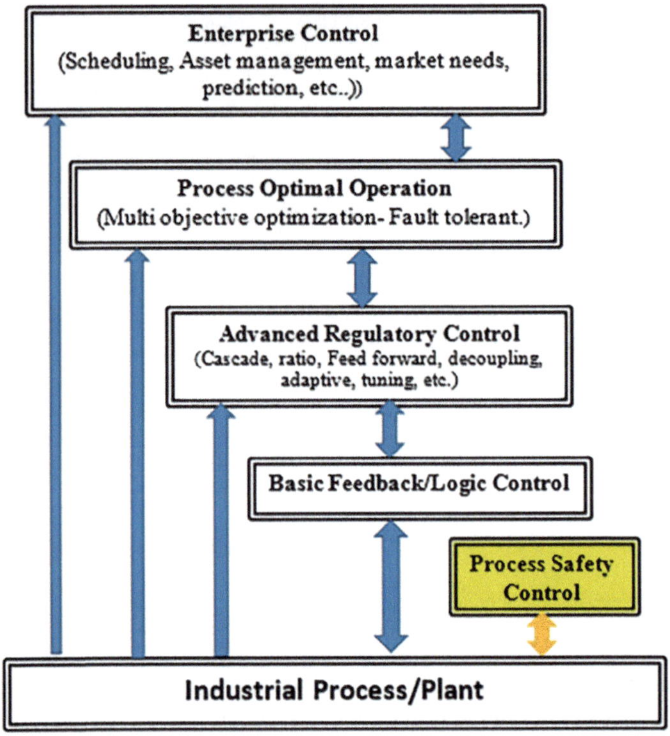


**Fig. 6.5** Control hierarchy in automation system

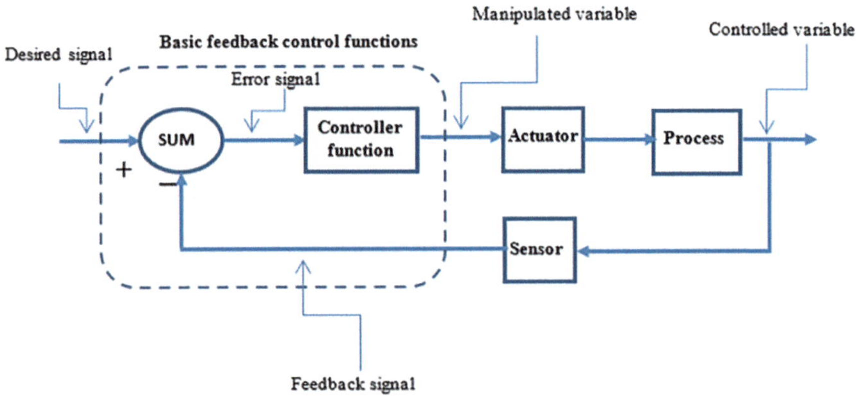


**Fig. 6.6** Basic feedback control loop

### 6.2.1 Automation and Cyber Physical System

The significant progress in Information Technology (IT) and the IoT are the main pivot of Industry 4.0 that introduced new automation frameworks, notably defined as Cyber-Physical Systems (CPS). CPS represent interactions among the physical, communication, and computation systems as illustrated in Fig. 6.7 [3–5].

A typical CPS could be divided into these components:

Sensor Network: The layer of sensors that collect data from the physical environment.
Actuator Network: This layer performs actions based on instructions,
Service Controller: This part represents the processing and decision-making layer based on the analysis of sensor data.
Cloud-Based Control: This layer involves remote monitoring, data storage, and further processing, often utilizing the computational power of the cloud.

In CPS, it is clear that the field devices, sensors and actuators, can interact through a communication network instead of connect each device directly to the controller.

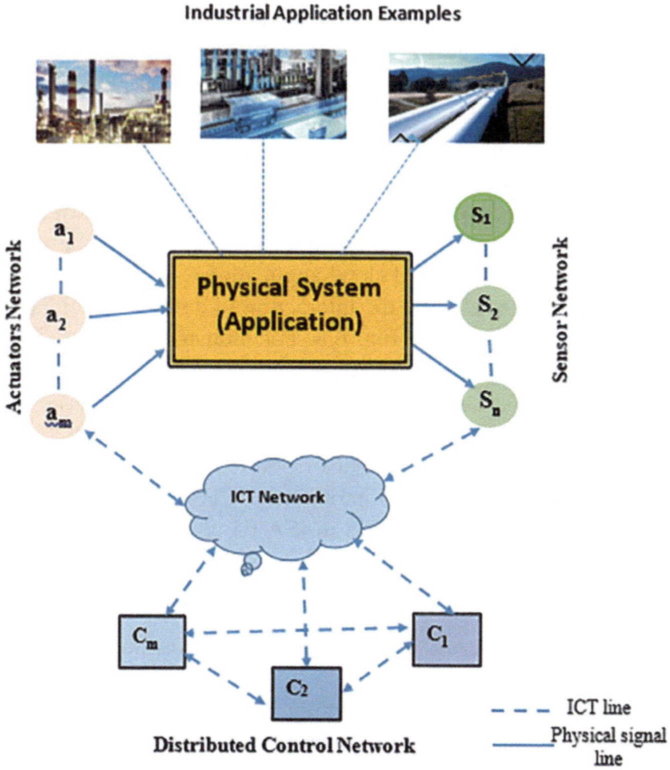


**Fig. 6.7** Typical cyber physical system

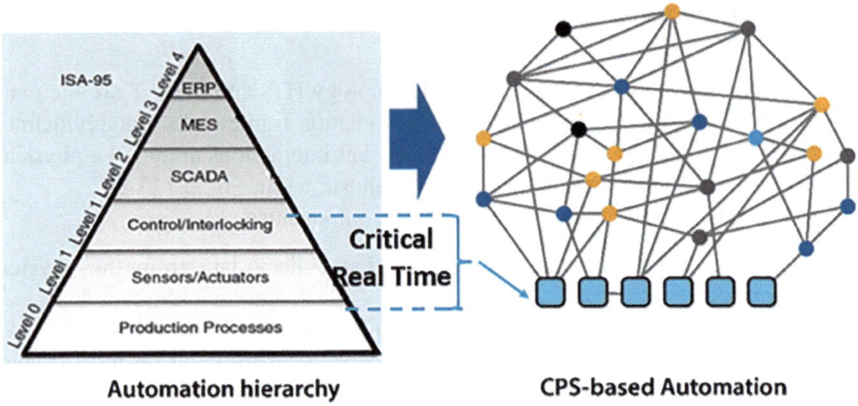


**Fig. 6.8** CPS-based automation system

Moreover, the controller collect data and send commend through communication network in a distributed manner that can be located in remote place instead of collect all in the same location of physical system (system to be controlled). Hence, some of the control functions, in Fig. 6.5, can be implemented remotely through cloud based service provider as illustrated in Fig. 6.8. The service controllers react as distributed control systems, seamlessly integrating physical processes with cloud-based applications and ICT networks. In industrial settings, CPS plays a pivotal role in enabling real-time monitoring, data analysis, and decision-making by harnessing sensor and actuator networks, service controllers, and cloud computing.

CPS eliminate the need to install all control functions and hardware locally on-site. Instead, they enable a distributed architecture where some control features are managed via service provider-based cloud applications.

Based on the definition of CPS, the interaction of automation hierarchy (ASNI/ISO 95) and CPS is illustrated in Fig. 6.8. For industrial application, the critical functions such as field devices level and basic regulatory controllers should be implemented near the physical system, while the other control function can be implemented through could base service providers.

The general CPS evolve to a concept of Service Oriented Architecture (SOA) [4–7]. The implementation of ISA 95 into SOA of CPS is illustrated in Fig. 6.9 [8]. It is clear that different control function in ISA 95 could be implemented through cloud-based service providers. The philosophy of SOA offers several advantages such as:

(1) Cost Efficiency: Reduces the need for extensive on-site hardware, lowering installation and maintenance costs.
(2) Scalability and Dynamic Configuration: Cloud-based control systems allow industries to scale their operations without significant infrastructure changes.
(3) Remote Access and Control: Operators can monitor and control industrial systems from anywhere, enhancing flexibility and efficiency.

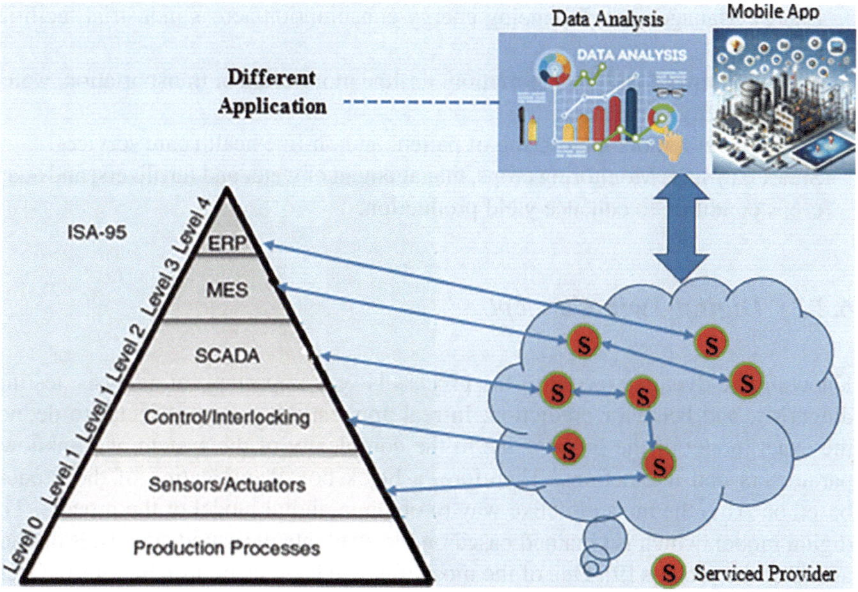


**Fig. 6.9** CPS and service oriented architecture (SOA)

(4) Real-Time Data Processing: The integration with cloud applications enables real-time data analysis, empowering predictive maintenance and timely decision-making in addition to building digital model (Digital Twin) of the.
(5) Resilience and Adaptability: Distributed control reduces dependency on local systems, enhancing system resilience and adaptability to changing demands.
(6) Less Human dependency: SOA utilize the expertise of Service provider without needs to have over-qualified employees and reduce the number operators;
(7) High automation features: SOA enable the utilization highly automated tasks such as condition monitoring and fault prognosis features.
(8) Digital Twin development: The availability of on-line data supports the development of identified digital model, which is called as Digital Twin (DT). DT has many applications in behavior prediction and control adaptation.

However, the main drawback of SOA is the security and cyber-attack that limit these options for noncritical application.

CPS could be implemented in different applications such as:

Smart Manufacturing: Monitoring production lines, optimizing workflows, and reducing downtime.
Process Automation: Controlling complex systems in industries like chemical production or power plants.
Predictive Maintenance: Using sensors and cloud analytics to anticipate equipment failures before they happen.

Energy Management: Balancing energy consumption across industrial facilities effectively.
Service Network and Transportation: on-line monitoring of transportation, water, oil, electricity networks.
Health care: remote monitoring of patient and on-line health care services.
Smart Farming: Monitoring crops, management of water and fertilizers, and adapt crops condition to enhance yield production.

### *6.2.2 Digital Twin Concept*

Knowing the dynamic model of the process is very important for analysis, testing, diagnosis, and behavior prediction. In real application, it is so difficult to deduce the exact model of the process due to the complexity of the system and unknown parameters and interactions. Therefore, a black box identification of the process based on AI is the most effective way to obtain a digital model of the process. The digital model, which is obtained based on the available measured signals, is defined as DT of the process [9]. One of the most effective benefits of utilizing SOA system is to build DT model by employing IoT system and cloud base service provider. DT model of the system can be generated based on the historical data analysis and AI implementation as illustrated in Fig. 6.10. DT model can be utilized in advanced application such as behavior prediction, virtual reality, big-data analysis, testing, state estimation, condition monitoring, tuning and adaptation, etc.

DT model has five dimension [10, 11] as illustrated in Fig. 6.11, named as: (1) Physical system; (2) Services; (3) Virtual model; (4) Digital Twin data; and (5) IOT connections between each pair of the 4 subsystems. Hence DT employ data from physical system and services to build virtual model employing AI to be used in prediction, control, diagnosis, etc.

## 6.3 Application of Industry 4 in WEF Nexus

Advanced automation based on industry 4 has many applications in WFE nexus such as agriculture 4 employing smart irrigation, solar pumping, crops health monitoring, etc. Some of this application will be addressed here. Since agriculture needs water for irrigation, and energy for water production and supplying, and agriculture residues and food are used to produce energy, it is a typical application of WEF nexus. Hence, the illustrated applications in this section will be focused on agriculture 4/Smart agriculture. Smart agriculture has the potential to significantly reduce input costs, improve crop production, and enhance agricultural sustainability and efficiency. The use of technology and data analytics, such as precision irrigation, remote sensing and machine learning, can optimize crop growth, minimize the unnecessary use of fertilizers and pesticides and improve water use efficiency [12, 13]. Use of AI and

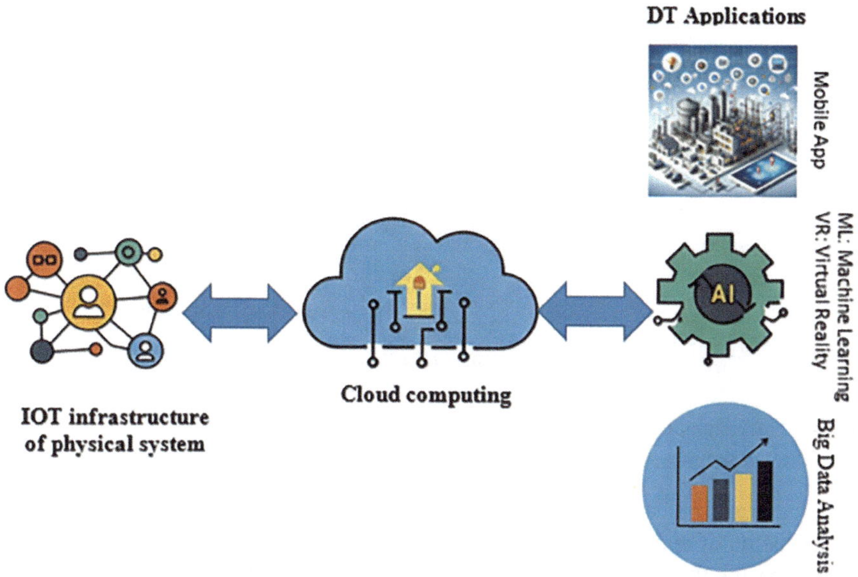


**Fig. 6.10** Industry 4 enabling DT

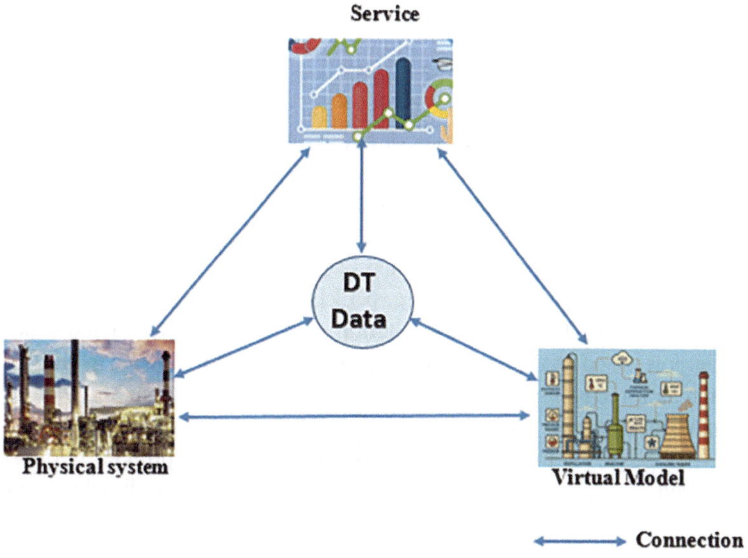


**Fig. 6.11** Five dimension DT model

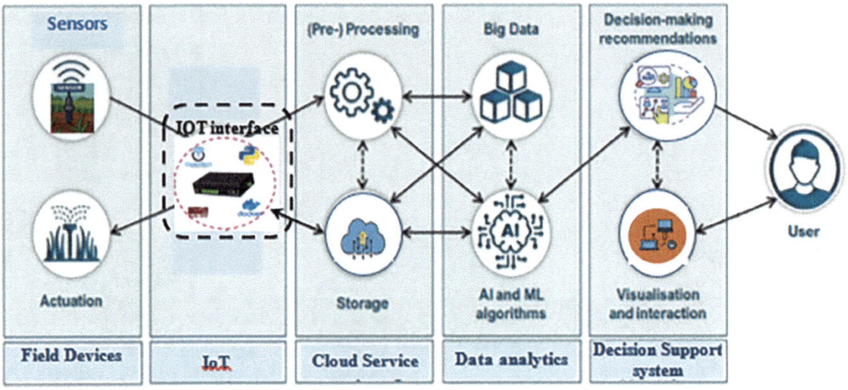


Fig. 6.12 Agriculture 4 concept [16]

IoT in cyber-physical farm management is the foundation of smart agriculture [14]. The benefits of applying automation in smart agriculture can be summarized into: (1) increasing the quantity of real-time data on the crop; (2) remote monitoring and control of farmers; (3) managing water and other natural resources; (4) improving livestock management; (5) accurate appraisal of soil and crops; and 6) increasing agricultural output [15].

Based on DT concept, Fig. 6.12 shows main technologies and connections of agriculture 4 [16]; which consist of: (1) Sensors and actuators to represent field level infrastructure according to the needs of the system; (2) IoT infrastructure that provides connectivity based on protocol and network for data communication; (3) Cloud computing for data storage and processing; (4) Data analytics including big data and AI and ML-based algorithms; and (5) Decision support system to provide data visualization, guidance functions, and user interaction. The domain of applications of Agriculture 4 covers different tasks such as monitoring, control, prediction and logistic that are summarized in Fig. 6.13 [16]. Each domain has several applications, for example, the predicting function in Agriculture 4.0 includes decision-making support based on the analysis of real-time and historical data to predict actual events such as forecasting weather conditions, crop development and yield estimation, animal development and market demand. Similarly, control applications cover the following tasks: (1) Irrigation management system; (2) optimal use of fertilization; (3) weed, pest and disease control; and (4) harvesting of all subsystems. The other application in monitoring and logistic are stated in Fig. 6.13.

## *6.3.1 Smart Irrigation System*

Irrigation is essential for agricultural productivity as it ensures that crops receive the necessary water to thrive. Irrigation has direct impact on plant growth influenced

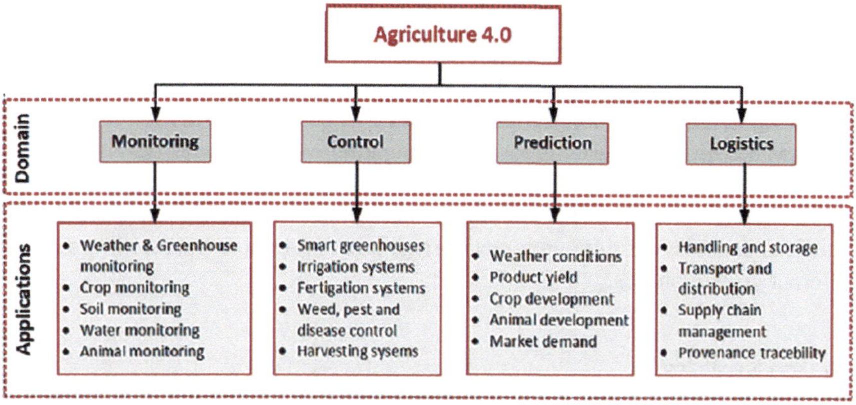


**Fig. 6.13** Domain of different applications in agriculture 4 [16]

by factors such as water availability, soil moisture dynamics, and external climatic conditions like temperature, rainfall, and wind. The relationship between irrigation and plant growth is multifaceted, particularly in water-limited regions where maintaining crop yield is crucial. To enhance irrigation systems, it is important to adapt them for different climatic conditions, utilizing smart and automated technologies that incorporate real-time data from soil moisture sensors, weather stations, and evapotranspiration models to optimize water use, reduce waste, and improve plant growth outcomes. Future research should focus on enhancing the efficiency of these systems under extreme weather conditions and developing sustainable irrigation practices to tackle global water scarcity challenges.

Implementing a smart agriculture system has become a necessity to improve performance. The integration of IoT and systems has also proven to be extremely valuable. IoT technology helps to reduce the overall cost of technology and allows for more efficient management and monitoring of irrigation processes. Wireless sensor networks (WSNs) also play a crucial role in real-time monitoring for precision farming and irrigation activities [17]. One example of smart agriculture in practice in Egypt is the use of precision irrigation systems, which use sensors to monitor soil moisture levels and adjust irrigation schedules accordingly. This technology has been shown to improve water use efficiency and reduce water waste [18]. Another example of smart agriculture in Egypt is the use of remote sensing technology for crop monitoring and management.

A case study of smart irrigation has been demonstrated in [19] as simplified in Fig. 6.14. The study aims to demonstrate that automated irrigation system reduces water and power consumption for enhancing the Water-Energy-Food nexus at field scale. The proposed system consists of: a) WSN nodes to measure soil moisture and local weather; b) data logger and IoT interface to collect WSN and controlled actuators; c) cloud-based services including irrigation decision support system and forecasting systems based on IoT interface to predict the water demand based on the weather condition and moisture of land and schedule water demand; d) onsite solar pump system equipped with optimal energy management strategy to operate the solar-powered automated irrigation system; e) Web services and AI interface to build DT of the farm system; and f) energy management and control systems for improving irrigation in more economical and energy-efficient way.

There are different systems for smart irrigation one of them is designed in [20] that called "AgriSens". "AgriSens" employs an IoT system according to a dynamic irrigation scheduling for efficient water management of irrigated crop fields. The AgriSens enables real-time, automatic irrigation, and remote manual irrigation for different growth phases.

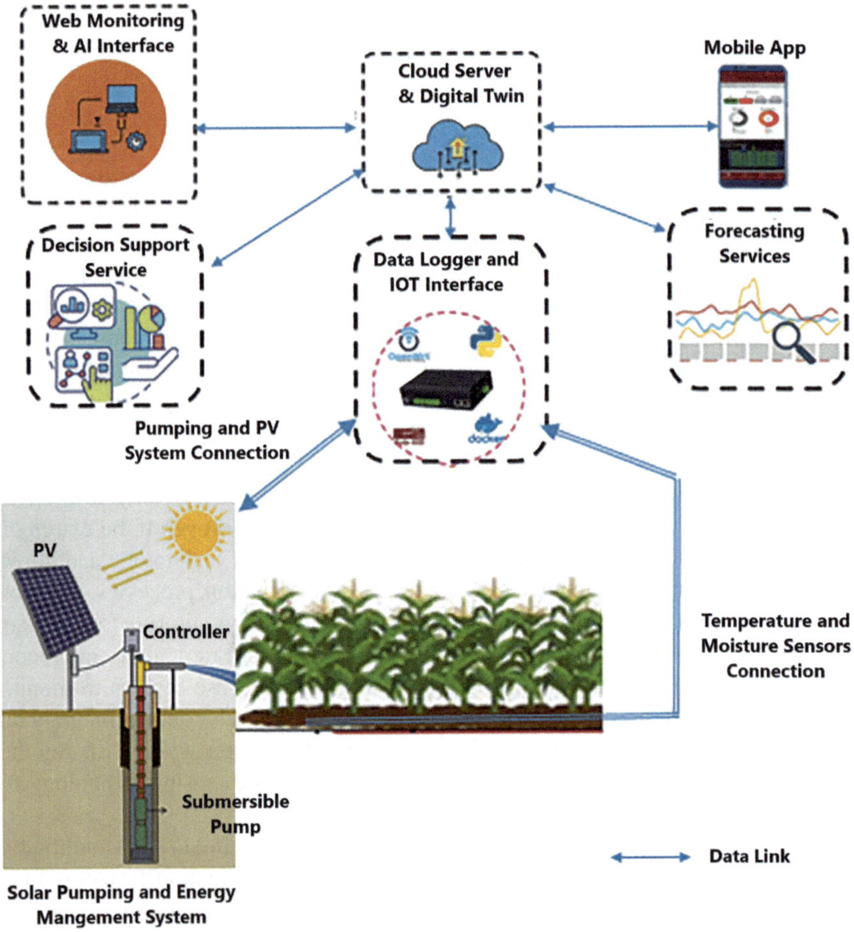


**Fig. 6.14** Smart irrigation employing solar pumping

### 6.3.2 Solar Pump System Condition Monitoring

Most of the smart irrigation system utilize solar pumping system especially in rural area as illustrate in the application in Fig. 6.14 [19]. Most of renewable energy resources, in particular PV pumping systems, are exiting in rural area or owned by nontechnical persons for particular purposes either on grid, islanded micro grid and off grid. The main objectives of PV owner are to enhance the operability, productivity, availability and reliability. The lake of expertise of the owner of the PV and the existence of PV, in most cases, far away from the city and services make the remote monitoring, and fault diagnosis and fast recovery to health condition are mandatory tasks. Hence, it is necessary to have an autonomous system based on IoT, AI and DT to monitor the pumping system, predict the generated power and supplied water for short and long period, perform fault diagnosis and fault tolerant of a PV pump system. A smart PV monitoring based on cloud computing is proposed in [21]. A proposal of smart condition monitoring and fault diagnosis system, capable of identifying the cause of power drop and has the capability to tolerate fault if possible, is illustrated in Fig. 6.15. The proposed system includes:

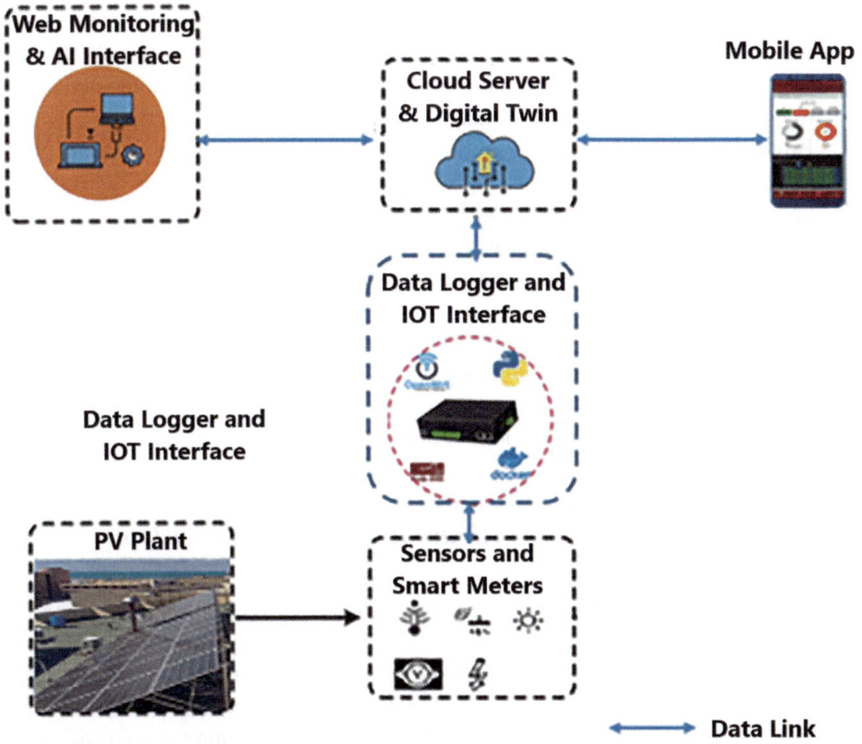


**Fig. 6.15** Smart PV pump system condition monitoring

1. PV plant integrated with solar pump for water irrigation.
2. IoT infrastructure including sensors and smart meters, data logger, access point, and cloud interface and services
3. Cloud based services covering DT and monitoring Platform for access control, monitoring, AI model, diagnosis and tolerant algorithm, etc.
4. Remote access and mobile application

The sensors and smart meter system is designed according to IEC 61724 standard for PV monitoring in addition to weather monitoring sensors, radiations, temperature and wind speed and direction [22, 23]. The collected information is analyzed using cloud based services to build the DT model of the PV pumping system. A decision support is developed based on DT model to schedule the operation, estimate the production of energy and irrigation water, regulate the control loop, enhance reliability and carry online condition monitoring of the system.

The proposed system, here, will support the PV owner with a remote monitoring system based on IoT technology not only to monitor the power generated but also has the capability to:

1. Evaluate the weather condition and determine the expected production for this condition
2. Evaluate the deviation between generated power and estimated one (simulated system) due to environmental conditions, such as soiling that can be handled by regular maintenance, cleaning, or resulting from any other sources that may require further actions, e.g., replacement of panels or similar activities.
3. Building up an AI and ML model based on the online measurement of PV system that is used to predict a priori when a maintenance cycle is warranted, and automated cleaning systems are required.
4. Use the AI to determine the root cause of energy deviation between the generated and estimated that may be due to dust, faulty string or PV panel, human error, bad contact, short circuit, etc....
5. Suggest the corrective actions to be taken in order to tolerate the PV behavior, either by updating the maintenance procedure and/or plan, isolating PV string, if necessary, etc.
6. Forecast the energy production for short time period.

### *6.3.3 IOT-Based Smart Fertigation Management System*

A framework for the IoT-based irrigation and fertilization system are proposed in [24] to make both long-term and short-term planning. The system structure is similar to Fig. 6.14 but it has local control system for short-term planning [24]. The proposed system employing IoT sensors, such as temperature, PH, Humidity, and camera, to monitor the crops growing and status and send this information through IoT infrastructure to decision making system based on AI that make short term planning of fertilizing through determining the optimal quantity and schedule. All these information and short-term planning decision support system interact with upper level decision support system to provide farm and crop information. The long term planning system employing DT model of the farm to plan for long term of the water and fertilize needs based on the crops growth period.

### *6.3.4 IoT Based Smart Livestock Farming (SLF)*

Smart Livestock Farming (SLF) is an important application of Agriculture 4 that supports real-time monitoring of productions, health, and welfare of livestock, and to ensure optimal yield, to increase the management capacity of animals. A variety of sensors and actuators and decision-making tools, and IoT technologies are used in SLF [25, 26]. Electronic identification systems such as ear tags, ruminal boluses, and sub-cutaneous radio-frequency identification; on-animal sensors like accelerometers, global positioning systems, and social activity loggers; and stationary management systems, milking parlor, and related technologies and flock management software are used in SLF [25]. Different sensors categories are split into three categories: (1) "At Cow" sensors, (2) "Near Cow" sensors, and (3) "From Cow" sensors [26] as simplified in Fig. 6.16. The proposed sensors "At COW" cover measuring acceleration, temperature, heart rate, and PH. The proposed sensors "Near COW" used to analyze video and sound information, as well as climate analysis, feed analysis, and GPS. The last group of sensors "From Cow" are employed to monitor products and wastes like milk, hair, saliva, sweat, nasal secretion, breath, and faces taken from the cow. All the sensors are interacted with the cloud and decision support system through IOT that support environmental control, disease early warning, precise feeding, and remote diagnosis.

## 6.4 Conclusion

The automation system evolution and cyber physical system have been introduced including the concept of DT and IoT implementation. The applications of CPS based automation, which depend mainly on IoT, cloud-based services, for WEF nexus in particular agriculture 4 are highlighted. Four case studies have been demonstrated including: (1) smart irrigation system as an example for Water-Food nexus; (2) smart solar pump condition monitoring as an example for Energy-Water nexus; (3) smart Fertigatiom management system; and (4) smart livestock farming. Based on the presented case studies, the CPS based automation, has many added values in terms of cost reduction, productivity enhancement, high availability and efficiency, scalability, real-monitoring for short- and long-term forecasting and scheduling and high quality of product/services. Hence, expanding the application of CPS based automation in the WEF nexus will overcome some of the interacted WEF challenges and SDGs of

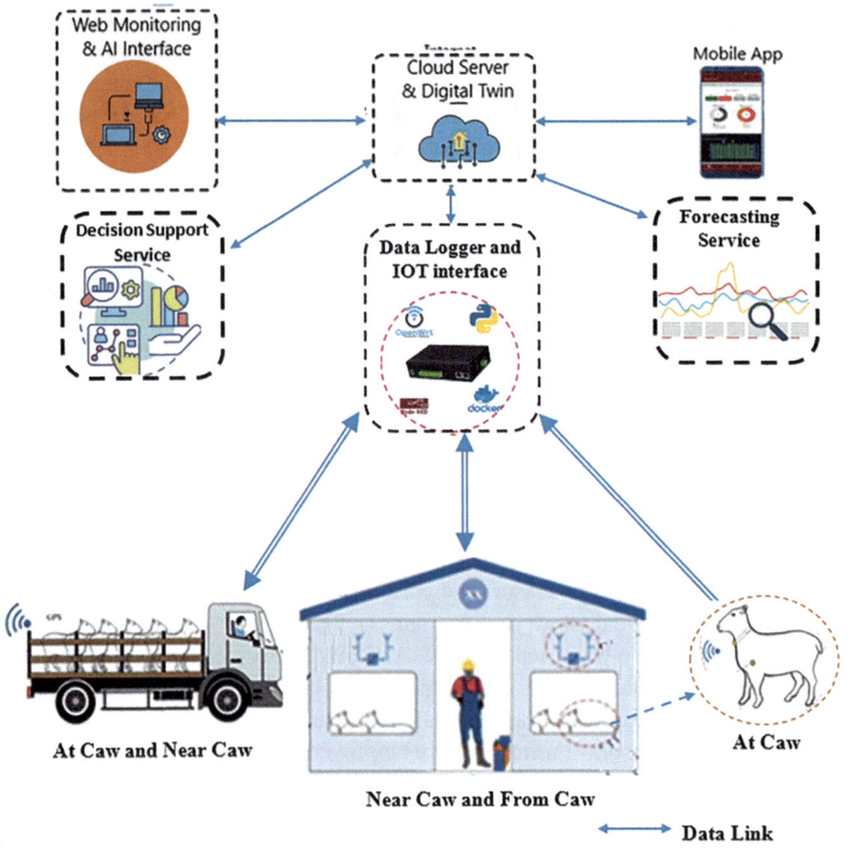


**Fig. 6.16** IOT based Smart Livestock Farming (SLF) [26]

united nation directly. Moreover, it will integrate all factors/players either direct or indirect to design a robust decision support system. For example, in agriculture, the crops production, yield logistics and market demand are very important players. So, CPS based automation application considering direct and indirect main players will be the focus of future work.

## References


1. Bieber N, Ker JH, Wang X, Triantafyllidis C, Van Dam KH, Koppelaar RHEM, Shah N (2018) Sustainable planning of the energy-water-food nexus using decision making tools. Energy Policy 113:584–607. https://doi.org/10.1016/j.enpol.2017.11.037
2. Liu Y, Ma X, Shu L, Hancke GP, Abu-Mahfouz AM (2021) From industry 4.0 to agriculture 4.0: current status, enabling technologies, and research challenges. IEEE Trans Indus Inform 17(6):4322–4334. https://doi.org/10.1109/TII.2020.3003910

3. Iserman R (2006) Fault-diagnosis systems: an introduction from fault diagnosis to fault tolerant. Springer-Verlag, Berlin Heidelberg
4. Hegazy T, Hefeeda M (2015) Industrial automation as a cloud service. IEEE Trans Parallel Distrib Syst 26(10):2750–2763
5. Leitão P, Colombo AW, Karnouskos S (2016) Industrial automation based on cyber-physical system technologies: prototype implementations and challenges. Comput Ind 81:11–25
6. Espí-beltran JV, Gilart-iglesias V, Ruiz-fernandez D (2017) Enabling distributed manufacturing resources through SOA: the REST approach. Rob Comput Integr Manuf 46:156–165
7. Jammes F, Karnouskos S, Bony B, Nappey P (2014) Industrial cloud based cyber physical systems: the IMC AESOP approach—chapter (promising technologies for SOA based industrial automation systems). Springer, pp 89–109
8. Bigheti JA, Fernandes MM, Risso SL, Arao DP, Godoy EP (2019) Controle de Processos baseado em Serviços para a Indústria 4.0. COBISA, Campinas, Spain
9. Aheleroff S, Xu X, Zhong RY, Lu Y (2021) Digital twin as a service (DTaaS) in industry 4.0: an architecture reference model. Adv Eng Inform 47:101225. https://doi.org/10.1016/j.aei.2020.101225
10. Tao F, Liu W, Zhang M, Hu T (2019) Five-dimension digital twin model and its ten applications. Comput Integr Manuf Syst CIMS 25(1):1–18. https://doi.org/10.13196/j.cims.2019.01.001
11. Uribe D, Durand C, Baudouin C, Bigot R (2022) Vers l'asservissement du pilotage en énergie d'une opération de forgeage: développement d'un métamodèle prédictif pour un jumeau numérique. In: 25th Congrès Français de Mécanique conference (CFM 2022), France
12. Amato F, Havel J, Gad AA, El-Zeiny MA (2015) Remotely sensed soil data analysis using artificial neural networks: a case study of El-Fayoum depression, Egypt. ISPRS Int J Geo Inform 4:677–696
13. Al-Metwally SAH, Hassan MK, Mourad MH (2020) Real time internet of things (IoT) based water quality management system. J Proc CIRP 91:478–485
14. Bacco M, Barsocchi P, Ferro E, Gotta A, Ruggeri M (2019) The digitization of agriculture: a survey of research activities on smart farming. J Array 3(4):00009
15. Said ME, Belal AA., Kotb Abd-Elmabod S, El-Shirbeny MA, Gad A, Zahran MB (2021) Smart farming for improving agricultural management. Egypt J Rem Sens Space Sci 24:971-981
16. Araújo SO, Peres RS, Barata J, Lidon F, Ramalho JC (2021) Characterising the agriculture 4.0 landscape-emerging trends, challenges and opportunities. Agronomy 11(667):1–37. https://doi.org/10.3390/agronomy11040667
17. Obaideen K, Yousef BAA, Al-Mallahi MN, Tan YC, Mahmoud M, Jaber H, Ramadan M (2022) An overview of smart irrigation systems using IoT. Energy Nexus 7:100124
18. Abd El-kader SM, Mohammad El-Basioni BM (2013) Precision farming solution in Egypt using the wireless sensor network technology. Egypt Inform J 14:221–233
19. Yadav N, Pattabiraman B, Tummuru RN, Kasiviswanathan C, Adeloye AJ, Sen S, Maurya M, Vijayalakshmanan S (2024) Toward improving water-energy-food nexus through dynamic energy management of solar powered automated irrigation system. Heliyon 10:e25359
20. Roy SK, Misra S, Raghuwanshi NS, Das SK (2021) AgriSens: IoT-based dynamic irrigation scheduling system for water management of irrigated crops. J IEEE Internet Things 8(6):5023–5030. https://doi.org/10.1109/JIOT.2020.3036126
21. Emamian M, Eskandari A, Nedaei A, Sizkouhi M, Lobato K, Milimonfared J, Aghaei M (2023) Smart monitoring of photovoltaic plants with cloud computing and IoT using machine learning. In: 40th European photovoltaic solar energy conference, Portugal, 20387-1-4. https://doi.org/10.4229/EUPVSEC2023/4CV.1.21
22. Klise K, Stein J, Cunningham J (2017) Application of IEC 61724 standards to analyze PV system performance in different climates. In: 44th IEEE photovoltaic specialists conference Washington, DC
23. IEC Standard for PV Monitoring (2021). IEC 61724-1
24. Lin N, Wang X, Zhang Y, Hu X, Ruan J (2020) Fertigation management for sustainable precision agriculture based on Internet of Things. J Clean Prod 277:124119. https://doi.org/10.1016/j.jclepro.2020.124119

25. Vaintrub MO, Levit H, Chincarini M, Fusaro I, Giammarco M, Vignola G (2021) Precision livestock farming, automats and new technologies: possible applications in extensive dairy sheep farming. J Animal 15:100143. https://doi.org/10.1016/j.animal.2020.100143
26. Zhang M, Wang X, Feng H, Huang Q, Xiao X, Zhang X (2021) Wearable internet of things enabled precision livestock farming in smart farms: a review of technical solutions for precise perception, biocompatibility, and sustainability monitoring. J Clean Prod 312:127712

# Chapter 7
# Capacity Building Towards Water Sustainability in Agriculture

**Daniela D'Agostino, Abdelouahid Fouial, Aymen Sawassi, and Roula Khadra**

**Abstract** This paper explores the critical role of capacity building in enhancing water management strategies within agricultural systems, and particularly the contributions made by the International Center for Advanced Mediterranean Agronomic Studies of Bari (CIHEAM Bari) in this area in the recent past. With increasing pressures on water resources due to climate change, population growth, and rising agricultural demands, effective water management is essential for ensuring sustainable agricultural productivity and food production. Capacity building, which involves improving skills, knowledge, and institutional frameworks, is pivotal in fostering more efficient and adaptive water use practices in agriculture. The study explores various approaches to capacity building, including training programs, technological innovations, and knowledge-sharing initiatives aimed at farmers, water managers, and policymakers. By strengthening local expertise and institutional capacities, these initiatives can promote more informed decision-making and the implementation of sustainable water management practices. The paper also examines the successes and challenges of capacity building efforts in water management. Through case studies and best practices, the study highlights the positive impacts of capacity building on improving water use efficiency, reducing waste, and enhancing agricultural productivity. The findings emphasize that investing in capacity building is essential for achieving long-term sustainability in agriculture, ensuring both food security and the preservation of valuable water resources.

D. D'Agostino (✉) · A. Fouial · A. Sawassi · R. Khadra
Mediterranean Agronomic Institute of Bari (CIHEAM BARI), Valenzano, Italy
e-mail: dagostino@iamb.it

A. Fouial
e-mail: fouial@iamb.it

A. Sawassi
e-mail: sawassi@iamb.it

R. Khadra
e-mail: khadra@iamb.it

 
A. Angelini et al. (eds.), *Security Enhancement for Climate Change Impacting Urban Resources - SECCURe*, NATO Science for Peace and Security Series C: Environmental Security, https://doi.org/10.1007/978-94-024-2345-7_7

**Keywords** Capacity building · Water management · Agriculture · Sustainability · CIHEAM Bari

## 7.1 Introduction

The importance and relevance of capacity building to sustainable development is featured in the SDG target 17.9 to "Enhance international support for implementing effective and targeted capacity building in developing countries to support national plans to implement all the sustainable development goals, including through North–South, South-South and triangular cooperation" [1]. The terms "capacity," "capacity building," and "capacity development" (hereafter referred to as "capacity building") are used widely across the development, health, agricultural, and environmental disciplines as an important step toward achieving the goals of a specific program [2–4]. Capacity building in water management is essential for ensuring the efficient and sustainable use of water resources in agricultural practices. As water scarcity becomes an increasingly pressing issue, improving the skills and knowledge of stakeholders involved in irrigation, from farmers to policymakers, is crucial for optimizing water use, enhancing crop yields, and preserving water supplies for future generations [5]. Effective capacity building in this context involves providing training on modern irrigation technologies, water-efficient farming practices, and strategies for mitigating the impacts of climate change on water availability [6]. Additionally, fostering collaboration among local communities, governmental bodies, and technical experts ensures that irrigation practices are adapted to local conditions and are implemented in a way that balances agricultural productivity with environmental sustainability [7]. Ultimately, the capacity building process empowers farmers and other stakeholders to make informed decisions, improving water management and contributing to the long-term sustainability of water resources in agricultural systems.

The sustainable management of water resources in agriculture is a global challenge that involves numerous actors and stakeholders at local, national, and international levels [8]. The ability of these entities to understand and implement effective water management strategies is a cornerstone for advancing territorial sustainable development [9]. Capacity building among different stakeholders, including governments, local communities, businesses, and civil society organizations, is therefore crucial in addressing the water challenges of our time [10]. Capacity building fosters communication and cooperation between these stakeholders by providing tools and platforms for collaboration. When stakeholders understand their roles and responsibilities, they can work together more effectively to address water challenges and ensure that solutions are well-coordinated and equitable [11].

Capacity building serves as a foundational element for addressing complex and interconnected challenges related to water resources. It ensures that stakeholders acquire the necessary knowledge and technical skills to manage water resources effectively. This includes understanding hydrological processes, water conservation techniques, pollution control, and climate change impacts on water availability. With better expertise, stakeholders can implement sustainable practices [12].

Water resources are constantly changing due to environmental, social, and economic factors. Building capacity enables stakeholders to adapt to these changes by enhancing their ability to monitor, assess, and respond to emerging challenges. With the right skills and tools, stakeholders can implement adaptive management strategies that can adjust to shifting conditions, ensuring that water resources are managed sustainably over the long term. For water management to be sustainable, it requires strong governance and sound policies [13]. Capacity building helps policy-makers understand the importance of water sustainability, the complexities of water systems, and the needs of various stakeholders. This enables them to craft better regulations, incentives, and policies that promote sustainable practices, protect water resources, and ensure equitable access to clean water for all. Local communities are often the first to feel the effects of water mismanagement, such as droughts or floods [14]. Capacity building empowers communities with the knowledge and tools to manage their own water resources, improving resilience and reducing dependence on external aid. Empowered communities can actively engage in water conservation efforts, advocate for better management practices, and create local solutions that align with their specific needs and contexts of the territory [15].

Sustainable water management relies on the long-term participation of well-equipped stakeholders. Capacity building strengthens institutions, equips future generations with the skills needed for sustainable practices, and fosters a culture of continuous improvement [16, 17]. This ensures that the principles of water sustainability are embedded in society, rather than being a temporary initiative. In summary, capacity building in water management is vital for the success of any effort aimed at ensuring the sustainability of the resource. By equipping stakeholders with the necessary knowledge, skills, and tools, a more resilient, equitable, and sustainable approach can be created to managing water resources.

The present work emphasizes the importance of capacity building in the context of sustainable water management in agriculture, as it is a bedrock for effective territorial sustainable development, exploring the contributions made by the CIHEAM Bari in such area in the recent past. One of the missions of CIHEAM Bari is to ensure the protection of water resources, enhance the resilience of territories, and promote the development that respects the environment and meets the needs of future generations through an integrated and participatory approach.

## 7.2 Challenges of the Agricultural Sector

The agricultural sector faces numerous challenges, particularly in regions that are experiencing increasing water scarcity, climate variability, and growing populations that hinder the efficiency and sustainability of the sector. One of the primary issues is the inefficient use of water in irrigation systems, which results in significant water losses and low agricultural productivity.

Another challenge is the degradation of water resources due to overexploitation, pollution, and poor management practices, which exacerbate water scarcity and lead to the depletion of aquifers [18].

Additionally, the impacts of climate change, including unpredictable rainfall patterns, extreme weather events, and rising temperatures, further complicate water management strategies, making it difficult to ensure consistent and reliable water supplies for irrigation [19]. Changes in temperature can also lead to shifts in growing seasons and the geographical distribution of certain crops, disrupting established agricultural practices. For example, in sub-Saharan Africa, prolonged droughts have led to crop failures and food insecurity, exacerbating poverty in rural areas [20].

Soil degradation is a significant threat to agricultural productivity. Practices such as overgrazing, monoculture farming, and the excessive use of chemical fertilizers and pesticides lead to soil erosion, loss of soil fertility and desertification. According to the FAO, approximately 33% of the Earth's soil is already degraded, which poses a serious risk to long-term agricultural productivity [21]. The degradation of soil not only reduces crop yields but also impacts water retention and increases the vulnerability of ecosystems to floods and droughts.

The lack of technical expertise, inadequate infrastructure, and limited financial resources also hinder the adoption of modern technologies which have the potential to transform agriculture but require substantial investment and maintenance. Smallholder farmers, who make up the majority of the agricultural workforce in many parts of the world, often lack access to advanced machinery, quality seeds, fertilizers, and digital tools that could boost productivity. Furthermore, limited investment in agricultural research and development (R&D) hinders innovation in crop breeding, pest management, and sustainable farming practices [22].

In many parts of the world, there is a growing shortage of agricultural labor as younger generations move to urban areas in search of better employment opportunities. This shift leads to an aging farming population, which reduces the overall productivity and innovation in the sector. In Europe and parts of Asia, this demographic trend is becoming increasingly problematic, as older farmers retire without a sufficient number of young people to replace them [23].

Finally, institutional and policy gaps, including weak governance and insufficient coordination among stakeholders, limit the effectiveness of water management policies and regulations [24].

Addressing these challenges requires integrated solutions that include technological innovation, better water management practices, strong institutional frameworks and capacity building, crucial in equipping farmers with the knowledge and tools needed to implement sustainable agricultural practices. Capacity building initiatives, such as training in climate-smart agriculture, financial literacy, and advanced farming technologies, can help farmers increase productivity, improve resource management, and adapt to changing environmental conditions [25]. Furthermore, strengthening institutional support and improving agricultural policies at the governmental and local levels can foster better governance and enable more effective policy implementation [26]. Through such interventions, agricultural resilience can be improved, contributing to greater food security and economic stability in rural communities.

## 7.3 Capacity Building Initiatives in Irrigation Water Management

Efficient irrigation water management plays a crucial role in ensuring the sustainability of agricultural production systems, especially in areas where water scarcity is a significant challenge. Given the increasing pressure on water resources, capacity building initiatives have become essential for enhancing the efficiency and sustainability of irrigation practices and for ensuring the long-term viability of water management.

A key component of capacity building in the agricultural sector is the dissemination of modern irrigation technologies and practices. Training programs that introduce farmers to these technologies and teach them how to implement and maintain them are vital in promoting water-efficient farming. Numerous funded initiatives from international organizations, development banks, and private-sector partnerships have contributed significantly to capacity building in irrigation water management. These initiatives not only focus on enhancing technical knowledge and skills but also address the broader challenges of water governance, policy development and sustainable agriculture. Moving forward, there is an increasing emphasis on integrating technology, promoting community engagement and enhancing resilience to climate change through capacity building programs. The FAO has been particularly active in organizing training sessions and workshops in developing countries, helping farmers adopt modern techniques that not only save water but also reduce costs and improve crop yields [18]. The FAO-TCP Program (Technical Cooperation Programme) provides funding for small to medium-scale capacity building projects in irrigation, focusing on improving water productivity, enhancing irrigation infrastructure, and disseminating efficient irrigation practices. The FAO's Africa's Green Revolution project focuses on building local capacities to implement improved irrigation and water management systems across Sub-Saharan Africa, by training farmers in efficient irrigation practices, such as drip irrigation and water harvesting techniques. The World Bank's Irrigation and Drainage Sector Strategy has funded initiatives across countries like Egypt, Morocco and India, which include training programs, farmer field schools, and participatory water management strategies. The International Fund for Agricultural Development (IFAD) funds numerous projects aimed at improving water management and agricultural productivity in developing countries. These projects frequently include a capacity building component to empower local communities in irrigation management. In countries like Ethiopia and Nepal, IFAD has funded projects that offer training in low-cost, water-saving irrigation technologies, aiming to increase food security and reduce poverty. The United Nations Development Programme (UNDP) provides funding for a range of water management projects, with a focus on capacity building and sustainable water use in agriculture. In particular, UNDP funds community-driven initiatives to enhance local capacity in water management, which includes improving irrigation practices and water-use efficiency at the community level and also supports educational initiatives to develop the technical and managerial skills required for effective irrigation water management. For example, the UNDP's Integrated Water Resources Management Program in Jordan focuses on improving water use in agriculture through capacity building programs for farmers, promoting water-saving technologies, and increasing awareness about sustainable water use.

Another critical area of capacity building in irrigation water management is the improvement of decision-making and governance frameworks. Effective water management requires a strong institutional foundation, clear policies, and coordination between various stakeholders, including farmers, irrigation agencies, and local governments. Building the capacity of policymakers and water management authorities ensures that they have the necessary understanding to create and implement policies that promote sustainable water use. For instance, the creation of water user associations (WUAs) has been a successful strategy in several regions. These associations empower local communities to manage their own water resources, fostering a sense of ownership and responsibility for the sustainable use of irrigation water [27]. Capacity building efforts in this context often involve training in institutional development, legal frameworks, and water allocation systems, which help stakeholders understand the complexities of water governance and improve their ability to make collaborative decisions.

Moreover, building capacity to address climate change adaptation is becoming increasingly important in irrigation water management. Programs like the Global Water Partnership's (GWP) "Water, Climate, and Development Program" have provided training and technical assistance to local communities, teaching them how to integrate climate change considerations into their irrigation practices [28]. These initiatives emphasize water-saving techniques, the adoption of drought-resistant crop varieties, and the development of water storage systems to better cope with climate-induced variability.

In addition to technical and policy aspects, community-level capacity development plays an essential role in ensuring that irrigation practices are sustainable and inclusive. Capacity building efforts that target rural communities, especially women and marginalized groups, are critical to achieving equitable and effective water management. Training programs aimed at enhancing the skills of community leaders, extension workers, and local technicians help ensure that the benefits of improved irrigation management are distributed fairly. By involving local stakeholders in decision-making and providing them with the necessary tools and knowledge, capacity development initiatives can foster more resilient and self-sufficient agricultural systems [5].

These initiatives, when properly implemented, can significantly contribute to the reduction of water waste, the improvement of agricultural productivity, and the protection of water resources for future generations. By focusing on technology adoption, institutional strengthening, climate adaptation, and community involvement, capacity building is a critical step towards achieving sustainable irrigation practices.

## 7.4 Tools and Instruments of Capacity Building

Effective capacity building in the context of water management in agriculture relies on a variety of tools and instruments that facilitate knowledge transfer, skill development and practical application.

Key tools include workshops, educational and training programs to acquire new insights, skills and attitude, by providing hands-on experience and theoretical knowledge. These programs often integrate demonstration projects, where new technologies or water-saving techniques are tested in real-world conditions, allowing participants to witness their effectiveness firsthand. Interactive methods that emphasize "learning by doing" are preferred, although classroom teaching is effective for a number of purposes. The type of the program needs to be tailor-made to the target groups (e.g. farmers, water managers, and policymakers) and the objectives to be reached. The format can be longer-term programs, short and focused workshops, exposure programs.

Decision support system (DSS) is another critical tool, providing data-driven insights to guide irrigation scheduling, water use efficiency, and resource allocation, helping to optimize water management decisions. DSS integrates diverse datasets, such as weather patterns, soil moisture, crop water needs and irrigation system performance, allowing users to optimize water use and improve agricultural productivity. Rana et al. [29] highlights how DSS can enhance adaptive capacity by simulating various water management scenarios, helping stakeholders prepare for climate variability and extreme weather events. Additionally, DSS empowers local institutions and communities by fostering knowledge-sharing and improving technical skills in water resource management, thus building long-term resilience against water scarcity and climate change [30]. In this way, DSS are integral in strengthening the capacity of individuals and organizations involved in water management and supporting sustainable agricultural practices.

Additionally, the use of Information and Communication Technology (ICT) tools such as mobile apps and online platforms enhances real-time data collection and dissemination, enabling users to access important water usage information and educational resources remotely. ICT tools enable stakeholders to monitor and manage water resources more effectively. According to Challinor et al. [31], ICT helps farmers and water managers access vital information about weather patterns, crop water needs, and soil moisture, leading to more precise irrigation scheduling and resource allocation. Additionally, Sarma et al. [32] highlight how ICT platforms facilitate knowledge exchange and training, allowing local communities, governments, and agricultural extension services to share best practices and innovative solutions for sustainable water use. By bridging information gaps and providing platforms for interactive learning, ICT empowers users at all levels to make informed decisions and build their technical and managerial skills. As noted by Ali and Elhassan [33], ICT tools also foster greater collaboration between farmers, researchers, and policymakers, ensuring a more coordinated and integrated approach to water management and enhancing resilience to climate change.

Participatory approaches and community engagement also play a vital role, with tools like focus groups and stakeholder meetings fostering collaboration and ensuring that local knowledge and needs are integrated into capacity building efforts [34].

Moreover, monitoring and evaluation (M&E) tools are essential for assessing the effectiveness of capacity building initiatives, ensuring that goals are met and allowing for the continuous adaptation of strategies [35, 36].

These tools collectively create a comprehensive framework for empowering individuals and communities, enabling them to manage agricultural water resources sustainably and effectively.

## 7.5 Capacity Building at CIHEAM Bari

### *7.5.1 Basic Concepts and Philosophy*

The CIHEAM Bari recognizes the urgent need to train high-level officers, technicians and future decision makers on water diplomacy framework to enhance their capacity to understanding and managing complexity, uncertainty and contingency when addressing complex water problems in the face of multiple and competing demands in the natural, societal and political domains. CIHEAM Bari encompasses a wide array of capacity building initiatives: training and education tailoring programs to address the needs of target groups; territorial partnership, ensuring inclusive participation processes; research and cooperation projects facilitating the exchange of findings, best practices, and innovative solutions.

The philosophy of training that applies to CIHEAM Bari programs is based on fostering the holistic development of individuals through structured learning experiences that enhance both theoretical knowledge and practical skills. It emphasizes a learner-centered approach, where the needs, interests, and potential of each participant are taken into account to create engaging, inclusive, and adaptive training programs. The goal is to promote critical thinking, problem-solving, and continuous learning, while equipping individuals with the tools and capabilities necessary to apply what they have learned in real-world contexts. This philosophy values lifelong learning, empowerment, and the development of competencies based on partnerships, networking and collaboration and on best practices in education and training using modern information communications technology, realizing that the trainees of today may be tomorrow leaders.

Thanks to the education and training initiatives active since 1962, CIHEAM Bari has developed the capacities of 14,000 students and trainees coming from over 100 different countries.

### 7.5.2 The Master Course

The Institutional Master's program in "Sustainable Water and Land Management in Agricultural Ecosystems" proposed by CIHEAM Bari plays a role in preparing the new generation towards professional and academic careers and enabling their effective contribution to the sustainable management of water and land resources, and to the socio-economic development thereof, in view of important challenges that include water scarcity, land degradation, demographic pressures and climate change.

A major focus is on the increase of cross-sectoral coherence between Sustainable Development Goals and the application of modern technologies and tools that integrate agronomic, engineering, environmental and socio-economic aspects of water and land management.

The scope of the Master program is to train young, graduated students to let them acquire the necessary skills to apply integrated approaches of water and land management for sustainable agriculture, under significant challenges of climate change, resource scarcity, environmental degradation and societal changes. The program presents the basic principles and the latest scientific and technological achievements in irrigation systems at farm and large-scale level, taking into consideration the application of innovative "green" management solutions.

The interdisciplinary nature of the Master program fosters a holistic perspective, allowing students to address water-related issues within broader environmental, economic, and social contexts. The quality of the educational program is supported by the positive impact of the master's degree attainment upon academic and professional career placement. Graduates of the Master program are well-prepared to be agents of development and to take on leadership roles in addressing water scarcity and promoting sustainable agricultural practices, making a significant impact on food security and environmental sustainability. It is estimated that 74% of CIHEAM Bari students enter the labor market straight away, finding positions in local ministries, universities and private sector.

### 7.5.3 The Projects

Research and cooperation projects are powerful tools for capacity building and training, as they foster collaborative learning and provide hands-on experience in addressing real-world challenges. These projects often bring together diverse stakeholders, such as academic institutions, government agencies, local communities and private sector partners, creating a rich environment for knowledge exchange and skill development. Through joint research efforts, participants not only gain specialized expertise but also learn how to work effectively in teams, negotiate differences and adapt to evolving circumstances. Additionally, research and cooperation projects create lasting impacts by strengthening the human and institutional capacity to address complex issues and contribute to development goals in an informed, collaborative manner.

CIHEAM Bari promotes projects that involve capacity building activities such as stakeholders' workshops, technical and field visits and training sessions, enabling participants to acquire practical skills and transfer knowledge to broader contexts. Over the last four years (from 2021 to 2024), CIHEAM Bari successfully offered 105 courses within research and cooperation projects, engaging a total of 1,630 participants. This impressive achievement highlights CIHEAM Bari commitment to fostering lifelong learning, providing individuals with valuable opportunities to grow, develop new skills, and expand their knowledge.

In this session is provided an overview of CIHEAM Bari contribution to the capacity development through the experience in 3 different projects, currently running on, that demonstrated to be a powerful and effective vehicle for the building of knowledge in the field of sustainable water management.

**WATDEV project.** The WATDEV (Climate Smart WATer Management and Sustainable DEVelopment for Food and Agriculture in East Africa) project has significantly advanced capacity building efforts in water resources management and agricultural best management practices (BMPs) across Kenya, Ethiopia, Sudan and Egypt. Through a structured approach combining training, stakeholder brokerage and technical knowledge transfer, WATDEV has empowered local actors, strengthening their ability to implement sustainable solutions tailored to regional challenges.

A key component of the project has been the development and implementation of targeted training programs aimed at enhancing the technical expertise of farmers, extension workers, policymakers, and water managers. These training initiatives have focused on modern irrigation techniques, soil moisture conservation, precision agriculture, and integrated water resource management, equipping participants with essential tools to optimize water use and improve agricultural productivity. The participatory nature of these programs has facilitated the adaptation of innovative techniques to local conditions, ensuring their long-term applicability and effectiveness. In addition to formal training, WATDEV has fostered four brokerage events in each of the partner countries to encourage collaboration between researchers, practitioners, and decision-makers. These events have served as knowledge-sharing platforms, enabling cross-sectoral dialogue on climate-smart agricultural practices, water governance strategies and sustainable land use planning. By bringing together diverse stakeholders, the project has not only improved communication among key actors but has also enhanced policy uptake and institutional coordination for the successful implementation of BMPs.

The impact of these efforts is evident in the enhanced adaptive capacity of local communities and institutions to respond to water-related challenges. Stakeholders trained under WATDEV have demonstrated improved decision-making abilities, adopting evidence-based water management practices that contribute to increased resilience against droughts, improved irrigation efficiency and better soil health. Additionally, the project's participatory approach has strengthened community engagement in resource governance, fostering a sense of ownership and responsibility for sustainable water and agricultural management.

By integrating scientific knowledge with traditional farming practices, WATDEV has ensured that its capacity building interventions are both practical and culturally relevant. The project's success highlights the critical role of technical training, institutional support, and multi-stakeholder collaboration in achieving long-term sustainability in water and agricultural systems. These initiatives underscore the need for continued investment in capacity building to address water security concerns and enhance agricultural resilience in arid and semi-arid regions.

**Water Knowledge project.** The Water Knowledge (WK) project aims to enhance professional expertise and foster a shared vision on water management in Egypt, Arab countries, and African Neighborhood countries. It seeks to improve integrated water management at the institutional level through capacity development, infrastructural interventions, network enhancement, and training quality improvement. By leveraging local expertise and reinforcing it through training and capacity building initiatives, the project provides valuable knowledge to beneficiaries and facilitates the establishment of networks for future collaborations. Study visits ensure the sharing of experiences and best practices. The initiative aligns with the National Egyptian Strategy 2030 for Sustainable Development and the priorities of the Italian Agency for Development Cooperation.

The project has undertaken several key activities, including upgrading selected laboratories and equipment at the Regional Training Center for Water Resources and Irrigation (RCTWS) in Egypt, to improve hands-on learning experiences and enhancing remote training and e-learning infrastructure. Comprehensive curricula and strategic plans have been created to guide training activities, and targeted training sessions have been conducted for institutional and technical staff, regional participants, and new hires to address specific needs and enhance expertise. Workshops on managing water resources across boundaries have been hosted to promote collaboration and best practice sharing, and study tours have been organized for trainees, teachers, lab technicians, and youth to learn about water management practices and establish future collaboration networks.

As a result of these activities, the project has delivered high-level training courses and technical visits to youth Egyptian employees, Egyptian officials from key ministries for water management, regional officials from neighboring countries, Egyptian trainers, RCTWS tutors and lab technicians, didactic officers, and official representatives. The project has achieved several key outcomes, including improved facilities and enhanced skills, promoted collaboration between Italian and Egyptian water management institutions, strengthened partnerships between key Egyptian ministries and international bodies, created a network of institutions contributing to high-quality training, and connected RCTWS with over 98 national and international trainers. Additionally, it has developed the capacities of 550 trainees from 31 African countries and promoted discussions on conflict management in shared water resources, exemplifying the principles of water and science diplomacy through scientific collaboration to address global water challenges and foster international cooperation.

**Sun4Water project.** In the framework of the program Sun4Water (Promoting climate friendly, water- and energy efficient solar-powered irrigation systems), CIHEAM Bari has developed and successfully implemented training of trainers (ToT) programs on solar-powered irrigation systems (SPIS) across five sub-Saharan African countries, namely: Ethiopia, Kenya, Malawi, Mali and Togo. So far, CIHEAM Bari has trained over 400 future trainers, who are now equipped to design and deliver a range of training sessions targeted at diverse stakeholder groups. These include farmers, extension service technicians, SPIS technology providers and installers, financial service providers, and policy advisors.

A key insight gained through the implementation of these programs is the importance of adapting training content to the specific needs of each stakeholder group. This targeted approach ensures that the training is both relevant and impactful, enhancing the capacity of local actors to support the adoption and sustainable use of SPIS technologies. By focusing on multiple target groups, the programs have fostered a complementary effect on the promotion of SPIS in these countries, acknowledging that involving all stakeholders is key to ensuring the positive impact of these systems. This contributes to broader goals such as improved access to irrigation water, increased agricultural productivity and the sustainable management of limited water resources. Moreover, the programs emphasize the necessity of equipping trainers with a comprehensive skill set that extends beyond technical expertise.

In addition to in-depth knowledge of SPIS planning, design, implementation, operation and maintenance, trainers are also provided with practical approaches for engaging local communities, navigating financial mechanisms and addressing policy and regulatory frameworks. To enhance learning outcomes, all training sessions incorporate field visits, allowing participants to gain firsthand insights into the practical aspects of SPIS deployment. These visits facilitate direct interactions with farmers, enabling trainers to understand real-world challenges and explore potential solutions collaboratively. This holistic and practical approach has proven instrumental in fostering an enabling environment for the effective integration of SPIS technology into agricultural practices, particularly for smallholder farmers, thereby promoting long-term sustainability.

All training sessions incorporated a pre-assessment to identify knowledge gaps related to SPIS, followed by a post-evaluation to measure learning outcomes and training effectiveness. The majority of participants highlighted the significance of the trainings, indicating substantial improvements in their understanding of all aspects of SPIS implementation, from technical design to long-term management. Additionally, participants recognized the potential positive impact of SPIS on farmers and rural communities, particularly in enhancing access to irrigation and improving agricultural productivity. At the same time, the training also provided critical insights into the potential negative environmental impacts of SPIS if not properly planned, designed, implemented and managed. The positive feedback from participants stresses the value of this comprehensive and context-sensitive training model in promoting the sustainable and responsible adoption of SPIS technologies.

## 7.6 Conclusions

Capacity building for irrigation water management has proven to be crucial in improving water-use efficiency, especially in water-scarce regions. Key lessons learned from various initiatives highlight the importance of integrating technical knowledge with local needs and practices. Effective programs must prioritize community involvement, ensuring that local farmers and water managers are actively engaged in the learning process. Training should cover not only technical aspects of irrigation systems but also water conservation techniques, climate resilience, the use of modern irrigation technologies, and alternative energy and water sources.

Strong partnerships among governments, NGOs and local communities are often a cornerstone of success, enabling a collaborative approach to sustainable water management. Evidence from countries like India and Egypt have shown that empowering stakeholders knowledge and tools results in more sustainable water use, better crop yields, and improved climate adaptation [37, 38]. Additionally, integrating data-driven methods and ICT tools into capacity building efforts is essential for long-term effectiveness. Capacity building is not a one-off intervention but a dynamic, ongoing process that evolves alongside changing environmental and socioeconomic conditions.

As science evolves to tackle new challenges, so too does the approach to disseminating scientific findings to the end-users, farmers, through the technical cadre that support the farming sector.

In summarizing CIHEAM Bari's capacity building activities and highlighting the numbers of trained beneficiaries, we aim to demonstrate the critical role of training in international development. Much of CIHEAM Bari's scientific impact in its mandate regions stems from its robust training programs. However, important research gaps and future challenges remain. There is a need to better tailor training programs to local agroecological contexts, evaluate the long-term effectiveness of capacity building efforts and enhance integration of indigenous knowledge with modern practices. Further research is also required to optimize digital training platforms, ensure equitable access to technology, and address gender and youth inclusion within training frameworks.

In light of global food security challenges and environmental stressors, the role of training and technology transfer in international development must be reinforced. CIHEAM Bari is committed to advancing its mission through revitalized capacity building, grounded in scientific innovation and responsive to future needs.

## References

1. United Nations (UN) (2015) Resolution adopted by the General Assembly on 25 September 2015
2. Bolger J (2000) Capacity development: why, what and how. Can Int Dev Agency Policy Branch 1(1):9
3. Eade D (1997) Capacity building: an approach to people-centered development. Oxfam, 263 pp
4. OECD (2006) The challenge of capacity development: working towards good practice. In: DAC guidelines and reference series, p 46
5. Molden D (2007) Water for food, water for life: a comprehensive assessment of water management in agriculture. International Water Management Institute
6. Pereira LS, Cordery I, Iacovides I (2009) Improved water use efficiency in irrigation: perspectives and challenges. Irrig Drain 58(1):5–14
7. FAO (2017) Water for sustainable food and agriculture: a report from the high-level panel of experts on food security and nutrition. Food and Agriculture Organization of the United Nations
8. Labadi S, Giliberto F, Rosetti I, Shetabi L, Yildirim E (2021) Heritage and the sustainable development goals: policy guidance for heritage and development actors. Int J Herit Stud
9. García L, Rodríguez D, Wijnen M, Pakulski I (eds) (2016) Earth observation for water resources management: current use and future opportunities for the water sector. World Bank Publications
10. Nagy K (2010) Hanna: transforming government and building the information society: challenges and opportunities for the developing world. Springer, New York, NY
11. Sigala M, Marinidis D (2012) E-democracy and web 2.0: a framework enabling DMOs to engage stakeholders in collaborative destination management. Tour Anal 17(2):105–120
12. Pahl-Wostl C (2009) A conceptual framework for analysing adaptive capacity and multi-level learning processes in resource governance regimes. Glob Environ Chang 19(3):354–365
13. Mullin M (2009) Governing the tap: special district governance and the new local politics of water. MIT Press
14. National Research Council (2013) Sustainability for the nation: Resource connections and governance linkages. The National Academies Press, Washington, D.C
15. Epstein PD, Coates PM, Wray LD, Swain D (2006) Results that matter: improving communities by engaging citizens, measuring performance, and getting things done. Wiley
16. Jiménez A, Saikia P, Giné R, Avello P, Leten J, Liss Lymer B, Schneider K, Ward R (2020) Unpacking water governance: a framework for practitioners. Water 12(3):827
17. Tropp H (2007) Water governance: trends and needs for new capacity development. Water Policy 9(S2):19–30
18. FAO (2017) The state of food and agriculture: climate change, agriculture, and food security. Food and Agriculture Organization of the United Nations
19. Kumar S, Singh R, Sharma D (2020) Climate change and irrigation in developing countries: challenges and solutions. Environ Sci Policy 108:123–133
20. FAO (2014) Climate change and food security: risks and responses. Food and Agriculture Organization of the United Nations
21. FAO (2015) The state of the World's soil resources. Food and Agriculture Organization of the United Nations
22. World Bank (2017) World development report 2017: agriculture for development. The World Bank
23. OECD (2020) The future of rural youth in developing countries: from dependence to the digital age. OECD Publishing
24. Rijsberman FR, Molden DJ, Murray-Rust D (2018) Water scarcity and irrigation in the 21st century: challenges and strategies. Water Resour Dev 34(1):1–15
25. Bene C, Wood RG, Newsham A, Davies M (2016) Resilience, poverty and development. Dev Policy Rev 34(1):27–51

26. World Bank (2012) Agriculture for development: world development report 2008. The World Bank
27. Rogers P, Hall AW (2003) Effective water governance. TEC background paper no. 7. Global Water Partnership
28. Global Water Partnership (GWP) (2011) Water, climate, and development programme. www.gwp.org
29. Rana MS, Khan AZ, Ahmed MS (2018) Role of decision support systems in irrigation management. J Irrig Drain Eng 144(4):04018001
30. Mekonnen MM, Hoekstra AY, Gerbens-Leenes PW (2021) Water productivity and decision support tools in irrigation systems: a global assessment. Agric Water Manag 246:106718
31. Challinor AJ, Ewert F, Arnold S (2018) Information and communication technologies for climate change adaptation and sustainable agriculture. Glob Environ Chang 53:10–20
32. Sarma KR, Vikas P, Singh P (2020) Role of ICT in enhancing agricultural water management and sustainability. J Agric Inform 11(3):45–59
33. Ali J, Elhassan H (2019) ICT for water management in agriculture: advancing rural development. Int J Water Resour Dev 35(2):345–358
34. D'Agostino D, Borg M, Hallett SH, Sakrabani RS, Thompson A, Papadimitriou L, Knox JW (2020) Multi-stakeholder analysis to improve agricultural water management policy and practice in Malta. Agric Water Manag 229:8
35. Khadra R, Sagardoy JA, Taha S, Lamaddalena N (2017) MONEVA—a monitoring & evaluation system to assess the performance of participatory irrigation management/irrigation management transfer programs in the Mediterranean Region. Water Resour Manage 32(1):123–140
36. Fisher R, Harris C, Dubois S (2019) Participatory approaches for sustainable water management: tools and techniques for capacity building. J Environ Manage 240:145–153
37. Bastakoti R, Pandit R, Sharma R (2017) Sustainable irrigation practices and capacity building for improving water use efficiency in the Asia-Pacific region. Asian Development Bank
38. FAO (2020) Capacity development for irrigation water management: lessons from field projects. Food and Agriculture Organization of the United Nations

# Part III
# Depletion of Resources and Circular Economy as a Strategic Route for the Rationalization of Resources in the Perspective of a Sustainable and Secure Grow

# Chapter 8
# New Challenges in Solar Energy Resource and Forecasting at Different Temporal and Spatial Scales

**Andreas Kazantzidis, Stavros-Andreas Logothetis, and Panagiotis Tzoumanikas**

**Abstract** The energy production of solar energy systems differs fundamentally from that of conventional energy sources. Solar energy availability is highly dependent on weather conditions, making it inherently variable. Various prediction approaches have been developed to address this variability, depending on the application and timeframe. Solar irradiance nowcasting and short-term forecasting are crucial for integrating solar plants into the electricity grid. Understanding the role of clouds and aerosols in these techniques is essential for enhancing their accuracy. This chapter provides an overview of the methods used to assess solar resources and predict solar radiation and energy across different spatial and temporal scales. Special emphasis is placed on short-term forecasting using all-sky imagers (ASIs), particularly for predicting rapid fluctuations in solar irradiance caused by cloud movements. Additionally, the integration of three-dimensional (3D) cloud fields into solar forecasting is highlighted as a key factor in improving the accuracy and reliability of solar energy predictions.

**Keywords** Solar energy · Resource · Forecast · All-sky imagers

A. Kazantzidis · S.-A. Logothetis (✉) · P. Tzoumanikas
Laboratory of Atmospheric Physics, Physics Department, University of Patras, Patras, Greece
e-mail: logothetis_s@ac.upatras.gr

A. Kazantzidis
e-mail: akaza@upatras.gr

P. Tzoumanikas
e-mail: tzumanik@ceid.upatras.gr



A. Angelini et al. (eds.), *Security Enhancement for Climate Change Impacting Urban Resources - SECCURe*, NATO Science for Peace and Security Series C: Environmental Security, https://doi.org/10.1007/978-94-024-2345-7_8

## 8.1 Solar Resource Data Are Important to Solar Power

Accurate and comprehensive solar resource data are crucial for the successful development, operation, and integration of solar energy systems. This data plays a vital role in various key aspects of solar power projects, including site selection, system design, performance forecasting, and grid management [1, 2].

Reliable solar resource data help developers identify the optimal locations for solar installations by analyzing the availability and variability of solar radiation. Detailed information on solar irradiance aids in assessing a site's potential energy output, directly impacting investment decisions and project feasibility. Additionally, a thorough understanding of local solar conditions is essential for designing efficient photovoltaic (PV) systems. Accurate solar irradiance data enable engineers to optimize the orientation, tilt, and configuration of solar panels to maximize energy capture [3].

Moreover, such data are essential for modeling the performance of solar energy systems and predicting energy yields by minimizing uncertainties in performance forecasts, supporting financial planning and risk assessment. Sengupta et al. [1, 2] demonstrate how high-quality solar data enhance energy yield predictions, which are crucial for securing financing and ensuring the economic viability of solar projects. Understanding the variability and predictability of solar resources is key to integrating solar power into the electrical grid, enable the development of strategies to manage fluctuations in solar power generation, ensuring grid stability and reliability.

Overall, solar resource data are indispensable across the entire lifecycle of solar energy projects. They underpin decisions in site selection, system design, performance modeling, grid integration, and technological advancement, thereby playing a pivotal role in the advancement of solar power as a cornerstone of sustainable energy solutions.

## 8.2 Spatial and Temporal Scale of Solar Energy Forecasting Methods

To cover the above-mentioned needs, solar energy forecasting methods are designed to predict solar irradiance and PV power output across various spatial and temporal scales. These scales are critical in selecting appropriate forecasting techniques and tools.

Temporal scales in solar forecasting range from seconds to days, influencing the choice of forecasting methods [4–8]:

- Nowcasting (Seconds to Minutes Ahead): This short-term forecasting focuses on predicting solar irradiance within the next few minutes. Techniques often involve ground-based sky imagers that capture real-time cloud movements to anticipate immediate changes in solar radiation [9–18].

- Short-Term Forecasting (Minutes to Hours Ahead): Covering time horizons up to several hours, these forecasts utilize satellite imagery to track cloud development and movement, providing data for predicting solar irradiance over broader areas [19–24].
- Day-Ahead Forecasting (Hours to Days Ahead): For predictions extending from several hours to a few days, numerical weather prediction (NWP) models are employed. These models simulate atmospheric conditions to forecast solar irradiance over larger regions [25, 26].

In solar forecasting, spatial scales range from localized areas to extensive regions:

- Point Forecasts: Targeting specific locations, such an individual PV installation, these forecasts require high-resolution data to accurately predict solar irradiance at a particular site.
- Regional Forecasts: Encompassing broader areas, these forecasts aggregate data from multiple sites to predict solar energy production across regions, aiding in grid management and energy distribution.

It is important to note that advanced forecasting methods incorporate both spatial and temporal data to enhance accuracy. These models analyze spatially distributed solar time-series data, capturing spatial and temporal correlations to improve forecast precision. Primarily based on machine learning approaches, they utilize data from multiple PV systems to model complex dependencies, enabling more accurate forecasts for multiple sites simultaneously [27].

## 8.3 Solar Resource and Forecasting from All-Sky Imagers (ASIs)

All-sky imagers (ASIs) are ground-based devices equipped with fisheye lenses that capture hemispherical images of the sky. These instruments have become essential in solar resource assessment and forecasting due to their ability to monitor cloud cover and dynamics in real-time [28].

ASIs facilitate the estimation of solar irradiance by analyzing sky images to determine cloud coverage and properties [29]. This information is crucial for assessing the solar energy potential of a specific location. The proposed methodologies include image processing techniques to extract features such as aerosol optical properties [30–34], cloud type [28, 35], optical thickness and coverage [36–38] from sky images, so that models can estimate the attenuation of solar radiation. Moreover, ASIs are important in short-term solar forecasting, particularly for predicting rapid fluctuations in solar irradiance caused by cloud movements. This can be achieved either by cloud motion vector analysis, i.e. by tracking the movement of clouds across consecutive sky images, so that algorithms can predict future cloud positions and their impact on solar irradiance or by machine learning models [39–41]. These approaches enable intra-hour forecasting, which is vital for grid management and energy storage optimization [39].

All these approaches, as well as hybrid ones, have shown promise in improving forecast accuracy. Additionally, establishing networks of ASIs enhances spatial coverage and forecasting capabilities. The Eye2Sky network in northwestern Germany exemplifies this approach, enabling accurate and high-resolution very short-term forecasts of solar irradiance [42].

In summary, highly advanced methodologies based on ASIs have been applied to solar resource assessment and forecasting. By utilizing image processing, cloud motion analysis, and machine learning, ASIs provide valuable data that improve the efficiency and reliability of solar energy systems.

However, these methodologies have certain limitations, including challenges in cloud identification, reduced performance over longer time scales, as well as operational and infrastructure constraints. For example, some cloud types, such as cirrus or mixed cloud formations, are difficult to detect using traditional image-processing techniques. Additionally, their complex dynamics can introduce significant uncertainties, even in short-term forecasts.

## 8.4 New Challenges in Solar Forecasting from ASIs

As solar power generation continues to grow rapidly worldwide, new challenges have emerged that demand more sophisticated forecasting techniques. These challenges include accurately predicting solar irradiance under dynamic weather conditions, integrating real-time data from advanced sensors and sky imagers and addressing the inherent variability and uncertainty of solar power generation. Advances in machine learning, cloud modeling, and radiative transfer simulations offer promising solutions, but further innovation is required to meet the increasing complexity of solar forecasting in diverse geographical and climatic contexts. This evolving landscape highlights the need for continued research and collaborative efforts to improve forecasting accuracy and reliability, paving the way for a more resilient and sustainable energy future.

Three-dimensional (3D) radiative transfer research in atmospheric sciences gained momentum in the early 1970s with pioneering studies by Mullamaa et al. [43] and McKee and Cox [44]. Over time, this research found its way into key radiative transfer texts, including works by Lenoble [45] and Marshak and Davis [46]. The motivation for developing 3D models lies in the complex relationship between radiation and the cloud field, influenced by cloud amount, shapes, and microphysical properties. This intricate interaction poses significant challenges for accurately simulating cloud systems and their climatic impacts within climate models. Notably, the Intergovernmental Panel on Climate Change (IPCC) AR-6 report highlights cloud radiative effects as a major factor contributing to the variation in climate sensitivity across different models.

Over the past decade, advancements in technology have led to the increased use of sky imaging systems for more frequent cloud detection and characterization compared to traditional meteorological observations. With the adoption of all-sky imagers, cloud detection has become an automated process, enabling the extraction of cloud properties such as cloud type and cover [28, 47, 48]. Simulating radiative transfer in clouds with both efficiency and accuracy presents a significant challenge, particularly when optical properties vary spatially. The high reflectivity of liquid droplets and ice crystals results in scenes dominated by multiple scattering, often involving thousands of photon-matter interactions. Many numerical methods for generating photorealistic images rely on Monte Carlo (MC) integration. Although these techniques have proven effective for volume rendering and naturally extend to multidimensional problems, applying MC rendering to atmospheric clouds encounters two key obstacles. First, the path space contributing significant transport is highly multidimensional, necessitating numerous samples to minimize variance in the estimates. Second, constructing individual path samples is computationally costly, even with advanced free-path sampling techniques [49, 50].

Deep learning neural networks represent a new and highly promising generation of neural models capable of delivering state-of-the-art accuracy, often surpassing human-level performance. These networks are composed of simple building blocks that, when appropriately combined, achieve remarkable results across a wide range of computational domains [51]. Recently, they have demonstrated the ability to accurately predict radiance distribution in artificial clouds under unoccluded distant lighting, even after an arbitrary number of scattering events [52, 53].

### *8.4.1 3D Reconstructed Cloud and Map Based Irradiance Fields from ASIs*

The integration of three-dimensional (3D) cloud fields into solar forecasting is of paramount importance for improving the accuracy and reliability of solar energy forecasting. Traditional solar forecasting methods often struggle to account for the complex spatial and temporal dynamics of cloud cover, which significantly influence solar irradiance on the Earth's surface. By representing clouds in three dimensions, 3D cloud fields provide a more realistic and detailed depiction of cloud structure, including variations in cloud height, thickness, and composition. This detailed representation allows for a better understanding of how clouds interact with solar radiation, accounting for phenomena such as cloud shading, scattering, and reflection, which are critical for precise solar irradiance estimation. Moreover, incorporating 3D cloud fields enhances the ability to predict the movement and evolution of clouds, enabling more accurate short-term solar forecasts, optimizing energy generation from solar panels, and improving grid management. As solar energy systems become increasingly central to the global energy mix, the ability to predict solar radiation with

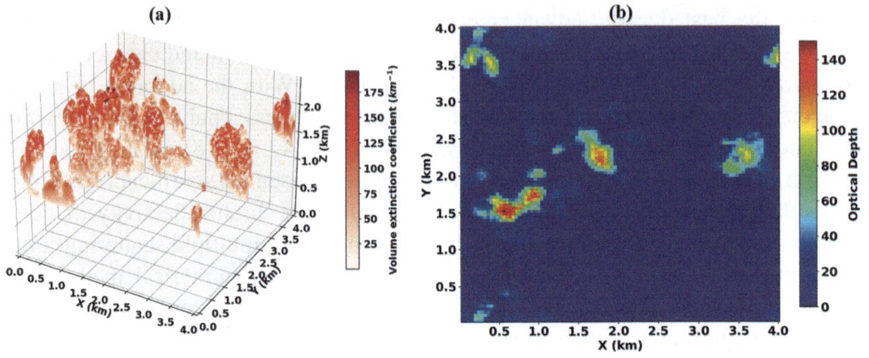


**Fig. 8.1** **a** 3D field of extinction coefficient and **b** 2D cloud optical depth for a reference cumulus cloud scene

higher fidelity using 3D cloud fields is essential for ensuring efficient, sustainable, and reliable energy production.

Figure 1a illustrates an idealized 3D cloud field designed for the calculation of radiance fields in atmospheric simulations. This domain spans 4×4 km on the x-y plane, representing a typical cumulus cloud structure. The cloud field depicted in Fig. 8.1 is highly detailed, capturing the spatial distribution and vertical variation in cloud properties, such as cloud density and topography, which are essential for accurately modeling radiative transfer in the atmosphere.

Figure 1b depicts the cloud optical depth derived from the reference cloud field presented in Fig. 1a. The cloud optical depth is a critical parameter in radiative transfer, as it quantifies the extent to which the cloud absorbs and scatters incoming solar radiation. The cloud optical depth values displayed in Fig. 1b provide insight into how the cloud's internal structure and composition influence its interaction with solar radiation, which is crucial for accurate solar irradiance predictions. Together, these figures demonstrate the use of detailed 3D cloud reconstructions and associated optical properties to enhance our understanding of radiative transfer processes in cloudy environments.

Figure 8.2 presents a reference solar irradiance output derived from 3D radiative transfer calculations, specifically for a solar zenith angle (SZA) of 30°. Figure 8.2 offers a detailed examination of how clouds influence the various components of solar irradiance, such as direct, diffuse, and global solar radiation. The simulation incorporates the complex interactions between solar radiation and the cloud field, including scattering, absorption, and reflection processes. The output emphasizes how clouds alter the distribution and intensity of solar radiation at different levels within the atmosphere and at the Earth's surface. Based on Fig. 8.2, it is apparent that the critical impact of clouds, especially at their edges, where the most pronounced effects on solar irradiance occur. Cloud edges are areas of significant interest because they are where solar radiation undergoes complex interactions between being partially transmitted, reflected, and scattered.

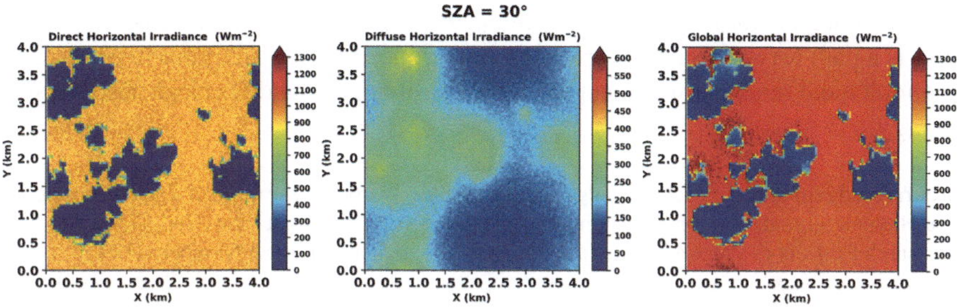


**Fig. 8.2** Direct, diffuse, and global horizontal irradiances for solar zenith angle of 30° for the real cumulus cloud scenes of Fig. 8.1 using 3D radiative transfer calculations

A powerful and promising approach for capturing the intricate 3D effects of clouds on solar radiative transfer involves the integration of several advanced techniques. This includes the combined use of 3D reconstructed cloud fields (such as Fig. 8.1) and radiance measurements derived from all-sky imagers (ASIs), simulated irradiance outputs from sophisticated 3D radiative transfer models, and the application of deep learning neural networks. Together, these components enable efficient and accurate rendering of irradiance within cloudy atmospheres, providing rapid and robust solutions for complex radiative transfer simulations.

## 8.5 Conclusions

The chapter provides a comprehensive overview of the evolving landscape of solar energy forecasting, emphasizing the critical role of accurate, real-time, and spatially resolved data in the efficient integration of solar power into modern energy systems. The discussion highlights how traditional forecasting methods are being enhanced by cutting-edge technologies, particularly through the deployment of all-sky imagers (ASIs) and the use of three-dimensional (3D) cloud modeling.

ASIs, with their ability to monitor and analyze cloud dynamics in real time, have emerged as indispensable tools in nowcasting and short-term forecasting. Their integration with advanced image processing and machine learning techniques has significantly improved forecast accuracy and temporal resolution. Moreover, the incorporation of 3D cloud fields into radiative transfer simulations represents a major leap forward, allowing for more precise estimations of solar irradiance by accounting for cloud structure, composition, and optical properties.

Despite these advancements, challenges remain in dealing with complex cloud formations, computational demands, and the scalability of forecasting models. However, the synergy between physical models, AI-driven analytics, and sensor networks points toward a future of robust, adaptive, and intelligent forecasting systems.

This chapter not only illustrates the state-of-the-art in solar forecasting but also charts a path forward, underscoring the importance of interdisciplinary research and technological innovation to meet the growing global demand for clean, reliable solar energy.

**Acknowledgements** We acknowledge support of this work by the project D3D supported by the Hellenic Foundation for Research and Innovation (H.F.R.I.) under the "2nd Call for H.F.R.I. Research Projects to support Faculty Members & Researchers" (Project Number: 4129).

## References

1. Sengupta M, Habte A, Wilbert S, Gueymard C, Remund J (2021) Best practices handbook for the collection and use of solar resource data for solar energy applications, 3rd edn. IEA PVPS TASK-16
2. Sengupta M, Habte A, Wilbert S, Gueymard C, Remund J, Lorenz E, Van Sark W, Jensen A (2024) Best practices handbook for the collection and use of solar resource data for solar energy applications, 4th edn. IEA PVPS TASK-16
3. Jiménez-Torres M, Rus-Casas C, Lemus-Zúiga L, Hontoria L (2017) The importance of accurate solar data for designing solar photovoltaic systems—CASE studies in Spain. Sustainability 9:247
4. Cesar LB, Silva RAE, Callejo MÁM, Cira C-I (2022) Review on spatio-temporal solar forecasting methods driven by in situ measurements or their combination with satellite and numerical weather prediction (NWP) estimates. Energies 15:4341
5. Yang D, Gu C, Dong Z, Jirutitijaroen P, Chen N, Walsh WM (2013) Solar irradiance forecasting using spatial-temporal covariance structures and time-forward kriging. Renew Energy 60:235–245
6. Boland J (2014) Spatial-temporal forecasting of solar radiation. Renew Energy 75:607–616
7. Lan H, Zhang C, Hong Y-Y, He Y, Wen S (2019) Day-ahead spatiotemporal solar irradiation forecasting using frequency-based hybrid principal component analysis and neural network. Appl Energy 247:389–402
8. Dambreville R, Blanc P, Chanussot J, Boldo D (2014) Very short term forecasting of the Global Horizontal Irradiance using a spatio-temporal autoregressive model. Renew Energy 72:291–300
9. Chow CW, Urquhart B, Lave M, Dominguez A, Kleissl J, Shields J, Washom B (2011) Intra-hour forecasting with a total sky imager at the UC San Diego solar energy testbed. Sol Energy 85:2881–2893
10. Fu C-L, Cheng H-Y (2013) Predicting solar irradiance with all-sky image features via regression. Sol Energy 97:537–550
11. Marquez R, Coimbra CFM (2012) Intra-hour DNI forecasting based on cloud tracking image analysis. Sol Energy 91:327–336
12. Caldas M, Alonso-Suárez R (2019) Very short-term solar irradiance forecast using all-sky imaging and real-time irradiance measurements. Renew Energy 143:1643–1658
13. Nouri B, Blum N, Wilbert S, Zarzalejo LF (2021) A hybrid solar irradiance nowcasting approach: combining all sky imager systems and persistence irradiance models for increased accuracy. Solar RRL 6
14. Fabel Y, Nouri B, Wilbert S, Blum N, Schnaus D, Triebel R, Zarzalejo LF, Ugedo E, Kowalski J, Pitz-Paal R (2023) Combining deep learning and physical models: a benchmark study on All-Sky Imager-Based solar nowcasting systems. Solar RRL 8

15. Kuhn P, Wilbert S, Schüler D, Prahl C, Haase T, Ramirez L, Zarzalejo L, Meyer A, Vuilleumier L, Blanc P, Dubrana J, Kazantzidis A, Schroedter-Homscheidt M, Hirsch T, Pitz-Paal R (2017) Validation of spatially resolved all sky imager derived DNI nowcasts. AIP Conf Proc 1850:140014
16. Logothetis S-A, Salamalikis V, Wilbert S, Remund J, Zarzalejo LF, Xie Y, Nouri B, Ntavelis E, Nou J, Hendrikx N, Visser L, Sengupta M, Pó M, Chauvin R, Grieu S, Blum N, Van Sark W, Kazantzidis A (2022) Benchmarking of solar irradiance nowcast performance derived from all-sky imagers. Renew Energy 199:246–261
17. Kazantzidis, A., Tzoumanikas, P., Blanc, P., Massip, P., Wilbert, S., and Ramirez-Santigosa, L.: Short-term forecasting based on all-sky cameras. Elsevier eBooks, 153–178 (2017).
18. Wang P, Van Westrhenen R, Meirink JF, Van Der Veen S, Knap W (2018) Surface solar radiation forecasts by advecting cloud physical properties derived from Meteosat Second Generation observations. Sol Energy 177:47–58
19. Straub N, Karalus S, Herzberg W, Lorenz E (2024) Satellite-based solar irradiance forecasting: replacing cloud motion vectors by deep learning, Solar RRL 8
20. Prasad AA, Kay M (2021) Prediction of solar power using near-real time satellite data. Energies 14:5865
21. Kallio-Myers V, Riihelä A, Lahtinen P, Lindfors A (2020) Global horizontal irradiance forecast for Finland based on geostationary weather satellite data. Sol Energy 198:68–80
22. Gallucci D, Romano F, Cersosimo A, Cimini D, Di Paola F, Gentile S, Geraldi E, Larosa S, Nilo ST, Ricciardelli E, Viggiano M (2018) Nowcasting Surface Solar Irradiance with AMESIS via Motion Vector Fields of MSG-SEVIRI Data. Remote Sensing 10:845
23. Chen S, Li C, Stull R, Li M (2024) Improved satellite-based intra-day solar forecasting with a chain of deep learning models. Energy Convers Manage 313:118598
24. Chen S, Li C, Xie Y, Li M (2023) Global and direct solar irradiance estimation using deep learning and selected spectral satellite images. Appl Energy 352:121979
25. Zhang G, Yang D, Galanis G, Androulakis E (2021) Solar forecasting with hourly updated numerical weather prediction. Renew Sustai Energy Rev 154:111768
26. Mathiesen P, Kleissl J (2011) Evaluation of numerical weather prediction for intra-day solar forecasting in the continental United States. Sol Energy 85:967–977
27. Simeunovic J, Schubnel B, Alet P-J, Carrillo RE (2021) Spatio-Temporal graph neural networks for Multi-Site PV power forecasting. IEEE Trans Sustain Energy 13:1210–1220
28. Kazantzidis A, Tzoumanikas P, Bais AF, Fotopoulos S, Economou G (2012) Cloud detection and classification with the use of whole-sky ground-based images. Atmos Res 113:80–88
29. Nouri B, Kuhn P, Wilbert S, Hanrieder N, Prahl C, Zarzalejo L, Kazantzidis A, Blanc P, Pitz-Paal R (2018) Cloud height and tracking accuracy of three all sky imager systems for individual clouds. Sol Energy 177:213–228
30. Logothetis S-A, Giannaklis C-P, Salamalikis V, Tzoumanikas P, Raptis P-I, Amiridis V, Eleftheratos K, Kazantzidis A (2023) Aerosol optical properties and type retrieval via machine learning and an All-Sky imager. Atmosphere 14:1266
31. Cazorla A, Shields JE, Karr ME, Olmo FJ, Burden A, Alados-Arboledas L (2009) Technical note: determination of aerosol optical properties by a calibrated sky imager. Atmos Chem Phys 9:6417–6427
32. Román R, Antuña-Sánchez JC, Cachorro VE, Toledano C, Torres B, Mateos D, Fuertes D, López C, González R, Lapionok T, Herreras-Giralda M, Dubovik O, De Frutos ÁM (2022) Retrieval of aerosol properties using relative radiance measurements from an all-sky camera. Atmos Meas Techn 15:407–433
33. Kazantzidis A, Tzoumanikas P, Nikitidou E, Salamalikis V, Wilbert S, Prahl C (2017) Application of simple all-sky imagers for the estimation of aerosol optical depth. In: AIP conference proceedings
34. Scarlatti F, Gómez-Amo JL, Valdelomar PC, Estellés V, Utrillas MP (2023) A machine learning approach to derive aerosol properties from all-sky camera imagery. Remote Sens 15:1676
35. Wacker S, Gröbner J, Zysset C, Diener L, Tzoumanikas P, Kazantzidis A, Vuilleumier L, Stöckli R, Nyeki S, Kämpfer N (2015) Cloud observations in Switzerland using hemispherical sky cameras. J Geophys Res Atmos 120:695–707

36. Andrianakos G, Tsourounis D, Oikonomou S, Kastaniotis D, Economou G, Kazantzidis A (2019) Sky Image forecasting with Generative Adversarial Networks for cloud coverage prediction. In: 2019 10th international conference on information, intelligence, systems and applications (IISA), pp 1–7
37. Fabel Y, Nouri B, Wilbert S, Blum N, Triebel R, Hasenbalg M, Kuhn P, Zarzalejo LF, Pitz-Paal R (2022) Applying self-supervised learning for semantic cloud segmentation of all-sky images. Atmos Meas Techn 15:797–809
38. Park S, Kim Y, Ferrier NJ, Collis SM, Sankaran R, Beckman PH (2021) Prediction of solar irradiance and photovoltaic solar energy product based on cloud coverage estimation using machine learning methods. Atmosphere 12:395
39. Logothetis S-A, Salamalikis V, Nouri B, Remund J, Zarzalejo LF, Xie Y, Wilbert S, Ntavelis E, Nou J, Hendrikx N, Visser L, Sengupta M, Pó M, Chauvin R, Grieu S, Blum N, Van Sark W, Kazantzidis A (2022) Solar irradiance ramp forecasting based on All-Sky imagers. Energies 15:6191
40. Nouri B, Fabel Y, Blum N, Schnaus D, Zarzalejo LF, Kazantzidis A, Wilbert S (2024) Ramp rate metric suitable for solar forecasting. Solar RRL 8
41. Chu Y, Pedro HTC, Li M, Coimbra CFM (2015) Real-time forecasting of solar irradiance ramps with smart image processing. Sol Energy 114:91–104
42. Blum NB, Wilbert S, Nouri B, Stührenberg J, Galeano JEL, Schmidt T, Heinemann D, Vogt T, Kazantzidis A, Pitz-Paal R (2022) Analyzing spatial variations of cloud attenuation by a network of All-Sky imagers. Remote Sens 14:5685
43. Mullamaa YR, Sulev MA, Poldmaa VK, Ohvril HA, Niylisk HJ, Allenov MI, Tchubakov LG, Kuusk AF (1975) Stochastic structure of cloud and radiation fields. Acad Sci Estonian SSR 192
44. McKee TB, Cox SK (1974) Scattering of visible radiation by finite clouds. J Atmos Sci 31:1885–1892
45. Lenoble J (1985) Radiative transfer in scattering and absorbing atmospheres: standard computational procedures
46. Marshak A, Davis A (2005) 3D radiative transfer in cloudy atmospheres
47. Tzoumanikas P, Nikitidou E, Bais AF, Kazantzidis A (2016) The effect of clouds on surface solar irradiance, based on data from an all-sky imaging system. Renew Energy 95:314–322
48. Hasenbalg M, Kuhn P, Wilbert S, Nouri B, Kazantzidis A (2020) Benchmarking of six cloud segmentation algorithms for ground-based all-sky imagers. Sol Energy 201:596–614
49. Kutz P, Habel R, Li YK, Novák J (2017) Spectral and decomposition tracking for rendering heterogeneous volumes. ACM Trans Graph 36:1–16
50. Szirmay-Kalos L, Georgiev I, Magdics M, Molnár B, Légrády D (2017) Unbiased light transport estimators for inhomogeneous participating media. Comput Graph Forum 36:9–19
51. LeCun Y, Bengio Y, Hinton G (2015) Deep learning. Nature 521:436–444
52. Kallweit S, Müller T, Mcwilliams B, Gross M, Novák J (2017) Deep scattering: rendering atmospheric clouds with radiance-predicting neural networks. arXiv (Cornell University) 36:231
53. De Parga CJ, Palomo SRG (2018) Efficient algorithms for real-time GPU volumetric cloud rendering with enhanced geometry. Symmetry 10:125

# Chapter 9
# Advancing the Bio-circular Economy: Application of Innovative Processes for Converting Renewable Biomass Resources into High-Value Products

**Jose A. P. Coelho, Maria Paula Robalo, Georgi St. Cholakov, and Roumiana P. Stateva**

**Abstract** Green extraction techniques employ eco-friendly solvents such as supercritical $CO_2$, deep eutectic solvents, and bio-based solvents to recover valuable compounds from natural sources like rice bran and algae. These methods replace toxic chemical solvents, ensuring a safer and more sustainable approach. Rice bran is rich in gamma oryzanol, tocopherols, and phytosterols, which have antioxidants and cholesterol-lowering properties. Traditional hexane-based extraction leaves harmful residues, whereas green solvents enhance oil yield, preserve nutrients, and minimize environmental impact. Algae species, available in millions, are an important biological resource of polyunsaturated fatty acids, carotenoids, lipids, proteins, and other valuable biochemicals, that can find application in the food, nutraceutical, pharmaceutical and cosmetic industries as well as for the production of biofuels. Conventional methods for the recovery of bioactives and secondary metabolites can not only degrade some of them, particularly those that are heat sensitive, but can harm the ecosystem. Green extraction, especially using supercritical $CO_2$, ensures

J. A. P. Coelho (✉) · M. P. Robalo
Instituto Superior de Engenharia de Lisboa, Instituto Politécnico de Lisboa, Lisbon, Portugal
e-mail: jcoelho@deq.isel.ipl.pt

Centro de Química Estrutural, Institute of Molecular Sciences, Instituto Superior Técnico, Universidade de Lisboa, Lisbon, Portugal

M. P. Robalo
e-mail: paula.robalo@isel.pt

G. St. Cholakov
University of Chemical Technology and Metallurgy, Sofia, Bulgaria
e-mail: cholakov@uctm.edu

R. P. Stateva
Institute of Chemical Engineering, Bulgarian Academy of Sciences, Sofia, Bulgaria
e-mail: roumiana.stateva@abv.bg



A. Angelini et al. (eds.), *Security Enhancement for Climate Change Impacting Urban Resources - SECCURe*, NATO Science for Peace and Security Series C: Environmental Security, https://doi.org/10.1007/978-94-024-2345-7_9

higher purity and efficiency while reducing energy consumption and waste production. It promotes healthier, high-quality products, reduces carbon footprint, and aligns with clean-label trends. It supports advancing the biocircular economy by efficiently utilizing agricultural and freshwater and marine algae byproducts. This innovative approach fosters sustainability in food, nutraceutical, and energy sectors, making it a crucial step toward eco-friendly industrial practices.

**Keywords** Sustainable bioresources · Advanced green techniques · Integrated biorefineries

## 9.1 Introduction

The bio-circular economy is an emerging framework that promotes sustainability by using renewable biomass resources to create high-value products. By incorporating principles of circularity and engineering, this approach aims to minimize waste, lessen reliance on fossil fuels, and enhance economic and environmental resilience.

According to EUROSTAT in 2021, global $CO_2$-emissions were estimated to amount to 36.82 billion tonnes, from which the EU produced around 8%. In the fourth quarter of 2023, EU economy greenhouse gas emissions were about 4.0% lower than in the same quarter of 2022. Still, the projections for global $CO_2$-emissions will increase by about 1.1% per year.

Nowadays, fossil resources account for over 80% of global energy production. Therefore, a rapid shift from fossil to renewable resources is urgently needed not in the near future but immediately by adopting on a large scale the idea of sustainability.

Biomass, derived from organic materials such as plants and/or organic waste, though it is a storehouse of a variety of valuable compounds with numerous applications, in many cases it is underutilized and its potential overlooked. However, biomass as a feedstock has a significant advantage over any other raw material since the cost of raw materials may comprise 30–70% of the total product cost. Thus, biomass transformation into bioproducts, platform chemicals, and biofuels can not only lower the overall production costs and reduce or even eliminate the waste treatment costs but also offer a multifaceted approach to addressing some of the most pressing global challenges, including climate change, human health and wellbeing, sustainable development, and waste management [1–3].

Using abundant but underutilized biomass aligns with the principles of the circular bioeconomy, which advocate that the waste product from one manufacturing process can serve as the raw material for another. This approach minimizes waste, carbon footprints, and greenhouse gas emissions. Those principles impose the circular bioeconomy as a more social and environmental economic model, which targets the development of a new generation of advanced processes integrated into zero-waste biorefineries, which can produce biochemicals and biobased materials, unlike oil refineries. An Integrated zero-waste biorefinery plays a major role in the transition to a circular bioeconomy. It is considered one of the best options for achieving carbon neutrality since it maximizes the value of the biomass feedstocks [2, 4].

A portfolio of advanced innovative technologies with limited application of conventional organic solvents must be used to realize the idea of a zero-waste biorefinery. This synergistic approach will help advance zero-waste biorefineries that epitomize the nexus between sustainability and circular bio economics, thus paving the way to a cleaner, more sustainable, and economically resilient future. Nonetheless, it should be pointed out that technologies for biomass conversion to achieve the zero-waste concept are still in their infancy [1, 5].

To outline the state-of-the-art of biomass conversion to non-energy and energy-related compounds, the challenges, hurdles, obstacles encountered, and possible ways of overcoming those applying advanced innovative techniques are presented using two biomass resources chosen as generic examples, namely rice bran and algae.

## 9.2 The Biomass Resources

### *9.2.1 Rice Bran*

According to forecasts, in January 2025, the most significant serial production in the world for 2024/25, by grain type, will be corn, wheat, and rice, totaling 1.21 billion tons, 0.79 billion tons, and 0.53 billion tons per year, respectively, as detailed on Statista's homepage [6].

Rice bran is the outer layer of the rice grain that is removed during milling. It is a highly nutritious byproduct with various uses in food, animal feed, and industrial applications.

The primary use of rice bran is to produce bran oil. Rice bran oil, with its unique composition containing about 23% saturated fatty acids, 44% oleic acid, and 30% linoleic acid, is a good source of beneficial fats, including unsaturated fats and omega-3 fatty acids. These compounds benefit heart health, making rice bran oil a heart-healthy dietary choice. Additionally, the oil is rich in protein, providing a plant-based protein source for vegetarian diets. Moreover, γ-oryzanol-enriched rice bran oil has been shown to decrease cholesterol and low-density lipoprotein cholesterol levels in hypercholesterolemic patients. This suggests that compounds in the rice bran oil may reduce cardiovascular disease risk factors, offering hope for improved heart health. Additionally, vitamin E and certain phenolic compounds in rice bran have been shown to have potentially beneficial effects, such as anticancer and antioxidant properties activity [7–11].

Rice bran has various industrial applications, including food products like cereals, bread, and biscuits; livestock and poultry feed; cosmetics for skincare and hair care; and biofuels, as rice bran oil is used in biodiesel and other products like wax and soap [11, 12].

These applications and uses are essential to environmental benefits and a circular economy. In what follows, we will present some findings on extracting rice bran oil (RBO) using green solvents, such as supercritical $CO_2$. Green solvent extraction of rice bran provides a sustainable, safe, and efficient alternative to traditional methods, ensuring improved health benefits, environmental protection, and economic advantages [13–15]. This approach aligns with the global shift toward clean and green technologies.

## 9.3 Sustainable Extraction Methods

Rice bran, a rice milling byproduct, is a rich source of oil, antioxidants, and nutrients. Its extraction must be sustainable to reduce environmental impact and enhance resource efficiency. Additionally, preserving rice bran before extraction is crucial for oil quality, shelf life, and waste reduction [8].

Rice bran has a shorter shelf-life than white rice due to free fatty acid production during storage [16]. Proper stabilization techniques improve extraction and enhance the economic and nutritional value of rice bran products [17].

From 2015 until the beginning of 2025, there were about 89 results in the literature related to the combination of rice bran and green extraction, whereas 63 were from the last 5 years. Thus, the process is undeniably a crucial component in the valorization of rice bran for extracting oil, polysaccharides, antioxidants, fibers, natural polymers, protein, sugars, and more.

Table 9.1 presents various methods, parameters, and key results from extracting rice bran over the past five years, though it is not exhaustive. Many other methods have been used to obtain various compounds from rice bran, which, despite claiming to be green, use solvents such as hexane or others whose toxicity is equally high.

Table 9.1 highlights a range of technologies, where supercritical fluids are utilized, highlighted, as well as others also considered green, like the ultrasound assisted, pressurized liquid extraction, deep eutectic solvent, microwave assisted extraction, or enzymatic hydrolysis for protein production.

While hexane extraction and cold-pressed extraction are the most common methods discussed in the literature, this analysis includes non-conventional solvent extraction, microwave-assisted extraction, ultrasound-assisted extraction, supercritical fluid extraction, enzyme-assisted aqueous extraction, and subcritical solvent extraction. A comprehensive summary of the main findings of these environmentally friendly techniques shows that significant yields of rice bran oil (RBO) and γ-oryzanol can be obtained without using hydrocarbon solvents.

The application of $scCO_2$ enables the production of high-quality RBO with a meaningful percentage of fatty acids, even when different varieties are analysed [27], as illustrated in Fig. 9.1, along with excellent antioxidant capacity.

Other non-conventional solvents, such as ethanol, isopropanol, citric acid, DES, and other preferred derivatives from bio-based agricultural biomass sources [29], are essential. Combining extraction techniques, where solvents are minimized, such as

**Table 9.1** Sustainable extraction methods, conditions, parameters of rice brain extracted, and main outcomes

| Extraction method | Optimized condition | Outcomes | References Year |
|---|---|---|---|
| Cold atmospheric plasma (CP) + ultrasound assisted (UAE) in hexane | CP,10 min + US, 20 min at 40% intensity | UAE and CP are eco-friendly alternatives for oil extraction, preserving natural compounds | [18] 2025 |
| Supercritical carbon dioxide ($scCO_2$) with a co-solvent and maceration (SPM) in ethanol-hexane mixtures | $scCO_2$ 200 bar, 40 °C, 90 min, versus SPM, 7 days | $scCO_2$ has a higher potential than SPM for extracting antioxidants, including γ-oryzanol and α-tocopherol content | [13] 2024 |
| Extrusion-stabilized rice bran underwent $scCO_2$ + enzymatic hydrolysis for protein production | 40 MPa, 60 °C, 20 kg/h flow rate for 80 min. Enzyms, 1% alcalase with 3% flavourzyme or 3% protamex | Extrusion stabilizes oil; $scCO_2$ achieves over 70% recovery while preserving bioactive. Sequential enzymatic hydrolysis produces proteins with improved functionality | [19] 2024 |
| Ultrasonic-assisted, 30 min, 35 kHz with a UAE output of 860 Watts. In %ethanol, temperatures, and pH | 57% ethanol, 50 °C and pH 3.1 | Maximum TPC content. Vanillic acid is the main compound. The extract's cytotoxic effects suggest cosmetic potential | [7] 2024 |
| Pressurized liquid extraction (PLE) uses 99.5% ethanol, compared with Soxhlet (SE), using petroleum ether as solvent. Changes in T and static time (St) | Maximum yields (≈ 27%) to an optimal temperature between 102 and 108 °C, for any static time | Oil offers higher yields and improved bioactive compounds, including phenolics and β-Sitosterol, essential for health. Offers superior oxidative stability and strong AChE inhibition with potential neuroprotective benefits | [20] 2024 |
| Deep eutectic solvent-based ultrasonic-assisted extraction (DES-UAE) | DES choline chloride/ ethylene glycol 1:2 ratio and 20% water. UAE at 37 kHz, 50 °C, 40 min | The total composition of the oil includes anthocyanin at 4.55 mg C3G/g dry weight (DW), phenolic compounds at 26.49 mg GAE/g DW, flavonoids at 6.57 mg QE/g DW, and an inhibition rate of DPPH radicals at 77.83% | [21] 2024 |

(continued)

**Table 9.1** (continued)

| Extraction method | Optimized condition | Outcomes | References Year |
|---|---|---|---|
| UAE and microwave assisted extraction (MAE) with a solvent ratio (v/v) citric acid 0.1 M/ ethanol | UAE at a temperature of 50 °C and frequency of 380 W MAE at 55 °C with citric acid at 0.1 M/ ethanol (50:50, v/v), with a mass ratio of solute to solvent of 1:24, g/mL | Both methods achieved similar anthocyanin recovery rates: 83.73% for UAE and 81.44% for MAE. UAE had a higher cyanidin-3-glucoside concentration at 2.41 mg/g compared to MAE's 2.11 mg/g. Both techniques are eco-friendly, earning a green certificate over 90 (A), with excellent EcoScale ratings above 75 | [22] 2024 |
| Hexane extraction (HE), cold-pressed extraction (CPM), subcritical liquefied dimethyl ether extraction (UBLDME), and $scCO_2$ | HE, at 50 °C, 1 h. CPM, three times at a temperature of oil flowing from the press cylinder lower than 50 °C SUBLDME, pressure range of 1.6–1.8 MPa, 70 °C for 60 min $scCO_2$ at 25 MPa, 50 °C, $CO_2$ flow rate of 38–150 L/h for 1 h | The yield of crude rice bran oil from hexane extraction was similar to subcritical dimethyl ether extraction. Cold-pressed extraction had the lowest yield but the highest γ-oryzanol content. Phytosterols and policosanol in crude rice bran oil via subcritical dimethyl ether extraction were significantly higher than other methods | [23] 2023 |
| Supercritical carbon dioxide ($scCO_2$) | Temperature 40 °C; pressure had no impact within the 300–400 bar range | Possible to obtain a good quality RBO, rich in γ-oryzanol and fatty acids, using low temperatures and pressures at a larger scale | [24] 2023 |
| UAE, at 20 kHz and 40% of amplitude, isopropanol as solvent. Compared with SE isopropanol and conventional hexane extraction | Maximum yield at at 90 °C and 1:9 bran to solvent ratio | SE and UAE yielded similar oils and γ-oryzanol, but UAE reduced extraction time significantly. LCA shows UAE has lower emissions impacting climate change and ozone depletion than SE | [25] 2022 |

(continued)

**Table 9.1** (continued)

| Extraction method | Optimized condition | Outcomes | References Year |
|---|---|---|---|
| Water and citric acid extraction. Water with temperature, continuous agitation at 100 rpm, pH adjusted to 4.50 with citric acid 50% | 90 min and 75 °C were chosen to improve protein purity | Water and citric acid effectively extract proteins from rice bran and can combine green technologies for new products | [14] 2022 |
| Supercritical carbon dioxide ($scCO_2$) | Pressure of 34.35 MPa, temperature of 339.5 K, and $scCO_2$ flow rate of $1.8 \times 10^{-3}$ kg/min | Yield, polyphenols, flavonoids, and fatty acid composition are similar to conventional methods but use a green solvent and lower time | [26] 2022 |
| Supercritical carbon dioxide ($scCO_2$) | $CO_2$ flow rate (0.5–1.5 L/min), pressure (20–60 MPa), and temperature (40–80 °C) | Three Japonica varieties—*Ariete, Euro,* and *Opale*—and five long-grain Indica varieties—*Minima, Ellebi, Sprint, Gladio,* and *Sirio*—were extracted. The oils showed excellent oxidation susceptibility parameters. *Ariete* and *Gladio* had superior DPPH and ABTS scavenging activities, while *Minima, Ellebi,* and *Sirio* exhibited the lowest. The cultivated *Indica Sirio* achieved the highest oil yield at 15.60%, compared to 11.48% for *Japonica Euro* | [27] 2021 |
| Ultrasound assisted (UAE) with acidified ethanol | Solid–liquid ratio of 1:17.46; ethanol concentration of 78.37%; UAE time of 55.23 min; and pH of 2.31. Power of 400 W and a at 24 kHz | The anthocyanin level was 5.80 mg/g, with paeoniflorin being the primary anthocyanin compound. The IC50 values for the DPPH radical scavenging and superoxide anion scavenging activities of the sample were 53.51 and 2.375 μg/ml, respectively | [28] 2020 |

MAE, UAE, or PLE, represents an increasingly significant contribution and option for advancing greener technologies.

Water as a solvent or enzyme in heated environments is crucial for extracting components like protein enrichment or fiber analysis. Furthermore, processes incorporated within the biorefinery concept that merge extraction methods—where the rice bran oil is obtained in the initial phase following the extraction of other bioactive compounds, proteins, and fibers—are gaining significance. [18, 19, 30, 31].

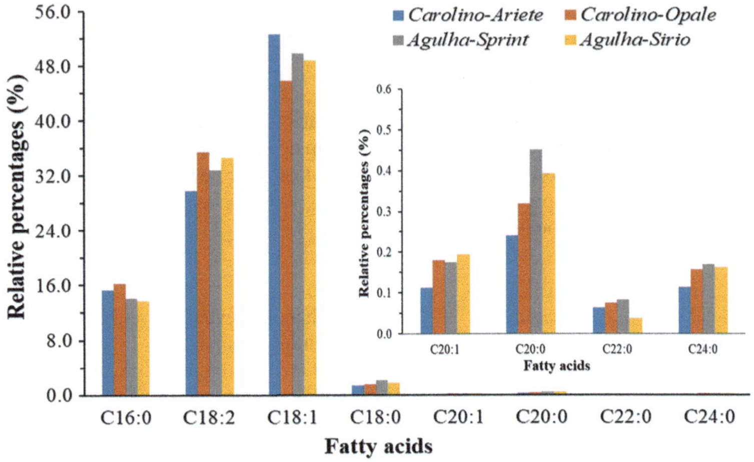


**Fig. 9.1** Relative percentage of the main fatty acids in lipids extracted by SFE at 40.0 MPa and 313.2 K, at flow rate of $CO_2$: $1.90 \times 10^{-3}$ kg $min^{-1}$, from rice bran (*O. sativa* L.) samples to four Portuguese cultivars [27]

## 9.3.1 Algae

**Algae importance as a resource**

Algae, as a biological resource, has drawn considerable interest in the last years. When compared to terrestrial plants, which have been used and overused greatly, algae have enormous commercial potential as the most suitable and sustainable feedstock for producing green energy since the whole process is carbon–neutral in nature. The overview of algae applications, however, shows that this feedstock is still neglected and highly underused as a resource [32].

If collated with land plants, algae have important advantages like: exist in millions compared to terrestrial plants; possess much higher biomass productivities (around 10–50 times higher); have higher $CO_2$ fixation rate; require arid or low quality agricultural land for their cultivation. Lastly but not least the diversity of compounds produced by algal species is estimated to be over 10 times greater than those produced by land plants.

Algae can be used as a feedstock to obtain two types of products—energy and non-energy—based on their potential usage.

With regard to the first group, algae have the potential to reduce the dependence on petroleum-based fuels and chemicals. The first generation of bioenergy strategies involved biofuel production based on sugar, starch, vegetable or animal oils using conventional technology, but these methods have been globally criticized because they deplete food resources. To circumvent this problem, the second generation of bioenergy uses non-edible or waste vegetable oils and agricultural wastes such as lumber, straw and leaves; however, the availability of these was less [33].

At present, microalgae are considered the most suitable and sustainable alternative feedstock for producing green energy not only because of the benefits outlined briefly above, but also because they contain a much higher lipid content per biomass than plants. Consequently, they were proclaimed as a revolutionary third generation resource for biofuels. [34]. Algal biodiesel is a carbon–neutral fuel, which means it assimilates about as much $CO_2$ during algal growth as it releases upon fuel combustion, and hence is an effective and sustainable response to climate change. However, biodiesel production from algae biomass requires release of lipids from their intracellular location, which should be done in the most energy-efficient and economical ways.

With regard to the second non-energy products group, microalgae biomass has been recognized to have great potential as a source of novel bioactive compounds with industrial as well as health promoting applications in human, animal and aquatic lives. They are rich in antioxidants, carbohydrates, lipids, proteins, starch, cellulose and polyunsaturated fatty acids (PUFAs), etc. that can find multiple uses as food additives and nutraceuticals, pharmaceuticals, pigments, fertilizers, energy crops, natural colorants and also as animal feed after oil extraction. Of particular importance are the natural bioactive compounds synthesized by the microalgae, which have great potential for commercialization in the near future due to their antioxidant, antiviral, antibacterial, antifungal, anti-inflammatory, antitumor, and antimalarial effects.

In order to maximize the conversion of algae biomass to high value products and reduce environmental pollution, caused by traditional oil-based chemical industries and to stimulate regional and local development, innovative cutting-edge solutions for its valorization adhering to the circular bioeconomy principles that contribute to Back to Earth Alternatives should be used.

The possible pathways for the sustainable conversion via different techniques of algae biomass are hereunder examined. Also, an analysis of the impact of the latter on the recovery of high value substances with potential applications in various industries, are illustrated on representatives from of two very different genera, namely *Scenedesmus obliquus* BGP and *Porphyridium cruentum*, used as generic examples.

***Scenedesmus obliquus* BGP**

Among the over 150,000 algal species identified worldwide, green unicellular eukaryotic algae (*Chlorophyceae*), particularly *Scenedemus* sp., were extensively studied due to their commercial and pharmaceutical value [35]. For example, recently, representatives of *Scenedesums* genus, namely *S. obliquus, S. dimorphus, S. quadricauda, S. protuberans*, have been a subject of an extensive research [36, 37].

Among those, *S. obliquus,* freshwater green algae, is of particular interest since it multiplies quickly, adapts easily to changing conditions, is known for its high growth rates, and suitability for semi-industrial and industrial cultivation. Also, the species are high in nutrition, synthesize bioactives (carotenoids, chlorophylls, mycosporin-like amino acids, polyphenols, etc.) and can be an excellent renewable, sustainable, and economical source for the food, nutra-, pharmaceutical, energy, etc. industries.

Yet, another interesting representative of the genus *Scenedesmus* is the recently isolated Bulgarian strain, which, after a careful taxonomic and morphological analyses, was identified as *Scenedesmus obliquus* and was accepted to be referred to as *Scenedesmus obliquus* BGP [38]. It has been shown that *S. obliquus* BGP is one of the top lipid producers among the *Scenedesmus* genus species reported in the literature. However, its potential as a renewable bioresource of value-added compounds with a wide range of applications was neglected and only very recently was examined and reported in [39].

***Porphyridium cruentum***

The algae of the genus *Porphyridium* are representatives of *Rhodophyta.* They are a very different group of marine, deep-sea red algae that are characterized by a pigment composition dissimilar to that of green algae. They accumulate in their cells pigments—phycobilioproteins, which are widely used in medicine, and are rich sources of carotenes, amino acids, exopolysaccharides, pigments, antioxidants, various microelements like Ca, Mg, Zn, and K, etc. [40]. *P. cruentum* synthesize several classes of essential compounds like long-chain polyunsaturated fatty acids (PUFA), e.g. omega-3 eicosapentaenoic (EPA, 20:5 ω-3), docosahexaenoic (DHA, 22:6), and omega- 6 arachidonic (AA, 20:4 ω-6) acids, which the human body is incapable of producing.

One of the main challenges for the sustainable transformation of any biomass, algae biomass including, to value added compounds is to realize extracts recovery without damaging them, particularly those which might comprise heat sensitive compounds and bioactives. In view of this, there is a need for mild, and advanced extraction techniques to overcome that hurdle.

*Techniques for extracts recovery*

Till present over 100 articles reporting application of innovative techniques like supercritical extraction (SCE) to recover carotenoids, chlorophylls, tocopherols, lipids, and fatty acids from microalgae, with a special section devoted to the genera *Scenedesmus* were published [41]. Still, the newly discovered *S. obliquus* BGP is under-examined by large extend, and the data obtained is very limited. Tsvetanova et al. [39] were the first to report the integral sustainable transformation of *S. obliquus* BGP to valuable substances applying SCE.

Extracts recovery from *P. cruentum* applying advanced methods like SCE, microwave and ultrasound assisted extractions and pressurized liquids was reported in [42–44].

Tsvetanova et al. [39] reported extracts recovery from the green algae *S. obliquus* BGP. and from the red algae *P. cruentum* applying neat supercritical $CO_2$ (sc$CO_2$) and sc$CO_2$ with a co-solvent. The efficiency of the techniques regarding yield was compared with that of traditional one- and two-step Soxhlet method used as a bench mark. In the one-step, ethanol was used as a solvent, while in the two step Soxhlet—firstly *n*-hexane, followed by ethanol on the residual matrix. *P. cruentum* yield achieved by of one-step Soxhlet though higher was still commensurable with that of *S. obiquus* BGP (26.5 vs. 23.6) %.

The species compositional difference is demonstrated when the yields of first step Soxhlet *n*-hexane was compared. *S. obliquus* BGP yield was over four times higher than that of *P. cruentum*, which was not surprising taking into consideration that the Bulgarian strain is one of the highest producers of lipids, while compositional analysis of *P. cruentum* showed a low content of lipids. The situation was reversed when the second-step Soxhlet with ethanol yields were compared—that of *P. cruentum* was over 1.7 times higher than that of *S. obliquus* BGP. The latter can be explained by the fact that marine deep see red algae are generally richer than fresh-water algae in long-chain polar PUFAs, the recovery of which is enhanced by the application of ethanol on the residual matrix.

The yield of $scCO_2$ extraction of *S. obliquus* BGP was quite low when compared to that of Soxhlet *n*-hexane and depending on the temperature and pressure applied, varied in the range (1.12–1.96) %. The extraction yield was increased considerably when 10% ethanol was added as a co-solvent. Thus, for *S. obliquus* BGP the highest yeild registered at 60 C was 12.29% which, however, was about two-times lower than one-step Soxhlet ethanol extraction.

*P. cruentum* was subjected only to $scCO_2$ + 10% ethanol extraction. Still, the original expectations that the addition of ethanol as a polar co-solvent would enhance the extraction recovery was not confirmed—the highest yield registered was 1.93% at $p = 400$ bar, over 13 times lower than Soxhlet one-step with ethanol.

*Extracts composition analyses*

The comparison of *S. obliquus* BGP with *P. cruentum* was further carried out on the basis of the following important characteristics of the extracts recovered: (i) fatty acid composition from FAME GC-FID; (ii) antioxidants, quali- and quantified by LC–MS/MS; (iii) total phenolic content and antioxidant activity, as well as based on a number of additional parameters and indices that allow outlining the two strains viability and potential applications.

The techniques and analytical methods applied to determine the extracts compositions, as well as their applications were outlined in details in [39], and are presented schematically on Fig. 9.2.

The results obtained marked the possible pathways for the conversion of the two algae species biomass via the methods examined and their impact on the recovery of bioactives, secondary metabolites and other high value biochemicals with potential applications in the food, nutra-, pharmaceutical, energy, etc. industries. For example, it was demonstrated that among the extracts of *P. cruentum* obtained the highest percentages of arachidonic acid (AA, 20:4 ω-6) were registered in the oils recovered by the first and second steps of Soxhlet. The same holds for the important eicosapentaenoic acid (EPA, 20:5 ω-3) but its amounts are over two times lower than those of AA.

Both those PUFAs were not even detected in the *S. obliquus* BGP oils. Consequently, the ability of *P. cruentum* to synthesize AA and EPA (the first in high amounts) defines the strain as a promising alternative to oils from fish and land-based plant sources, since both acids are used to enrich functional foods.

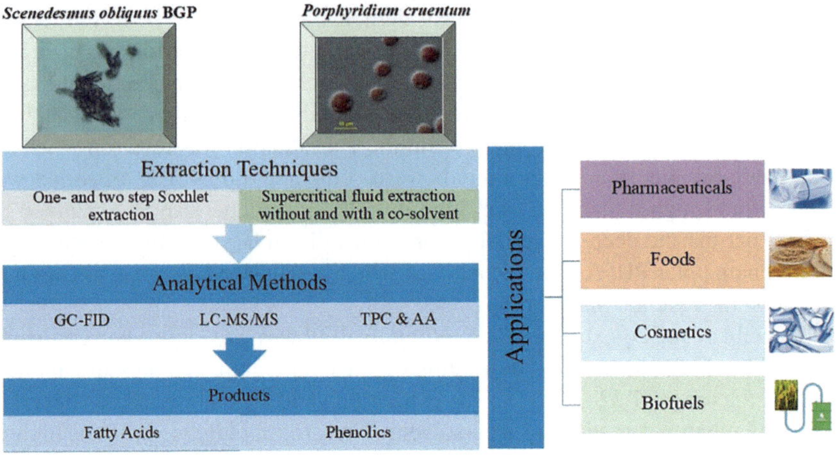


**Fig. 9.2** Extraction techniques and analytical methods applied to quali-quantify the extracts

The two strains perspectives as a sustainable source of bioactives was further revealed by examining the following indices:

- OSI, which indicates whether an oil is stable and can be used for the production of biodiesel without the addition of any antioxidants to enhance the stability during the OSI period, up to which the oil/biodiesel quality would remain unchanged and biodiesel must be entirely utilized for engine operation;
- h/H which is represented as the Σ hypocholesterolemic fatty acids/Σ hypercholesterolemic fatty acids ratio, which is related to cholesterol metabolism;
- IA, index of atherogenicity, which characterizes the atherogenic potential of fatty acids, and which is considered to be more adequate indicator when compared to the PUFA/SFA ratio, which according to the World Health Organization (WHO) recommendations for a "balanced diet" should be above 0.4, in order to avoid the risk of developing cardiovascular and other chronic diseases. Still, recently, PUFA/SFA ratio is being considered too general and unsuitable for assessing the atherogenicity of food products.

Regarding OSI, according to American Society for Testing and Materials standard oils can be classified as best, moderate, and poor. The best oils are characterized by an OSI $\geq$ 3 h. Both *S. obliquus* BGP and *P. cruentum* oils are classified as "best oils" since their OSI indices are in the range 3.89–3.90, and hence can be used for biodiesel production without any antioxidants.

Regarding h/H indices, it is accepted that nutritionally, higher h/H values are considered more beneficial for human health. Though the analysis of the fatty acids and polyphenols in *P. cruentum* oil indicates its superior potential for food and pharmaceutical applications compared to the green algae, still the h/H indices of both microalgal species oils are higher than or in the worst case commensurable with the range of h/H indices calculated for various fish species.

The atherogenicity index reflects the LDL to HDL ratio in the blood plasma. IA of *S. obliquus* BGP oil obtained with $scCO_2$ + 10% ethanol was lower than the IAs of most red and brown seaweeds. Consequently, it can be used as an additive to foods or products that can help prevent plaque accumulation and reduce levels of total cholesterol and "bad" cholesterol.

## 9.4 Bio-refineries Challenges

The one-feed multiproduct zero-waste biorefinery is considered at present as the best option to carbon-neutrality, and is expected to play a major role in the transition to circular bio-economy. A biorefinery can integrate any type of biomass feedstock and effective advanced processes to achieve the complete utilization of biomass/biowaste by creating multiple product streams.

Still, to maximize the value of the biomass feedstocks and consequently realize that green leap towards sustainability there are some bottlenecks and challenges inherent to solvent extraction problems, for example: those connected with the application of high pressures; the realization of efficient separation of different fractions without compromising and harming the other fractions; quality of the products—non-energy related being of particular importance, greenhouse gas (GHGs) e greenhouse gas (GHGs) emission and their impact on biodiversity, etc. Those issues are closely interrelated with the ability to model, design and optimize the future smart biorefineries that involve extraction and separation processes. That, in turn, requires availability of robust tools that must incorporate a framework for the reliable and efficient modelling of the thermodynamics of bio-based systems (a "thermodynamic modelling framework, TMF"). It should be based on up-front scientific knowledge, taking into consideration the importance of a priori evaluation of alternative processes that can address the presently existing controversies concerning the production and use of either biofuels or bioactives, before a decision to invest capital is taken.

As discussed in details in [45], in order for a TMF to be used as a powerful instrument it should include three elements; (i) a library of the available thermophysical properties of pure compounds and a variety of methods for estimating the missing ones; (ii) thermodynamic models for mixture properties; and (iii) model-independent routines, namely methods, algorithms, and numerical techniques, for performing stability analysis and solving the phase and chemical equilibrium relations.

Hierarchically, the first element of the TMF though situated on the lowest level, is of utmost importance. The information about the thermophysical properties of the constituents of the mixture under consideration is vital for the performance of both Elements 2 and 3. Thus, in cases when for certain pure compounds of interest to biofuels and/or food additives, nutra- and pharmaceutical production there are no experimental values of their thermophysical properties (either because the measurements are expensive or because cannot be realized) still such information must be provided, applying the available methods for properties estimation.

The values estimated can be qualified as "hypothetical", and are used further by elements 2 and 3 to realize the phase equilibria predictions and calculations. Moreover, the performance of the two modules can be regarded as a test for the reliability of the thermo-physical properties estimated as the uncertainties associated with the estimated values might have a considerable effect on the quality of the calculations and could lead, in certain cases, even to erroneous conclusions.

Though the availability of methods for properties estimation is considerable, still there is not a generally accepted criterion or a recipe how to evaluate beforehand the reliability of an estimated value of a given pure compound property. This is a serious impediment since even a minor uncertainty in a property value can influence negatively the quality of the phase equilibria predictions and calculations results, distort the design of unit operations based on phase equilibria and mislead the whole process of comparison of the alternatives generated as shown recently in [46].

For example, with regard to the recovery of energy related compounds, experimental data for the critical properties are available for bio-diesel methyl esters with up to 23 carbon atoms, but triglycerides of fatty acids with more than 14 carbon atoms in the ester chains decompose when measured. Hence, the missing properties have to be estimated.

The influence of the uncertainties in properties values can seriously influence the rigorousness of the phase behavior predictions and calculations in particular those performed at higher pressures. As demonstrated by Sovova and Stateva [47] on the example of the system triolein + $CO_2$, when some of the triolein critical parameters estimated with the in-built methods of commercial process simulators were applied to calculate the vapor–liquid (VL) equilibria of the system, either the extent of the VL region was erroneously predicted or even the system was identified as homogeneous at very low values of the pressure.

With regard to non-energy related compounds recovery, algae biomass contains a wide spectrum of high added value substances, for many of which the thermophysical properties are not known. For example, the genus *Porphyridium* synthesizes the nutritionally important PUFAs eicosapentaenoic acid, and arachidonic acid, the total amount of which can surpass 40% of the total fatty acids in certain strains. Still for AA, the only properties measured experimentally are its boiling and melting points, and vapor pressure, while for EPA—no data what-so-ever is available.

Yet another important example is the keto-carotenoid astaxanthin, for which only the melting point is available. Consequently, all missing required properties have to be estimated. The task can be further complicated by the fact that, as discussed by Sovova and Stateva [48], in the literature then there were no estimations of astaxanthin critical properties values reported, with which to compare the hypothetical values estimated by them.

Finally, taking into consideration that recently there was a shift in the biorefinery paradigm, namely from biomass-to-energy/biofuels to biomass-to-biochemicals in a biorefinery systems framework [49], it is important to note that in order to be economically efficient a biorefinery should be focused either to: low-value, high-volume products (like biofuels) or to high-value, low-volume biochemicals (like biopharmaceuticals, nutraceuticals, etc.).

Therefore, biorefineries for energy related products can be profitable since for example bioethanol costs just 0.5 $/L, but the global bioethanol market is estimated to be 140 billion L in 2022, with a compound annual growth rate of 7.6% from 2016 to 2022, and projected to reach $ 114.7 billion L in 2027. Biodiesel costs 1.59 Euros/L, but Europe biodiesel market size is valued at $25.1 billion in 2022, and is estimated to reach $34.0 billion by 2032.

On the other hand, commercial market price of astaxanthin varies from $2000–$7000 $kg^{-1}$ depending on the producer, purity, etc. Still, to the best of our knowledge, there is just one "the most vertically integrated astaxanthin producer in world" from *Haematococcus pluvialis* Cyanotech –Hawaii, situated in Kona, Hawaii [50]. Details are not disclosed aside from the fact that the process includes several stages from "the reddening system" open ponds, where *H. pluvialis* cultures turn from green to red by stress, to the in-house solvent free custom $CO_2$ extraction facility, operating up to 1000 bar, encapsulation, etc.

In conclusion, the advancement of bio-circular economy through biomass transformation can be achieved not by evading challenges but by embracing and addressing them. This can be accomplished by: introducing innovative, advanced methods and technologies that exhibit low environmental impact within integrated zero-waste biorefineries; applying dependable, robust, and efficient design and optimization methodologies; conducting a comprehensive examination of the advantages and trade-offs associated with the designed biorefinery flowsheets to gain deeper insights into their potential and limitations; and lastly by realizing Back to Earth alternatives by implementing the "cradle to cradle" concept.

**Acknowledgements** The authors acknowledge with gratitude the funding from the European Union's Horizon 2020 research and innovation programme under the Marie Sklodowska-Curie grant agreement No 778168. JAPC, and MPR also thank LA/P/0056/2020 (IMS DOI:https://doi.org/10.54499/LA/P/0056/2020), CQE UIDB/00100/2020 and UIDP/00100/2020 (https://doi.org/10.54499/UIDP/00100/2020) funded by national funds through FCT/MCTES (PIDDAC).

## References

1. Solarte-Toro J, Cardona Alzate C (2023) Sustainability of biorefineries: challenges and perspectives. Energies 16(9):1–24
2. Sarwer A, Hussain M, Al-Muhtaseb A, Inayat A, Rafiq S, Khurram M, Ul-Haq N, Shah N, Alaud Din A, Ahmad I, Jamil F (2022) Suitability of biofuels production on commercial scale from various feedstocks: a critical review. ChemBioEng Rev 9(5):423–441
3. Introduction to Chemicals from Biomass, 2nd edn. Wiley, West Sussex
4. Bamigbade G, Oyelami O, Babalola O, Adewolu A, Omemu A, Ogunsanya T, Sanusi J, Daniel O (2025) An updated comprehensive review on waste valorization: Informetric analysis, current insights and future perspectives on cereal waste and byproduct utilization for sustainable industrial applications. Bioresour Technol 418:1–23
5. da Rosa R, Ferreira V, Barroso T, Castro L, Forster-Carneiro T (2025) Corn biomass as a feedstock for biorefinery: a bibliometric analysis of research on bioenergy, biofuel and value-added products. Biofuel Bioprod Biorefin 19:206–230

6. Grain production worldwide by type 2024/25. Statista Homepage. https://www.statista.com/statistics/263977/world-grain-production-by-type/. Accessed 29 Jan 2025
7. Win H, Teethaisong Y, Ketudat-Cairns J, Ketudat-Cairns M (2024) Optimizing extraction conditions for isolation of antioxidant compounds from black rice bran using ultrasonic-assisted extraction and response surface methodology. J Food Sci 90(1):1–16
8. Tan BS, Norhaizan M, Chan L (2023) Rice bran: from waste to nutritious food ingredients. Nutrients 15(11):1–25
9. Kumar V, Singh P, Sharma A (2023) Recent trends of potential medicinal uses of rice bran with its composition and bibliographic analysis. eFood 4(3):1–21
10. Eng H, Mohd Rozalli N (2022) Rice bran and its constituents: introduction and potential food. Int J Food Sci Technol 57(7):4041–4051
11. Punia S, Kumar M, Siroha A, Purewal S (2021) Rice Bran Oil: Emerging Trends In Extraction, Health Benefit, And Its Industrial Application. Rice Sci 28(3):217–232
12. Friedman M (2013) Rice brans, rice bran oils, and rice hulls: Composition, food and industrial uses, and bioactivities in humans, animals, and cells. J Agric Food Chem 61(45):10626–10641
13. Okonogi S, Riangjanapatee P, Sasarom M, Rodwattanagul S, Chantadee T (2024) Potential of supercritical carbon dioxide extraction for high recovery of bioactive compounds and antioxidant activities of pigmented rice bran. Appl Food Res 4(2):1–7
14. Rajchman M, Gutiérrez M, Curutchet A, Cozzano C (2022) Green extraction of rice bran proteins and full revaluation of the remaining by-products. Cereal Chem 99(1):184–193
15. More P, Jambrak A, Arya S (2022) Green, environment-friendly and sustainable techniques for extraction of food bioactive compounds and waste valorization. Trends Food Sci Technol 128:296–315
16. Irakli M, Katsantonis D, Ward P (2017) Rice Bran: a promising natural antioxidant component in breadmaking. J Nutraceut Food Sci 2(3:14):1–3
17. Borresen JP, Ryan EP (2014) Chapter 22: Rice bran: a food ingredient with global public health opportunities. In: Wheat and rice in disease prevention and health: benefits, risks and mechanisms of whole grains in health promotion. Academic Press: Cambridge, MA
18. Balegh S, Saadati P, Safiaghdam M, Fakhari A, Razavi R (2025) Effect of cold atmospheric plasma pre-treatment and ultrasound-assisted extraction on bioactive compounds and chemical quality of rice bran oil. LWT Food Sci Technol 218(117451):1–9
19. Ruttarattanamongkol K, Chittrakorn S, Tochampa W (2024) Rice bran valorization: Oil extraction via extrusion-SC-$CO_2$ processing and development of protein. Appl Food Res 4:1–13
20. Echenique J, Alvarez-Rivera G, Luna V, Antonio A, Mazalli M, Ibañez E, Cifuentes A, Oliveira A (2024) Pressurized liquid extraction with ethanol in an intermittent process for rice bran oil: evaluation of process variables on the content of ß-sitosterol and phenolic compounds, antioxidant capacity, acetylcholinesterase inhibitory activity, and oil quality. LWT Food Sci Technol 207(116650):1–9
21. Ratanasongtham P, Bunmusik W, Luangkamin S, Mahatheeranont S, Suttiarporn P (2024) Optimizing green approach to enhanced antioxidants from Thai pigmented rice bran using deep eutectic solvent-based ultrasonic-assisted extraction. Heliyon 10(1–14):e23525
22. Leonarski E, Kuasnei M, dos Santos E, Benvenutti L, Moraes P, Cesca K, de Oliveira D, Zielinski A (2024) Ultrasound and microwave-assisted extractions as green and efficient approaches to recover anthocyanin from black rice bran. Biomass Convers Biorefin 15:7251–7264
23. Wongwaiwech D, Kamchonemenukool S, Ho C, Li S, Majai N, Rungrat T, Sujipuli K, Pan M, Weerawatanakorn M (2023) Bioactives from crude rice bran oils extracted using green. Molecules 28:1–20
24. Fraterrigo Garofalo S, Cavallini N, Destefano R, Micera M, Cavagnero C, Botto A, Savorani F, Tommasi T, Fino D (2023) Carbon dioxide extraction of rice bran oil and γ-oryzanol using multi-factorial design of experiment. Waste Biomass Valoriz 14:3327–3337
25. Fraterrigo Garofalo S, Demichelis F, Mancini G, Tommasi T, Fino D (2022) Conventional and ultrasound-assisted extraction of rice bran oil with isopropanol as solvent. Sustain Chem Pharm 29:1–10

26. Coelho J, Robalo M, Fernandes I, Stateva R (2022) Optimization of oil recovery from japonica Luna rice bran by supercritical carbon dioxide applying design of experiments: characterization of the oil and mass transfer modeling. ChemEng 6(4):1–16
27. Pinto T, Coelho J, Pires B, Neng N, Nogueira J, Bordado J, Sardinha J (2021) Supercritical carbon dioxide extraction, antioxidant activity, and fatty acid composition of bran oil from rice varieties cultivated in Portugal. Separations 8(8):1–18
28. Wang Y, Zhao L, Zhang R, Yang X, Sun Y, Shi L, Xue P (2021) Optimization of ultrasound-assisted extraction by response surface methodology, antioxidant capacity, and tyrosinase inhibitory activity of anthocyanins from red rice bran. Food Nutrit Sci 8(2):921–932
29. Chemat F, Vian MA, Cravotto G (2012) Green extraction of natural products: concept and principles. Int J Mol Sci 13:8615–8627
30. Hou M, Wu J, Miao J, Zeng H, Liao J, Hang S (2024) Combination of fermentation and enzymolysis enhances bioactive components and function of de-oiled rice bran. J Sci Food Agric 104:9514–9523
31. Herrero M (2023) Towards green extraction of bioactive natural compounds. Anal Bioanal Chem 416(9):1–9
32. Adeniyi OM, Azimov U, Burluka A (2018) Algae biofuels: current status and future applications. Renew Sust Energ Rev 90:316–335
33. Trivedi J, Aila M, Bangwal DP, Kaul S, Garg MO (2015) Algae based biorefinery—how to make sense? Renew Sust Energ Rev 47:295–307
34. Dong T, Knoshaug P, Lieve RD, Laurens ML, Van S, Philip W, Pienkos T, Nagleet N (2016) Combined algal processing: a novel integrated biorefinery process to produce algal biofuels and bioproducts. Algal Res 19:316–323
35. Ishaq A, Matias-Peralta H, Basri H (2016) Bioactive compounds from green microalga—Scenedesmus and its potential applications: a brief review. Pertanika J Trop Agric Sci 39(1):1–16
36. Idrissi Abdelkhalek ELA, Mohamed B, Mohammed AM, Lotfi A (2016) Growth performance and biochemical composition of nineteen microalgae collected from different Moroccan reservoirs. Mediterr Mar Sci 17(1):323–332
37. Ganesan R, Manigandan S, Samuel MS, Shanmuganathan R, Brindhadevi K, Chi NTL, Duc PA, Pugazhendh A (2020) A review on prospective production of biofuel from microalgae. Biotechnol Rep 27
38. Vasileva I, Boyadzhieva S, Kalotova G, Ivanova J, Kabaivanova L, Naydenova G, Jordanova M, Yankov D, Stateva RP (2021) A new Bulgarian strain of *Scenedesmus* sp.—identification, growth, biochemical composition, and oil recovery. Bulg Chem Commun 53:105–116
39. Tsvetanova FV, Boyadzhieva SS, Coelho JAP, Yankov DS, Stateva RP (2024) Sustainable transformation of two algal species of different genera to high-value chemicals and bioproducts. Molecules 29(156):1–19
40. Pignolet O, Jubeau S, Vaca-Garcia C, Michaud P (2013) Highly valuable microalgae: biochemical and topological aspects. J Ind Microbiol Biotechnol 40:781–796
41. Tzima S, Georgiopoulou I, Louli V, Magoulas K (2023) Recent advances in supercritical $CO_2$ extraction of pigments, lipids and bioactive compounds from microalgae. Molecules 28(3):1410
42. Feller R, Matos A, Mazzutti S, Moecke E, Tres M, Derner R, Oliveira J, Junior A (2018) Polyunsaturated ω-3 and ω-6 fatty acids, total carotenoids and antioxidant activity of three marine microalgae extracts obtained by supercritical $CO_2$ and subcritical n-butane. J Supercrit Fluids 133:437–443
43. Gallego R, Martínez M, Cifuentes A, Ibáñez E, Herrero M (2019) Development of a green downstream process for the valorization of *Porphyridium cruentum* biomass. Molecules 24(8):1564
44. Ardiles P, Cerezal-Mezquita P, Salinas-Fuentes F, Órdenes D, Renato G, Ruiz-Domínguez MC (2020) Biochemical composition and phycoerythrin extraction from red microalgae: a comparative study using green extraction technologies. Processes 8(12):1–16
45. Stateva RP, Cholakov GS (2020) Chapter 4: challenges in the modelling of thermodynamic proper-ties and phase equi-librium calculations for biofuels process design. In: Adrian B-P, Rangaiah GP (eds) Process systems engineering for biofuels development. Wiley

46. Tronci S, Garau G, Stateva RP, Cholakov G, Wakeham WA, Errico M (2023) Analysis of hybrid separation schemes for levulinic acid separation by process intensification and assessment of thermophysical properties impact. Sep Purif Technol 310:1–8
47. Sovova H, Stateva RP (2011) Supercritical fluid extraction from vegetable materials. Rev Chem Eng 27:79–156
48. Sovova H, Stateva RP (2019) New developments in the modelling of carotenoids extraction from microalgae with supercritical CO2. J Supercrit Fluids 148:93–103
49. Giuliano A (2023) The transition of scientific research from biomass-to-energy/biofuels to biomass-to-biochemicals in a biorefinery systems framework. Energies 16(5):1–4
50. Cyanotech Corporation. Home page, https://www.cyanotech.com. Accessed 13 Jan 2025

# Chapter 10
# Food Waste Prevention and Valorization

**Gerasimos Lyberatos**

**Abstract** It is estimated that one third of the food produced worldwide is lost or wasted with significant social, economic and environmental consequences. Food loss prevention and waste reduction is the best option from a hierarchical point of view. However, food surplus is always generated, and this is best managed by redistribution and upcycling practices. The food that cannot be saved or utilized for human or animal consumption ends up as food waste. Food waste represents one of the most challenging environmental problems in our times. While its disposal to landfills is responsible for the emissions of potent greenhouse gases to the atmosphere, such as methane, if separately collected, it may be exploited to produce compost, biofuels and other added value products. Source separated and collected Household Food Waste may be directly led to anaerobic digestion or composting facilities for the production of biogas and/or compost. An interesting possibility is to dry and shred the food waste collected at the municipality level, generating a homogeneous, dry, biomass, called FORBI (Food Residue Biomass) that emits no bad odors, may be stored for prolonged periods of time without deterioration and used as a potential feedstock for alternative processes, such as the production of biogas (biomethane, biohydrogen and/or hythane), compost, alternative fuel for the cement industry, pellets for heating, activated carbon for water purification, animal feed etc. These possibilities are discussed in the light of the experiences gained from relevant research programs in two Municipalities, Halandri and Andros.

**Keywords** Food waste · Valorization · Waste prevention

---

G. Lyberatos (✉)
School of Chemical Engineering, National Technical University of Athens, Athens, Greece
e-mail: lyberatos@chemeng.ntua.gr

Institute of Chemical Engineering Sciences, FORTH, Patras, Greece



A. Angelini et al. (eds.), *Security Enhancement for Climate Change Impacting Urban Resources - SECCURe*, NATO Science for Peace and Security Series C: Environmental Security, https://doi.org/10.1007/978-94-024-2345-7_10

## 10.1 Introduction

According to the Food and Agriculture Organization of the United Nations every year 1.6 billion tons of «primary product equivalents» of food are wasted [1]. This corresponds to a carbon footprint of 3.3 billion tons of $CO_2$ equivalent of Green House Gases (GHG) released to the atmosphere, while 250,000 $m^3$ of water are used to produce food that is lost or wasted and 1.4 billion hectares of land—28 percent of the world's agricultural area—are used to produce food that is wasted.

Food is lost or wasted along the entire value chain [2]:

- During or immediately after harvesting on the farm
- After leaving the farm for handling, storage, and transport
- During industrial or domestic processing and/or packaging
- During distribution to markets, including at wholesale and retail markets
- In the home or business of the consumer, including restaurants and caterers

Food loss and wasting in the EU originates as follows [3]:

- 30% from the production and processing sectors
- 5% from wholesale and retail sectors
- 12% from food services
- 53% from households

As seen in Fig. 10.1 the biodegradable fraction of the municipal solid waste in a typical Greek Municipality (the Municipality of Halandri located in the suburbs of Athens with an approximate population of 100.000 inhabitants) is as high as 47%, most of which is food waste [4].

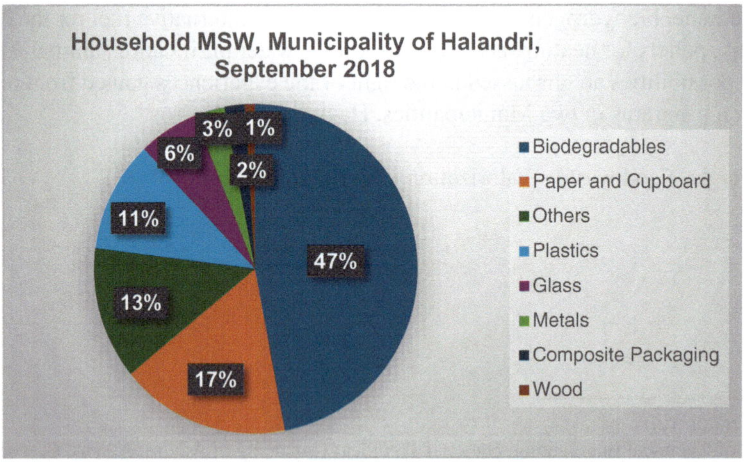


**Fig. 10.1** Composition of Household municipal solid waste (MSW) in the municipality of Halandri

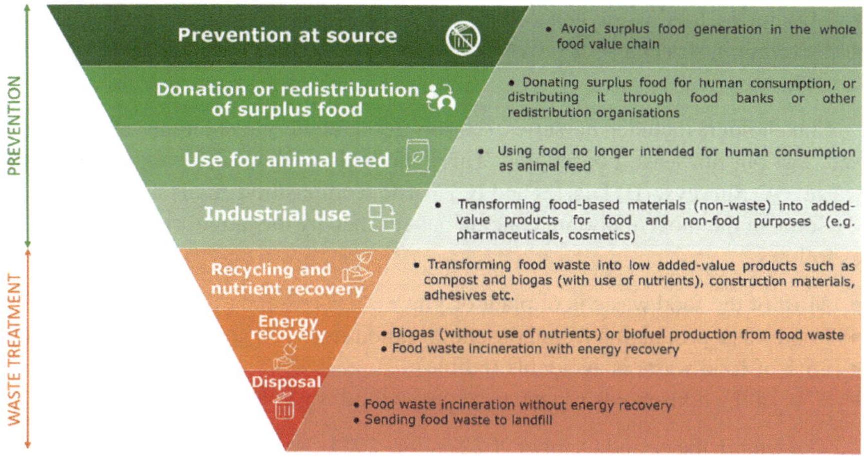


**Fig. 10.2** Food Waste Hierarchy [5]

Figure 10.2 presents the hierarchy for the management of food residues [5]. According to this hierarchy, prevention at the source is the most preferable solution. This is followed by reuse, preferably for human consumption and, if not possible, for animal consumption. However, no matter what the efforts and measures taken for prevention of food loss and reuse of food via redistribution and upcycling, some food wasting will be unavoidable, and its best use is then to valorize it for the production of chemicals, biofuels and other materials such as compost. The least desirable options are incineration and eventually landfilling, a practice that should be avoided as it leads to greenhouse gas emissions.

Following this hierarchy, the Food Waste Related Goals that should be set at Municipality level are then (a) Reduce food loss by prevention and reuse and (b) Divert the unavoidable food waste from landfilling by collection at the source and valorization for the production of biofuels, energy and compost. In the sequel, three examples of proper food waste management are presented following precisely the presented rationale.

## 10.2 Food Loss and Waste Prevention in the Municipality of Halandri

The Municipality of Halandri via three EU funded projects (FoodRUs, ToNoWaste and Food Connections) [6–8] founded in October 2022 a Food Loss and Waste Prevention Unit aiming at:

- Monitoring food waste generation
- Developing strategies for the reduction of food loss and food waste from farm to fork.

Stakeholders included nurseries, schools, food retailers, restaurants and cafeterias and, last but not least, households. Three methods were used for monitoring food waste: (A) Surveys, (B) Diaries and (C) Waste Composition Analysis.

Some of the most important conclusions reached by the monitoring campaign are as follows:

- Most of the food waste is of plant origin
- Food waste production from households amounts to 250 g/capita/d on the average
- 25% of food waste is avoidable (edible)
- 67% of the households separately dispose food waste to the brown bin

The Unit in the period October 2022—June 2024 carried out many actions aiming at reducing food wasting, such as:

- Campaigns with brochures to sensitize the general population
- Zero waste cooking events
- Living labs for reusing food leftovers
- Living labs for food upcycling (use of spent brewery grains to produce cookies)
- Diversion of end-of-day open market leftovers to the Municipality Social Grocery
- Events in schools and nurseries

## 10.3 Food Waste Collection at the Source and Valorization for the Production of Biofuels and Compost

In the framework of Waste4Think, a Horizon 2020 project [9], the Municipality of Halandri with the help of the National Technical University of Athens (NTUA) developed in the period "(2016–1020)" separate collection and valorization of Household Food Waste (HFW) at pilot scale.

Food waste was collected at the source from approximately 800 citizens (250 households) who were provided with (a) portable 30 L brown bins, (b) compostable bags, (c) instructions of what to dispose in the brown bin and (d) a key to a locked nearby 120 L bin, so that it would be accessible only by informed citizens. The latter is particularly important as it secured high purity of the collected waste (exceeding 98%).

The food waste was collected twice a week and taken to a developed dryer/shredder unit in the Municipality, where it was dried and shredded producing what we called a food-residue biomass (FORBI) with the following advantages:

- It has 1/4 to 1/5 the weight of food waste, implying reduced transportation costs
- It has low-moisture and may be stored for prolonged periods of time without deterioration
- It is homogeneous
- It does not emit odors
- It may be used for producing fuels, energy and other products

In a typical month, 4021 kg of household food waste (HFW) were collected and 1006 kg of FORBI were produced upon drying. The HFW weight was thus reduced by 77%. Figure 10.3 presents the management scheme of HFW in the Municipality. Table 10.1 presents the main characteristics of FORBI as determined at NTUA.

FORBI was then examined at NTUA for possible valorization in 8 different ways:

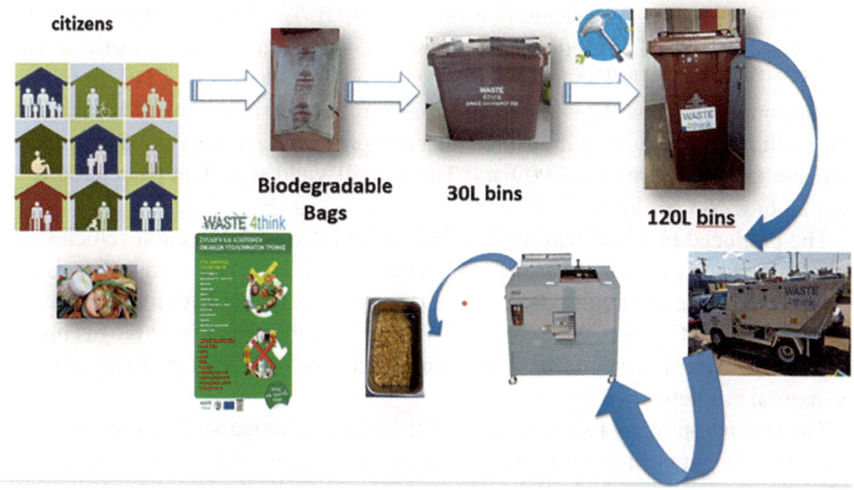


**Fig. 10.3** The food waste management scheme as developed in the Municipality of Halandri, Greece

**Table 10.1** Compositional analysis of FORBI (food residue biomass)

| Component | %, w/w, dry basis |
|---|---|
| Proteins | 13.7 ± 0.4 |
| Lipids | 12.3 ± 0.1 |
| Extractives (mainly sugars) | 27.3 ± 1.7 |
| Starch | 10.7 ± 0.1 |
| Pectins | 3.3 ± 0.8 |
| Cellulose | 10.3 ± 0.1 |
| Hemicellulose | 11.3 ± 0.2 |
| Total lignin | 6.8 ± 0.1 |
| Ash | 7.2 ± 0.3 |

1. Gaseous Biofuels (Methane, Hydrogen, Hythane)
2. Liquid Biofuels (Bioethanol)
3. Compost
4. Solid biofuels (pellets)
5. Alternative Fuel (AF) for the cement industry
6. Direct production of electricity (microbial fuel cell technology)
7. Adsorbent
8. Animal Feed.

Among the 8 alternatives, the most promising ones based on technology readiness level (TRL) and the yields obtained in the trials were: (a) biogas production, (b) use as fuel in the cement industry and (c) production of compost. A 4 $m^3$ methanogenic reactor was constructed and operated at NTUA under mesophilic conditions (35–40 °C) at a hydraulic retention time (HRT) of 40 days. The feed contained a suspension of FORBI with a concentration of 15g FORBI/L. The mean biogas productivity was 2.5 $m^3$/d. The methane content was 65%, so that the mean methane yield was 492 L/kg FORBI. The produced biogas was upgraded to bioCNG (biological origin compacted natural gas) using an amine scrubber for removing $CO_2$ and $H_2S$, a silica column for removing moisture and a compressor that compressed the purified methane (> 99% purity) to 200 bars. The overall yield was then calculated to be 0.27 kg bioCNG per kg of FORBI.

The produced bioCNG was used to fuel two solid waste collection vehicles, one diesel-based and one petrol-based (converted to operate on natural gas). This is then a perfect example of circular economy, as the vehicles that collect food waste were fueled by bioCNG generated from them. It should be mentioned that biogas may be alternately (a) used in a combined heat and power process (CHP) or (b) injected into the natural gas network.

The second option consists of using FORBI as an alternative fuel for the cement industry. The replacement of fossil fuels in the cement industry is highly desirable as it leads to: (a) Reduction of energy costs, (b) Saving of natural resources and (c) Reduction of $CO_2$ emissions. There are many examples of alternative fuels such as automotive tires, used oils, sludge, refinery sludge, meat meal, agricultural residues, RDF (Refuse Derived Fuel), SRF (Solid Recovered Fuel). To examine the suitability of FORBI as an alternative fuel EN 15.359 classification [10], presented in Table 10.2 was used. In this classification net calorific value, chlorine content and mercury content are determined, and the candidate waste is classified depending on the value determined in 5 possible classes, with class 1 being best. FORBI proved to be Class 3 in NCV (18 MJ/Kg), Class 2 in Cl (0.5% and Class 1 in Hg (0.006mg/MJ). Consequently, it may be used as it is as an alternative fuel for the cement industry.

Testing of FORBI as a alternative fuel took place in TITAN, the biggest cement company in Greece. 8.3 tons of FORBI stored over a period of 3 years replaced pet coke as a fuel. The material was very well preserved and had no undesirable odors. It was found to be excellent from an operational point of view.

Finally, in order to test the suitability of FORBI for composting, it was mixed with prunings in 1:1 proportion and a total of 4 tons of material were windrow composted.

**Table 10.2** Fuel classification according to EN 15.359

| Property | Unit | Class | | | | |
|---|---|---|---|---|---|---|
| | | 1 | 2 | 3 | 4 | 5 |
| NCV | MJ/kg | $\geq 25$ | $\geq 20$ | $\geq 15$ | $\geq 10$ | $\geq 3$ |
| Cl | %weight (dry basis) | $\leq 0.2$ | $\leq 0.6$ | $\leq 1.0$ | $\leq 1.5$ | $\leq 3$ |
| Hg | mg/MJ | $\leq 0.02$ | $\leq 0.03$ | $\leq 0.08$ | $\leq 0.15$ | $\leq 0.50$ |

It was found that in approximately 40 days complete composting was achieved. Consequently, FORBI may be used as a compost enhancer, since its readily degradable organic content helps speed up significantly the composting process of prunings, that would otherwise take months to achieve. The generated compost proved to be excellent and was bagged and returned to the participating citizens as a reward for their participation in the project.

From all the above it is clear that FORBI is an excellent material and its production from source-sorted food waste presents a very sustainable alternative. Details for the valorization processes developed in the framework of the WASTE4THINK project may be found in the references [11–20].

## 10.4 Food Waste Collection at the Source and Composting in an Island Community (the Greek Island of Andros)

The NTUA team participated as a partner in the Euro-Mediterranean project SIRCLES (Supporting Circular Economy Opportunities for Employment and Social Inclusion) [21]. Through this project, separate collection of food waste from households and HORECA (hotels, restaurants and cafeterias) was organized in the island community Korthi, Andros, following the model developed in Halandri.

As shown in Fig. 10.4, there is significant seasonal variation in the quantities of collected food waste. The highest amount was collected in mid-August and the lowest in December. Composting food waste for such quantities in a decentralized island community appears to be the best possible option for handling food waste. A significant challenge in this case is the inability to use typical windrow composting, because of the very strong winds that prevail during the summer months. Because of this, a system of closed 1 $m^3$ composting tanks was considered most suitable.

The feed to the composters consisted of a mixture containing approximately 2/3 of food waste and 1/3 prunings and leaves) as well as recycled compost (the course part after sieving) in more or less equal amounts. The mean feed composition is presented in Fig. 10.5.

Approximately after one month of operation in a composter (with turning and moisture addition as required) the mixture was transferred to a second composter and after one more month to a third one, as shown in Fig. 10..6. Following this,

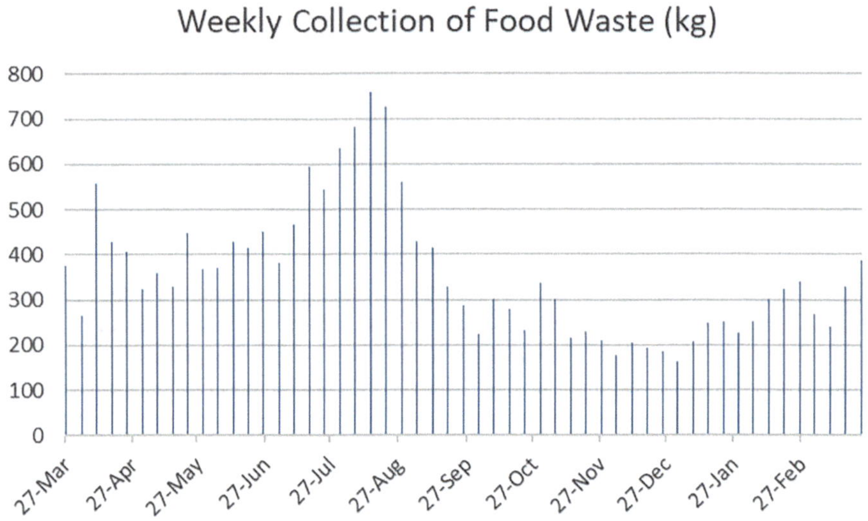


**Fig. 10.4** Variation of the collected food waste in Korthi, Andros [22]

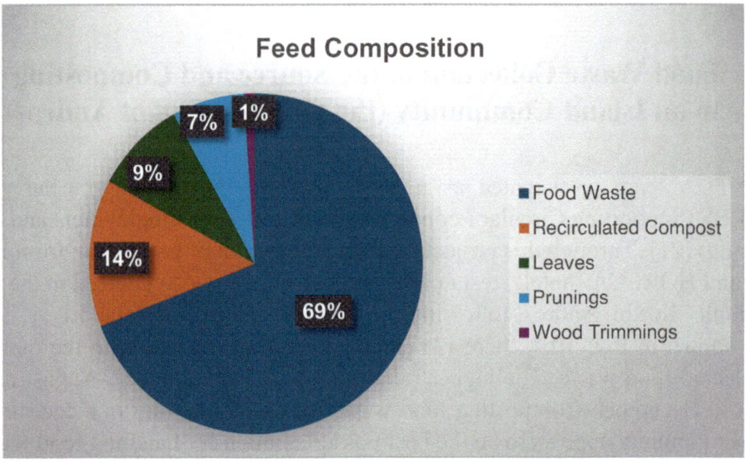


**Fig. 10.5** Mean composition of feed to the composters [22]

the mixture was sieved with approximately half of the compost (the fine part) being bagged as a final product and half (the course part) recirculated in the feed.

Figure 10.7 shows a typical temperature evolution of the composting process. In 12 months, the unit received 19 tons of biowaste, while 9 tons of compost were produced and donated to the community. Details may be found in [22].

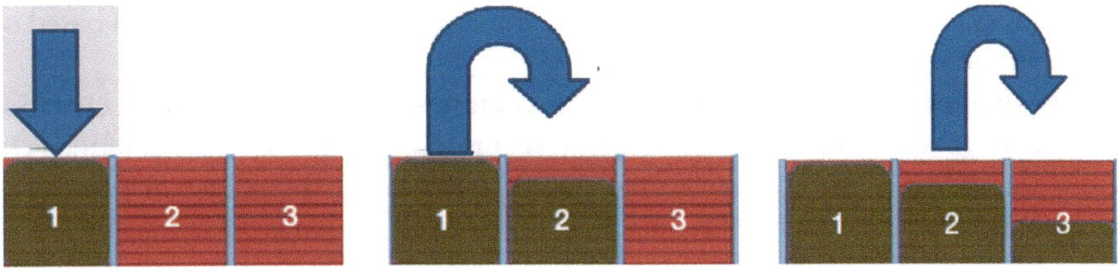


**Fig. 10.6** Scheme of composter operation

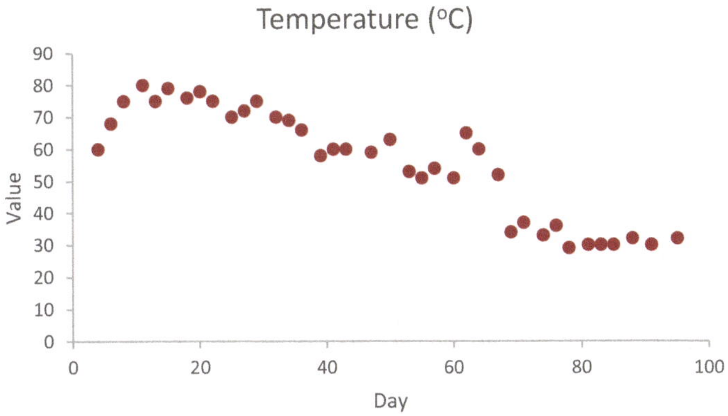


**Fig. 10.7** Typical temperature evolution during composting [22]

## 10.5 Conclusions

Following the food wasting hierarchy, the most important issue is prevention of food wasting. This is optimally handled at Municipality level as demonstrated by the Municipality of Halandri. No matter how effective the preventive measures are, it is unavoidable that food waste will be generated. If mixed with commingled waste and led to landfilling, an important contribution to greenhouse gas emissions results, while an important potential source of materials and/or energy is lost. What should be then done in order to properly collect and valorize the food waste which is generated at municipal level? The various alternatives were presented in this work. It is the author's belief that the best option depends on the community size. In any case, however, separate collection of food waste at the source is the only option if food waste is to be valorized.

For relatively large communities there are many options for the valorization of food waste. A significant aspect in this perspective is the generation of an intermediate product, FORBI, as presented in this work. FORBI may then be valorized in various ways, the most promising of which are the production of biogas via anaerobic digestion, the use of FORBI as a compost enhancer, or the direct use of FORBI as an alternative fuel for the cement industry. Which of these alternatives or which mix of them should be selected depends on the situation and should be decided based on a proper technoeconomic analysis.

For smaller communities, however, direct composting of the source-collected food waste, along with proper bulking material seems to be the most viable solution. The example of Korthi, Andros, proved the viability of this approach, especially for island communities.

## References


1. Food and Agriculture Organization (FAO): Food Wastage Footprint:Impacts on Natural Resources. FAO, Paris (2013)
2. Food and Agriculture Organization (FAO): Global food losses and food waste—extent, causes and prevention, Rome (2011)
3. Stenmarck Å, Jensen C, Quested T, Moates G (2016) Estimates of European food waste levels. https://doi.org/10.13140/RG.2.1.4658.4721
4. Mathioudakis D, Papadopoulou K, Lytras GM, Pavlopoulos C, Niakas S, Filippou K, Melanitou E, Lekkas DF, Lyberatos G (2021) A detailed characterization of household municipal solid waste. Waste Biomass Valorization 12:2945–2957
5. De Laurentis V, Casonato C, Mancini L, Garcia Herrero L, Valenzano A, Sala S (2024) Building evidence on food waste prevention interventions, Publications Office of the European Union, Luxembourg
6. Horizon 2020, Grant Agreement No 101000617, Circular Solutions for Resilient Food Systems [FOODRUS]
7. Horizon Europe, Grant Agreement No 101059849, Towards a New Zero Food Waste Mindset Based On Holistic Assessment [ToNoWaste]
8. Erasmus+Food Connections project (ID: 2022-2-DE02-KA210-ADU-000098503)
9. Horizon 2020, Grant Agreement No 688995. Moving towards Life Cycle Thinking by integrating Advanced Waste Management Systems [WASTE4THINK]
10. EN 15359:2011 Solid recovered fuels—Specifications and classes (2011)
11. Antonopoulou G, Ntaikou I, Pastore C, di Bitonto L, Bebelis S, Lyberatos G (2019) An overall perspective for the energetic valorization of household food waste using microbial fuel cell technology of its extract, coupled with anaerobic digestion of the solid residue. Appl Energy 242:1064–1073
12. Lytras G, Lytras C, Mathioudakis D, Papadopoulou K, Lyberatos G (2021) Food waste valorization based on anaerobic digestion. Waste Biomass Valorization 12:1677–1697
13. Mathioudakis D, Karageorgis P, Papadopoulou K, Astrup TF, Lyberatos G (2022) Environmental and economic assessment of alternative food waste management scenarios. Sustainability 14(15), art. no 9634
14. Mathioudakis D, Karageorgis P, Papadopoulou K, Lyberatos G (2021) LCA and LCC of dried and shredded food waste as an alternative fuel for the cement industry. Waste Manag Res
15. Mathioudakis D, Michalopoulos I, Kalogeropoulos K, Papadopoulou K, Lyberatos G (2021) Anaerobic digestion of dried/shredded food waste in a periodic anaerobic baffled reactor. Water Sci Technol, Water Sci Technol 84(2):420–430
16. Michalopoulos I, Lytras GM, Mathioudakis D, Lytras C, Goumenos A, Zacharopoulos I, Papadopoulou K, Lyberatos G (2020) Hydrogen and methane production from food residue biomass product (FORBI). Waste Biomass Valorization 11(5):1647–1655
17. Papanikola K, Papadopoulou K, Tsiliyannis C, Fotinopoulou I, Katsiampoulas A, Chalarakis E, Georgiopoulou M, Rontogianni V, Michalopoulos I, Mathioudakis D, Lytras GM, Lyberatos G (2019) Food residue biomass product as an alternative fuel for the cement industry. Environ Sci Pollut Res 26(35):35555–35564

18. Tremouli A, KamperidisT, Lyberatos G (2021) Comparative study of different operation modes of microbial fuel cells treating food residue biomass. Molecules 26(13), art. no. 3987
19. Tremouli A, Karydogiannis I, Pandis PK, Papadopoulou K, Argirusis C, Stathopoulos VN, Lyberatos G (2019) Bioelectricity production from fermentable household waste extract using a single chamber microbial fuel cell. Energy Procedia 161:2–9
20. Tremouli A, Pandis PK, Karydogiannis I, Stathopoulos VN, Argirusis C, Lyberatos G (2019) Operation and electro(chemical) characterization of a microbial fuel cell stack fed with fermentable household waste extract. Global NEST J 21(2):253–257
21. European Union ENI CBC Mediterranean Sea Basin Programme, SIRCLES "Supporting Circular Economy Opportunities for Employment and Social Inclusion" (Project Number: B_A.3.1_0157_SIRCLES)
22. Lytras C, Lyberatos V, Lytras G, Papadopoulou K, Vlysidis A, Lyberatos G (2024) Development of a model composting process for food waste in an Island community and use of machine learning models to predict its performance. Waste Biomass Valorization. https://doi.org/10.1007/s12649-024-02697-9

# Chapter 11
# Urban Biodegradable Waste: Available Resources for Sustainable Circular Cities

**Camilla M. Braguglia, Agata Gallipoli, Barbara Tonanzi, Michela Sbicego, Simona Crognale, Stefania Angelini, Francesca Angelini, Daniele Montecchio, Carlo Pastore, Simona Rossetti, and Andrea Gianico**

**Abstract** With over half the world's population currently residing in cities, and projections indicating a rise to two-thirds by 2050 (with 90% of this growth in low- and middle-income countries), urban centers face escalating demands for food, water, and energy. Simultaneously, urban populations generate increasing amounts of waste, notably organic waste and municipal wastewater. The Urban Nexus approach offers an action-oriented framework within the circular bioeconomy, positioning urban biodegradable waste as a valuable resource for creating sustainable, innovative cities and addressing waste management and ecosystem protection challenges. This paper presents applied research conducted at lab and semi-pilot scale, focusing on the anaerobic microbial conversion of urban waste (food waste, sewage sludge, and compostable bioplastics) into valuable by-products, including marketable carboxylic acids, not disregarding methane production. A preliminary economic assessment for the valorization of food waste and sludge from the city of Milan, Italy, is also presented.

**Keywords** Urban biorefinery · Feedstock pretreatments · Anaerobic conversion processes

C. M. Braguglia (✉) · A. Gallipoli · B. Tonanzi · M. Sbicego · S. Crognale · S. Angelini · F. Angelini · D. Montecchio · S. Rossetti · A. Gianico
Water Research Institute, National Research Council, Montelibretti, Rome, Italy
e-mail: camilla.braguglia@irsa.cnr.it

M. Sbicego
University of Rome La Sapienza, Rome, Italy

C. Pastore
Water Research Institute, National Research Council, Bari, Italy



A. Angelini et al. (eds.), *Security Enhancement for Climate Change Impacting Urban Resources - SECCURe*, NATO Science for Peace and Security Series C: Environmental Security, https://doi.org/10.1007/978-94-024-2345-7_11

## 11.1 Introduction

Urban areas are the epicenters of human activity, innovation, and economic development. However, they also concentrate environmental challenges like pollution, resource depletion (e.g. water, soil, and energy), waste production and climate change [1]. The challenge is to foster the convergence of these complex issues, where human and social capital are central determinant of resource productivity and sustainability, nourished by our research and innovation. The urban territory cannot be considered simply a land to be conquered, a resource to be exploited. On the contrary, we should see the city as an arena where competition unfolds, a crucible where innovation is forged—or stifled—depending on how it fosters a creative environment. Recognizing the critical need for sustainable urban development in an innovation-led context, knowledge cities serve as valuable hubs for increasing citizen skills and knowledge, fostering intellectual development [2], and ultimately contributing to sustainability, high-quality life conditions, and ecosystem services. This trajectory also generates new economic opportunities within the circular economy.

Human health, well-being, quality of life, the environment, economic efficiency, and the climate all depend significantly on how we manage waste [3]. Cities are responsible for more than 70% of human-induced greenhouse gas emissions and 50% of waste generated [4]. Over the years, there has been a constant increase in municipal waste generated per capita in the EU Member States (517 kg/capita in 2020) although per capita generation values are characterised by high variability at country level [5]. At the same time, the recycling rate for municipal waste (one of Eurostat indicators measuring the progress of circular economy policies and part of the EU Sustainable Development Goals set of indicators) that includes material recycling, composting and anaerobic digestion, was only 48.6% in 2020 in the EU27. Besides, cities are projected to consume around 90 billion tons of materials annually by 2050 [6] following the current trend (compared to 40 billion tons annually in 2010). This contributes to ecological degradation and makes urban ecosystems more susceptible to harm, with profound consequences for their structure and function. To foster urban development and societal transitions, the European Commission proposed that cities transform into major circular bioeconomy hubs [7]. This transformation is intended to improve livability and resilience through better resource utilization, reduced emissions, and the fulfillment of climate goals. Fast-tracking the circular and bio-economies offers indeed a powerful tool to lessen human impact on natural resources and capital. It achieves this by protecting and restoring biodiversity, while also reducing reliance on imported strategic materials. The circular economy aims, in fact, to use resources smarter, not harder. Recycling, symbiosis, and using by-products are all ways to implement the European waste hierarchy, that prioritize preventing waste, then reusing, recycling, and finally, disposal. In particular, the bioeconomy is founded on two main pillars: the utilization of renewable natural bioresources and the valorization of organic waste and by-products. It is inherently dependent on natural capital, and to be regenerative, it must employ biological resources in a manner that is compatible with their renewability. So, sustainability and circularity are beacon of the bioeconomy, aiming at the renewal of urban sectors, the protection of the environment and biodiversity, while helping to support welfare, employment, and economy of knowledge.

In this context, urban biodegradable waste represents a significant untapped biological resource with the potential to address natural resource depletion and promote a more sustainable and circular economy. According to the last ISPRA report [8], the annual production of sludge in Italy was 3.2 Million tons (at an average concentration of 23% of dry matter) with an estimated future production of 4 Million tons [9]. As regards the organic fraction of municipal solid waste, the production in Italy in 2023 was 7.5 million tons with a 70% of food waste fraction [5], the most relevant stream separately collected from household and HORECA sector, which includes hotels, restaurants and catering companies. And, in a safe circular economy approach, we cannot overlook the increasing amount of commercial compostable bioplastics typically associated with food waste, namely flexible bioplastic items (as shopper bags made of starch-based bioplastics) and rigid compostable items (as polylactic acid cutlery) managed by the organic waste treatment plants. A major ambition is to fully harness the vast resource of biodegradable urban waste to diminish our reliance on fossil fuels and other resources, thereby bolstering the circular bioeconomy. This necessitates the implementation of innovative green technologies for the sustainable creation of industrial bio-based products, biomaterials, and renewable energy, which will drive environmental protection, conserve resources, and stimulate urban economic growth. The potential of the knowledge economy to transform urban territories into hubs of resource recovery will be explored by promoting an integrated and interdisciplinary approach that will pave the way to profound change in cities, opening also new business frameworks based on an innovative circular approach to urban waste treatment (i.e. 'green jobs' as those that arise from recycling, ecosystem restoration, etc.).

## 11.2 The Urban Nexus Approach

While many nations still manage urban resource use and development through uncoordinated sectoral departments (water, energy, agriculture, food), realizing significant progress on community priorities and resource challenges at all levels demands a shift towards minimizing trade-offs and maximizing synergies across these areas. The urban nexus approach provides a framework for this shift by examining the interdependencies among urban sectors, including waste management. It involves a process of identifying, understanding, and acting on these interrelationships to unlock synergies and reduce conflicts. Crucially, it guides stakeholders on integrating management to improve resource efficiency and on viewing waste and wastewater as potential resources.

When specifically applied to waste management, the nexus approach offers a powerful tool for fundamentally changing waste from a liability into a valuable resource.

Key Strategies for Sustainable Waste Management within the Urban Nexus are

(a) Waste Reduction and Prevention: as (i) reduction at source by minimizing for example plastic packaging, (ii) implementing efficient waste collection and sorting systems to maximize recycling rates, (iii) diverting organic waste from landfills to produce compost for agriculture and urban green spaces.
(b) Waste-to-Energy: (i) incineration, (ii) anaerobic digestion, (iii) gasification.
(c) Material Recovery and Recycling: (i) establishing efficient recycling facilities to recover valuable materials like paper, plastic, glass, and metal, ii) transforming waste into new products.
(d) Wastewater Treatment and Reuse: (i) Advanced Wastewater Treatment, (ii) novel sustainable sludge Treatment.
(e) Landfill Management and Reclamation: (i) landfill Gas Recovery, (ii) Landfill Mining, (iii) Landfill Reclamation.

In this context anaerobic conversion processes offer most great opportunities, namely: waste reduction, decontamination and resource recovery (valuable products, biogas and digestate).

## 11.3 Anaerobic Digestion: A Multi-commodities Microbial Driven Biorefinery

Anaerobic digestion (AD) represents a widely applied biological technology for the stabilization of diverse organic waste materials, such as sludge, the organic component of municipal solid waste, and agro-industrial byproducts. Among various conversion methods, biological AD is gaining traction due to its economic advantages and effective carbon recovery in the form of renewable biogas fuel. At its core, AD is a series of biological reactions carried out by microorganisms to decompose organic matter in an oxygen-free environment. The microbial community involved in AD is complex and requires carefully balanced conditions for its various bacterial groups to function optimally [10]. This process unfolds in four sequential stages: hydrolysis, acidogenesis, acetogenesis, and methanogenesis. The initial stage, hydrolysis, involves hydrolytic and fermentative bacteria breaking down complex, insoluble organic polymers into simpler, soluble molecules. Subsequently, acidogenic bacteria convert these simpler molecules into compounds like $CO_2$, $H_2$, $NH_4$, and organic acids. These organic acids then undergo further transformation by acetogenic bacteria into acetic acid, along with additional $CO_2$, $H_2$, and $NH_4$. The final step, methanogenesis, is carried out by hydrogenotrophic and acetotrophic archaea, which convert these end products into the valuable biogas components, methane and carbon dioxide [11]. While AD offers significant benefits like methane production, digestate stabilization, and waste reduction, its application can be limited by the slow degradation kinetics of the initial hydrolysis step. This often becomes the rate-limiting factor, particularly when digesting complex, particulate substrates such as sludge or food waste. To overcome this, pre-treatment methods like sludge sonication and thermal hydrolysis can be employed to enhance anaerobic conversion. Our group's extensive research has investigated the applicability and technical–economic feasibility of these strategies. Through long-term, semi-continuous operation of large bioreactors under varying organic loading rates and temperature regimes, we have demonstrated the critical need for carefully optimizing the disintegration degree, energy input, and operational conditions (e.g., ADM1 modeling shows that longer HRTs diminish the methane gain from pretreatment). Furthermore, we have correlated these factors with microbial consortia activity and functions [12, 13]. The adaptability of anaerobic digestion (AD) systems, capable of processing diverse organic feedstocks and generating various by-products, solidifies its role as an environmentally sound and economically efficient treatment method within the circular economy [10, 14, 15]. However, the limited economic value of biogas and digestate is driving the development of alternative strategies to produce higher-value end-products. Consequently, there is increasing interest in generating bulk chemicals from mixed organic waste [16–19]. Carboxylic acids such as acetate, propionate, butyrate, and valerate, which are intermediates in the AD process, have broad applications in the chemical industry, serving as bulk industrial chemicals, precursors for bioplastics [20], and in biological nutrient removal.

### *11.3.1 The Urban Biorefinery Concept*

The use of abundant, readily available, organic waste streams (typically downgraded within cities or in their vicinities) has moved the frontiers of organic waste management and wastewater treatment towards more ambitious and articulated targets that may be fulfilled by the implementation of the urban waste biorefinery concept [21, 22]. Biorefineries are in fact the central technology in bioeconomy endeavor. Operating in a similar way to petrochemical refineries, they employ a range of techniques—such as feedstock pretreatments, primary refining, secondary refinery, conversion step, purification, product conditioning and recovery and re-use of by-product streams to transform organic waste into resources. This implies: (1) Reduction of investments costs; (2) Closing the cycle of materials (nutrients, organic matter, etc.); (3) Zero waste to be managed; (4) Increase in profitability. Despite their potential, many biorefinery conversion processes are still not fully optimized and are often in early development stages [23]. This results in cost and productivity challenges that require optimization. The upcoming European legislation, such as the Green Deal, will be a primary catalyst for the deployment of urban biorefineries, driven by the need to reduce fossil fuel dependence, the functional advantages of bio-based products, and increasing awareness of the impacts of fossil resource utilization.

Urban organic waste is now recognized as a readily available, renewable, and low-cost non-food feedstock [21, 22]. Utilizing these abundant waste streams can alleviate the demand for limited raw materials. This perspective has shifted the focus of organic waste management and wastewater treatment plants towards more ambitious and comprehensive goals, potentially achievable through the implementation of the waste biorefinery concept.

### *11.3.2 A Novel Biorefinery Platform Transforming Food Waste into Valuable Chemicals and Biogas Through Thermal Pretreatment Integration*

This innovative biorefinery platform proposes an integrated approach featuring thermal pretreatment to significantly enhance the yield of fermentable organic carbon from food waste (FW). Simultaneously, this pretreatment aims to improve the feedstock quality for the production of high-value carboxylic acids from the liquid fraction and biomethane from the solid residue. Notably, lignocellulosic waste like food waste requires pretreatment to release its components and improve microbial accessibility [24–27]. Reactor operating conditions (temperature, pH, hydraulic retention time, organic loading rate) significantly influence the yield and composition of volatile fatty acids (VFAs) in the fermentation broth. Among these, pH appears critical, with near-neutral conditions (5.5–6) generally favored for maximizing VFA production and increasing the proportion of longer-chain fatty acids [28, 29], likely due to the inhibitory effect of undissociated acids on microbial growth. Therefore, pH control during fermentation, despite its added cost, is a key strategy. The novelty of this study lies in leveraging thermal pretreatment coupled with solid–liquid separation to maximize carboxylic acid yield through controlled hydrolysis and acidogenesis. The resulting liquid fraction, rich in solubilized organics, is then fermented into commercially valuable medium-chain carboxylates. Concurrently, the solid residue offers the potential to enhance energy recovery via anaerobic digestion (AD).

The composition of the food wasted in households varies from region to region of the world. In Europe, the most wasted food categories are bread, fresh fruit, fresh vegetables, milk, and yogurt as reported in several studies [30]. Typical EU household food waste composition is reported in Fig. 11.1 (mean values calculated by the authors).

Kitchen FW is composed by around 60% of vegetables and fruit, 10% pasta and bread, dairy products (including eggs) and a low fraction (around 5–10%) of meat and fish residues. At IRSA-CNR we carried out in the last ten years research activity on canteen food waste with this fixed composition. The analyzed food waste exhibited, as expected, a high water content and a dominant organic fraction (comprising 95% of total solids), pointing to its suitability for anaerobic processes. Soluble carbohydrates constituted 18% of the total COD, likely due to the high water content and the physical pretreatment effect of the preliminary grinding. Additionally, the significant presence of vegetable waste resulted in a considerable lignocellulosic recalcitrant fraction, measured at 22% of the total COD [18].

The mild thermal pretreatment applied (132 °C, 3.4 bar, 20 min) proved efficient in solubilizing organic matter, transferring around 28% of the total COD, primarily

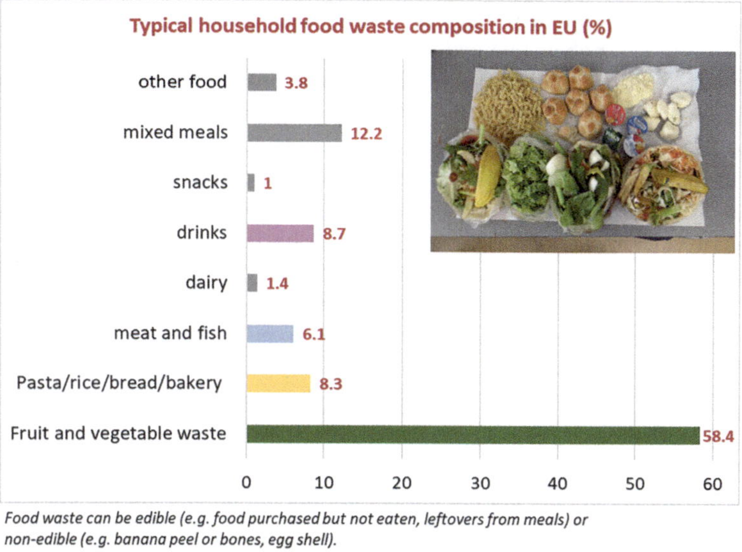


**Fig. 11.1** Typical European household food waste composition

starch, into the liquid phase, a result corroborated by previous studies [31, 32]. This process led to a decrease in starch within the pretreated solid and a concurrent rise in soluble carbohydrates. The integration of this pretreatment with solid–liquid separation significantly enhanced the extraction of organic components into the liquid fraction (Fig. 11.2), resulting in a 34% reduction in the total COD of the solid residue compared to untreated material. Notably, even this gentle thermal treatment achieved a slight breakdown of the recalcitrant lignocellulosic structure.

**Fig. 11.2** The liquid extract and the solid residue obtained after FW pretreatment and centrifugation

Anaerobic 3L fermenters, operating in semi-continuous mode with controlled pH at 6 (optimal for chain elongation to medium-chain fatty acids), and fed with the extract of raw or pretreated food waste were monitored in terms of VFAs composition, production and yield. The extract's high organic content, particularly bioavailable sugars, effectively drove primary fermentation, resulting in the production of short-chain fatty acids (acetic and butyric) and lactate. This lactate then facilitated chain elongation (CE) to produce caproic acid, a six-carbon medium-chain fatty acid. Notably, lactate was produced quickly after each feeding and subsequently fully utilized [28]. The fermentation process exhibited a high acidification degree of around 90% ($gCOD_{VFA}\ g^{-1}COD_{sol}*100$), and the conversion efficiency of soluble COD to caproate reached a maximum average of 31 ± 1% ($gCOD_{caproate}\ g^{-1}COD_{sol}*100$) at the end of the experiment. This caproate yield represents a slight improvement over the 26% achieved by Contreras-Davila et al. [33] in their FW fermentation without pH control. The overall average VFA production yield was 0.37 $g_{VFA}\ g^{-1}COD_{fed}$, and the maximum caproic acid concentration of approximately 8 g $L^{-1}$ was attained during the final feeding cycle. The microbiome composed by lactate-producing bacteria (i.e., Corynebacterium, Lactobacillus, and Olsenella) ensured the availability of electron donor for CE process operated by Clostridiaceae, and Caproiciproducens. However, this high caproate concentration likely led to the accumulation of toxic undissociated acids, negatively impacting microbiome activity and causing a decline in the chain-elongation process. By removing caproate from the fermentation broth the CE process restored, assuring higher yields (0.42 gVFA $g^{-1}COD_{fed}$, with 27 ± 2% of caproate specificity), besides process stability. The presence of a downstream phase of in situ removal and extraction of VFAs from the system was confirmed to be necessary for improving the performances of the CE process, in terms of productivity and stability.

The solid residues obtained after thermal pretreatment and solid–liquid separation showed an enriched soluble COD fraction, which enhanced methane production (0.33 $Nm^3CH4\ kg^{-1}VS_{fed}$) relative to untreated residues (0.29 $Nm^3CH4\ kg^{-1}VS_{fed}$). However, this benefit was accompanied by a heightened risk of acidification and methanogenesis failure when the hydraulic retention time (HRT) was short (20 days). Increasing the HRT to 40 days eliminated this instability, resulting in similar methane production (0.33 $Nm^3CH4\ kg^{-1}VS_{fed}$) for both pretreated and untreated residues. In the reactor processing pretreated residue at the longer HRT, the successful interaction between hydrogenotrophic methanogens and syntrophic bacteria ensured pH stability and prevented acidification [34].

### *11.3.3 Green and Efficient Conversion of Urban Organic Waste into Marketable Caproic Acid Without Chemicals Addition*

The significant potential of volatile fatty acids derived from urban waste is highlighted by their inclusion in the EU's top 20 innovative bio-based products, based on their development, innovation level, and market prospects [35]. To advance the application of fermentation/chain elongation (CE) in real-world anaerobic treatment systems, we established several experimental goals. Firstly, we aimed to provide the initial demonstration of stable caproate production from urban waste without the need for external electron donor addition or buffering agents. Secondly, we sought to assess the influence of different substrate feeding regimes on caproate yield. Finally, we intended to characterize the dynamic changes within the reactor's microbial community that corresponded with periods of increased caproate production.

While there are a number of studies and full-scale applications of co-digestion of sludge and organic waste for biogas production, there is limited literature on carboxylic acids production through co-fermentation [17, 36]. The stringent pH control (5.5–6) required for the industrialization of the chain elongation process for medium-chain carboxylic acid production presents a considerable challenge due to the large quantities of acid/base needed, an aspect lacking comprehensive investigation in current research [37]. To address this gap, we evaluated the capacity of anaerobic fermentation of mixed urban organic waste to produce caproic acid without external buffering. This strategy leveraged the high fermentative potential of food waste in conjunction with the inherent buffering capabilities of sludge.

A 5 Liter mesophilic bioreactor, inoculated with anaerobic fermentative biomass, was operated in semi-continuous mode at OLR 10 gVS $L^{-1}d^{-1}$ and HRT 4d by feeding a mixture of raw FW and pretreated WAS (test #1) or a pretreated mix of FW and WAS (test#2), two times per week, on Monday and on Thursday. The organic waste mix was composed by FW/WAS = 60/40 with respect to VS content. These long-term (around three months each) tests revealed a lactate-based CE pathway, since lactate was produced within the first hours after each feeding and then rapidly removed, while ethanol concentration remained always very low. From the 4$^{th}$ HRT the concentration of caproic acid remained stable with an average concentration of 3 ± 0.3 g COD/L until the end of the operation (test #1) and was the first most abundant carboxylic acid in the fermentation broth, with a high selectivity of 41 ± 1%, followed by acetic acid with 35% (see Fig. 11.3). The conversion efficiency of the soluble COD of the organic waste into caproic acid was 23 ± 1%, with an additional 10% converted into caprylic acid (C8), deriving from the elongation of caproic itself.

Thermal pretreatment of the feedstock favored, also the solubilization of proteins further converted, during the fermentation, into ammonium bicarbonate, buffering the system. The alkalinity of the system improved the buffering capacity by mitigating changes in pH caused by the rapid lactate accumulation so to foster chain elongation avoiding the addition of artificial pH control chemicals. Moreover, the thermal pretreatment of sludge resulted in the inactivation of endogenous anaerobic microorganisms present within the floc matrix. This inactivation served to prevent the metabolic conversion of VFAs to methane, while simultaneously contributing to the sanitation of the feedstock.

Preliminary results revealed a strongly selected microbial community mainly composed by members of phyla Actinobacteria (e.g., Olsenella) and Firmicutes (e.g., Caproiciproducens, Eubacterium nodatum, Pseudoramibacter) responsible for

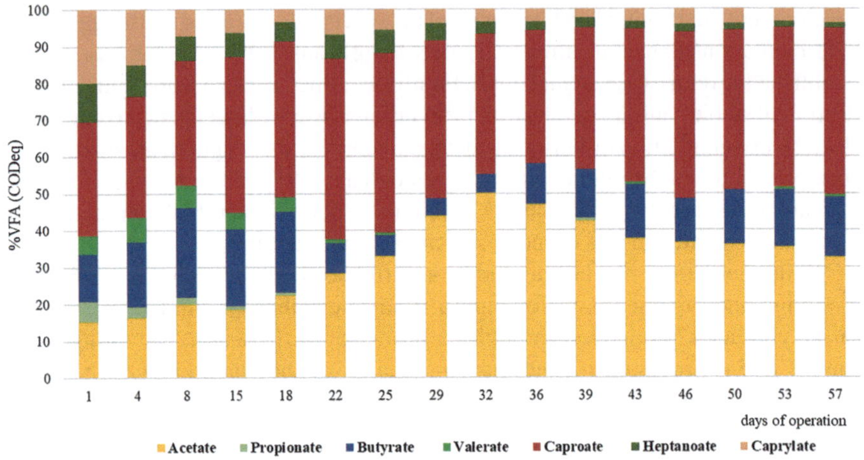


**Fig. 11.3** Carboxylic acids concentration selectivity (% in equivalent of COD) after each feeding cycle (test #1)

the production of lactate and subsequent elongation to caproic acid. Figure 11.4 shows the details of the partitioning and conversion of the soluble carbon of precursors (carbohydrates and proteins) and products (VFAs and lactate) during the first 24 h after the feeding and at the end of the cycle during the operation of test #2 with thermal pretreated mixture of food waste and sludge. The homo- and hetero-lactic fermentation of readily available sugars of the feedstock promoted the in situ lactate production, and within the first 6h, about 80% of the initial carbohydrates load was mainly converted to lactic acid, that increased up to 46.2 mmolC $L^{-1}$ (1.4 g $L^{-1}$), impacting on the system pH value (that reached the minimum of 4.9). The VFAs concentration remained almost stable during this first phase. Afterwards, in the following 48h, lactate was efficiently converted into butyrate and caproate with natural pH buffering. At the end of the cycle acidogenic metabolites represent almost 79% of the final soluble COD concentration, while the remaining fraction was due to the protein content. It is worthy to note that the net production of caproate during the cycle was 46 $\pm$ 0.2 mmolC $L^{-1}$ (Fig. 11.4). The thermal pretreatment of the entire mix (FW + WAS) promoted the caproate conversion efficiency that reached 42 and 16% with respect to the soluble and total COD fed.

Given its wide-ranging applications as a green antibiotic, food additive, lubricant, and surfactant [19, 37], caproate stands out as a particularly attractive product within the carboxylic acid platform. The economic success of its production hinges on efficient production and purification processes. Consequently, a preliminary techno-economic analysis was performed to evaluate the potential profitability of utilizing an urban biorefinery approach to convert mixed urban organic waste into caproate [38].

Considering an estimated food waste production of 100 kg/inhabitant/year (household + retail + horeca sector), a city like Milan (Italy) with a population of 3.1 Mio

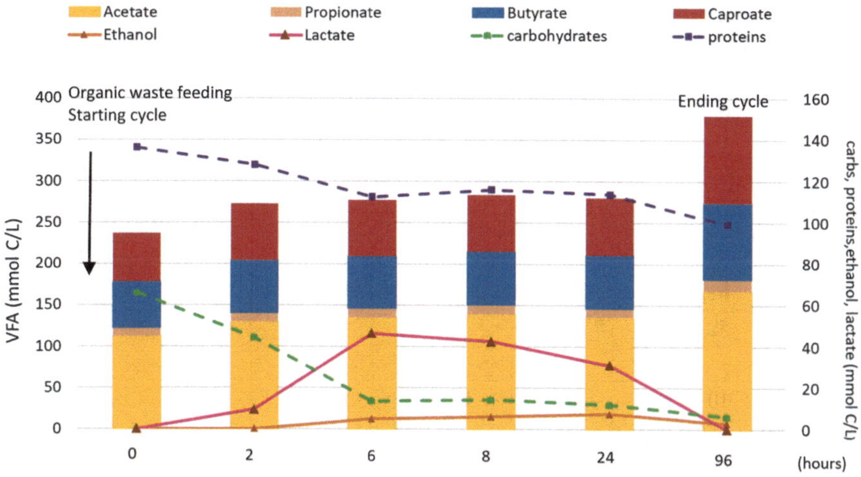


**Fig. 11.4** Carbon distribution between substrates and products during the fermentation-CE cycle (test #2)

could collect around is 310,000 tons (from household and restaurants). As regards sludge, the estimated production is around 77,000 ton of dry solids [8].

Table 11.1 shows the amount of annual organic substance (ton of Volatile Solids per year), based on typical food waste and sludge characterization, to be converted into methane (Scenario 1, with conventional Anaerobic Digestion) or into caproate (from mixed FW and WAS) as in the proposed biorefinery scheme (Scenario 2). The exceeding amount of sludge, not used for caproate production, could be digested to produce methane (Scenario 2). The economic model considered revenue generation from the sale of caproic acid, priced at 3000 €/t, and biomethane, at 600 €/t. On the cost side, the analysis factored in 500 €/t for the recovery and purification of caproate from the fermentation broth [38], and 0.11 €/Nm$^3$ for biogas upgrading [39].

The potential for increased revenues (from 8.3 to 10.8 million EUR) through the production of platform chemicals from mixed urban waste highlights the promise of this approach. However, realizing successful industrial commercialization of the proposed biorefinery necessitates a detailed cost–benefit analysis. Operating expenses, product selling prices, and yields represent key variables that could negatively impact the economic viability of waste valorization scenarios [38]. A comprehensive assessment must consider the full-scale costs of integrating thermal pretreatment in relation to revenues from caproate and/or methane sales, renewable energy subsidies, savings in waste disposal, and the estimated job creation [19]. A significant industrial advantage could be realized through carboxylic acid production at uncontrolled pH, potentially eliminating the 14% of process costs associated with the alkaline agents needed for pH stabilization [40].

While the adoption of sustainable waste management practices aimed at reduction and valorization offers numerous benefits for the sustainability of our circular cities,

**Table 11.1** Annual estimates for caproate and methane production utilizing urban organic waste from Milan, Italy

| | VS amount (t/year) | Methane production ($Nm^3$/year) | Caproate production (t/year) | Income value (EUR million) | Cost for recovery and purification (EUR million) | Revenue (EUR million) |
|---|---|---|---|---|---|---|
| Scenario 1 | | | | | | |
| Food waste | 62,000 | 18′600′000 | | 11.2 | 2.9 | 8.3 |
| Sludge | 52,000 | 7′800′000 | | | | |
| Scenario 2 | | | | | | |
| FW + Sludge (60/40) | 103,300 | | 4′130 | 12.4 | 2.1 | 10.3 |
| Sludge | 10,670 | 1,600,500 | | 0.68 | 0.18 | 0.5 |

Note of Table 11.1: Methane yield: 300 $Nm^3/tVS_{FW}$; 150 $Nm^3/tVS_{sludge}$ (from Tonanzi et al. [11]); caproate yield: 0.04kg/kgVSmix (from this study); Density $CH_4$: 0.708 kg/$m^3$

a thorough and accurate assessment of the environmental trade-offs within these systems remains a complex scientific challenge [41].

### *11.3.4 New Life for Post-Consumer Compostable Bioplastics Associated to Urban Food Waste*

In our consumer society, many consumer products on the market are made of non-biodegradable plastics deriving from fossil fuels. The mismanagement of end-of-life plastics leads to their accumulation and persistence in the environment, resulting in increased greenhouse gas emissions, pollution, and detrimental effects on soil and water quality, as well as biodiversity in both land and marine habitats [42]. Statistical data indicates that the packaging sector is the primary consumer of plastics (circa 40%) and the source of approximately 60% of post-consumer plastic waste within the EU, a substantial amount of which is intended for single use [43]. The disposal pathways for these post-consumer plastics, as also packaging materials, include recycling, incineration, landfilling, unregulated disposal, and environmental littering. The latter route has significant ecological consequences, exacerbated by the uncontrolled generation of micro- and nano-plastics [44]. Consequently, the environmental impact of seemingly minor littering events is substantial and often underestimated, with ecotoxicity being an area of infrequent monitoring.

One sound approach to address this challenge is to explore the application potential of bio-based and compostable plastic products, and innovative technological solutions to drastically reduce the uncontrolled emissions of persistent toxic pollutants as microplastics. A growing number of bio-based materials and products, several of them being marketed as biodegradable in specific environments and/or compostable, have been emerging over the last few years in our daily lives to offer more sustainable alternatives. Bio-based/bio-attributed plastics account for 1.5% of the world plastics production, estimated at around 391 Mt in 2021 [45].

Compostable bioplastics as tableware and cups are made of Polylactic acid (PLA), deriving from condensation reaction of lactide, while shopper and bags for food waste collection are made of thermoplastic starch–based plastics (blended with biodegradable polyesters as PBAT) and both account for the largest market share around (18.7%/each). PLA degrades in the environment in 6–24 months depending on size and shape polymer grade, Mater-Bi degrades in soil and less in marine environment.

Furthermore, biodegradation does not facilitate material recovery for reintroduction into the plastics lifecycle, leading to a loss of the inherent value within the polymer's structure. Consequently, the focus is shifting from biodegradation towards strategies for recycling and upcycling bioplastics. Innovative recycling and recovery technologies offer significant economic potential by transforming post-use bioplastics into their original components. These can then be continuously reintegrated into supply chains as feedstocks for new bioplastics, chemicals, manufacturing raw materials, and lower-emission transportation fuels, reducing the reliance on virgin resources. The potential of thermo(chemical) pretreatments combined with advanced acidogenic fermentation with chain elongation process to convert back thermoplastic starch- and PLA-based material into lactate and medium chain carboxylic acid is a new recycling route that we started to explore.

Preliminary results of the extraction of starch from bio-based compostable bags (SBB) by means of thermal pretreatment (autoclave for 20 min) were promising. Bio-based plastics (starch-based bag and PLA cup) were pretreated in the form of small pieces (not pulverized/powdered as often reported in literature), and once thermally pretreated, a size reduction was observed, together with the conglomeration and corrugation of the SBB surface. The disintegration degree of SBB was very high, namely 40%, while for PLA was only 2%. In the study of Mu et al. [46], the hydrothermal pretreatment at 120 °C on pulverized PLA solubilized 14.9% of PLA plastic into liquid phase, while the solubilization rate increased to 91.9% at 200 °C. The extraction of the organic material from SBB into the aqueous eluate after pretreatment was composed by $40 \pm 2\%$ of carbohydrates (expressed in COD), representing more than 90% of the total carbohydrate component detected in the SBB matrix. On the contrary, the mild pretreatment of PLA did not promote the lactate extraction/solubilization, as experienced elsewhere [47]. The addition of NaOH (up to pH 12) to the water before thermal pretreatment (time 180 min) promoted the depolymerization of PLA and 35% of the material was extracted as lactate.

Next steps will be the use of these eluates rich in lactate and starch as precursors to produce marketable carboxylic acids through fermentation and chain elongation, while the residual material will proceed to organic recycling with food waste to produce compost.

## 11.4 Conclusions

In conclusion, the future of integrated urban biorefineries is bright, driven by technological innovation, and a growing focus on circularity. They represent a crucial module in transitioning towards a more sustainable, low-carbon, and resource-efficient future for cities and urban settlemnts. While biofuels remain important, the economic viability of urban biorefineries will increasingly rely on the production of a wide range of high-value biochemicals (e.g., platform chemicals, organic acids), bioplastics (e.g., PHAs), biomaterials, and biofertilizers. This multi-product approach will create additional revenue streams. At the same time robust policy and regulatory frameworks are vital to drive the development and adoption of urban biorefineries. This includes waste management directives promoting valorization of bio-waste, biofuel mandates, and incentives for bio-based products, but also policies that promote urban-industrial symbiosis, where waste from one industry becomes feedstock for another, will facilitate the growth of integrated biorefineries.

## References


1. Wiedmann T, Allen C (2021) City footprints and SDGs provide untapped potential for assessing city sustainability. Nat Commun 12:3758
2. López-Ruiz VR, Alfaro-Navarro JL, Nevado-Peña D (2014) Knowledge-city index construction: an intellectual capital perspective. Expert Syst Appl 41(12):5560–5572
3. Lu M, Zhou C, Wang C et al (2024) Worldwide scaling of waste generation in urban systems. Nat Cities 1:126–135
4. IPCC (2022)Summary for policymakers. In: Climate change 2022: mitigation of climate change. Contribution of WG III to the sixth assessment report of the intergovernmental panel on climate change
5. ISPRA, Rapporto Rifiuti Urbani, Rapporti 406/2024 ISBN 978-88-448-1240-9
6. International Resource Panel (IRP) (2018) The weight of cities: resource requirements of future urbanization. In: Swilling M, Hajer M, Baynes T et al (eds) A report by the international resource panel. united nations environment programme, Nairobi, Kenya
7. European Commission (2018) A sustainable bioeconomy for Europe: strengthening the connection between economy, society and the environment
8. ISPRA, Rapporto Rifiuti Speciali Rapporti 402/2024 ISBN: 978-88-448-1219-5
9. The European House—Ambrosetti, Community Valore Acqua per l'Italia, Libro Bianco Valore Acqua per l'Italia (2023)
10. Braguglia CM, Gallipoli A, Gianico A, Pagliaccia P (2018) Anaerobic bioconversion of food waste into energy: a critical review. Bioresource Technol 248, Part A, 37–56
11. Tonanzi B, Gallipoli A, Gianico A, Montecchio D, Pagliaccia P, Rossetti S, Braguglia CM (2021) Elucidating the key factors in semicontinuous anaerobic digestion of urban biowaste: the crucial role of sludge addition in process stability, microbial community enrichment and methane production. Renewable Energy 179:272–284
12. Gagliano MC, Braguglia CM, Gianico A, Mininni G, Nakamura K, Rossetti S (2015) Thermophilic anaerobic digestion of thermal pretreated sludge: Role of microbial community structure and correlation with process performances. Water Res 68:498–509
13. Montecchio D, Astals S, Di Castro V, Gallipoli A, Gianico A, Pagliaccia P, Piemonte V, Rossetti S, Tonanzi B, Braguglia CM (2019) Anaerobic co-digestion of food waste and waste activated

sludge: ADM1 modelling and microbial analysis to gain insights into the two substrates' synergistic effects. Waste Manage 97:27–37
14. Carmona-Cabello M, Garcia IL, Leiva-Candia D, Dorado MP (2018) Valorization of food waste based on its composition through the concept of biorefinery. Curr Opin Green Sustain Chem 14:67–79
15. Kuruti K, Begum S, Ahuja S, Rao Anupoju G, Juntupally S, Gandu B, Kumar Ahuja D (2017) Exploitation of rapid acidification phenomena of food waste in reducing the hydraulic retention time (HRT) of high rate anaerobic digester without conceding on biogas yield. Bioresour Technol 226:65–72
16. Owusu-Agyeman I, Balachandran S, Plaza E, Cetecioglu Z (2021) Co-fermentation of municipal waste streams: effects of pretreatment methods on volatile fatty acids production. Biomass Bioenerg 145:105950
17. Fernando-Foncillas C, Varrone C (2021) Effect of reactor operating conditions on carboxylate production and chain elongation from co-fermented sludge and food waste. J Clean Prod 292:126009
18. Gianico A, Gallipoli A, Gazzola G, Pastore C, Tonanzi B, Braguglia CM (2021) A novel cascade biorefinery approach to transform food waste into valuable chemicals and biogas through thermal pretreatment integration. Biores Technol 338:125517
19. Tonanzi B, Gallipoli A, Frugis A, Gianico A, Lazzazzara M, Angelini S, Cecchini G, Braguglia CM (2024) Bio-based production of medium-chain carboxylic acids from food waste and sludge without chemical addition: the pivotal role of mix ratio and pretreatment. J Clean Prod 436:140560
20. Valentino F, Moretto G, Gottardo M, Pavan P, Bolzonella D, Majone M (2019) Novel routes for urban bio-waste management: a combined acidic fermentation and anaerobic digestion process for platform chemicals and biogas production. J Clean Prod 220:368–375
21. Ma Y, Liu Y (2019) Turning food waste to energy and resources towards a great environmental and economic sustainability: an innovative integrated biological approach. Biotechnol Adv 37:107414
22. Alibardi L, Astrup TF et al (2020) Organic waste biorefineries: looking towards implementation. Waste Manage 114:274–286
23. IEA Bioenergy—Global biorefinery status report: task 42 biorefining in a circular economy (2022)
24. Loow YL, Wu TY, Jahim JM, Mohammad AW, Teoh WH (2016) Typical conversion of lignocellulosic biomass into reducing sugars using dilute acid hydrolysis and alkaline pretreatment. Cellulose 23(3):1491–1520
25. Strazzera G, Battista F, Herrero Garcia N, Frison N, Bolzonella D (2018) Volatile fatty acids production from food wastes for biorefinery platforms: a review. J Environ Manage 226:278–288
26. Gallipoli A, Gianico A, Crognale S, Rossetti S, Mazzeo L, Piemonte V, Masi M, Braguglia CM (2021) 3-routes platform for recovery of high value products, energy and bio-fertilizer from urban biowaste: the revenue project. Detritus 24–30
27. Pagliaccia P, Gallipoli A, Gianico A, Gironi F, Montecchio D, Pastore C, di Bitonto L, Braguglia CM (2019) Variability of food waste chemical composition: impact of thermal pre-treatment on lignocellulosic matrix and anaerobic biodegradability. J Environ Manage 236:100–107
28. Crognale S, Braguglia CM, Gallipoli A, Gianico A, Rossetti S, Montecchio D (2021) Direct conversion of food waste extract into Caproate: metagenomics assessment of chain elongation process. Microorganisms 9:327
29. Gottardo M, Crognale S, Tonanzi B, Rossetti S, Annibale LD, Dosta J, Valentino F (2024) Volatile fatty acid production from hydrolyzed sewage sludge : effect of hydraulic retention time and insight into thermophilic microbial community. Biomass Conv Bioref 14:14921–14932
30. Grant F, Di Veroli JN, Rossi L (2023) Characterization of household food waste in Italy: three year comparative assessment and evaluation of seasonality effects. Waste Manage 164:171–180
31. Cesaro A, Belgiorno V (2014) Pretreatment methods to improve anaerobic biodegradability of organic municipal solid waste fractions. Chem Eng J 240:24–37

32. Mahmoodi K, Karimi K, Taherzadeh MJ (2018) Hydrothermal processing as pretreatment for efficient production of ethanol and biogas from municipal solid waste. Bioresour Technol 261:166–175
33. Contreras-Davila CA, Carrion VJ, Vonk VR, Buisman CNJ, Strik D (2020) Consecutive lactate formation and chain elongation to reduce exogenous chemicals input in repeated-batch food waste fermentation. Water Res 169:115215
34. Gallipoli A, Angelini F, Angelini S, Braguglia CM, Montecchio D, Tonanzi B, Gianico A (2024) Thermally enhanced solid-liquid separation process in food waste biorefinery: modelling the anaerobic digestion of solid residues. Front Bioeng Biotechnol 12:1343396
35. European Commission: Directorate-General for Research and Innovation, Top 20 innovative bio-based products—Task 3 of "Study on Support to R&I Policy in the Area of Bio-based Products and Services, Publications Office (2019)
36. Perez-Esteban N, Vinardell S, Vidal-Antich C, Peña-Picola S, Chimenos JM, Peces M, Dosta J, Astals S (2022) Potential of anaerobic co-fermentation in wastewater treatments plants: a review. Sci. of tot. Environm. 813:152498
37. Zhu X, Huang H, He Y, Wang X, Jia J, Feng X (2022) A preliminary study on the feasibility of industrialization for n-caproic acid recovery from food wastewater: from lab to pilot. Bioresour Technol 366:128154
38. Arhin SG, Cesaro A, Di Capua F, Esposito G (2023) Acidogenic fermentation of food waste to generate electron acceptors and donors towards medium-chain carboxylic acids production. J Environ Manage 348:119379
39. Danish Energy Agency: Technology data for renewable fuels (2017)
40. Lopez-Gomez JP, Latorre-Sanchez M, Unger P, Schneider R, Coll Lozano C, Venus J (2019) Assessing the organic fraction of municipal solid wastes for the production of lactic acid. Biochem Eng J 150:107251
41. Nordahl SL, Devkota JP, Amirebrahimi J, Smith SJ, Breunig HM, Preble CV, Satchwell et al (2020) Life-cycle greenhouse gas emissions and human health trade-offs of organic waste management strategies. Environ Sci Technol 54(15):9200–9209
42. European Environmental Agency (EEA) (2023) From source to sea—the untold story of marine litter, Web Report https://www.eea.europa.eu/publications/european-marine-litter-assessment/from-source-to-sea-the
43. European Economic and Social Committee, Recycling food packaging and food waste in plastics revolution https://doi.org/10.2864/154293 (2020)
44. United Nations Environment Programme (UNEP) Drowning in Plastics—Marine Litter and Plastic Waste Vital Graphics (2021)
45. Nova-Institute (2022) Data for bio-based structural polymers, preliminary estimations for 2021
46. Mu L, Zhang L, Ma J et al (2021) Enhanced biomethanization of waste polylactic acid plastic by mild hydrothermal pretreatment: Taguchi orthogonal optimization and kinetics modeling. Waste Manage 126:585–596
47. Marchelli F, Mattonai M, Ferrentino R et al (2024) Fostering bioplastics circularity through hydrothermal treatments: degradation behavior and products. ACS Sustain Chem Eng 12(24):9257–9267

# Chapter 12
# Waste Streams Potential for Resilience Through Eco-Industrial Parks for Arid Regions

**Dhabia M. Al-Mohannadi**

**Abstract** Arid regions face acute resource challenges driven by water scarcity, limited arable land, and growing populations. This work investigates how process integration and eco-industrial park frameworks can systematically convert waste streams—such as biosolids and plastics—into valuable resources, thereby enhancing resource resilience and sustainability. Technologies including anaerobic digestion, pyrolysis, and gasification are evaluated for their techno-economic performance and greenhouse gas implications, demonstrating strong potential to generate energy, recover materials, and reduce reliance on landfills. A resource integration modeling framework optimizes multi-process networks, maximizing profit and circularity metrics while meeting environmental targets. Comparative case studies on industrial biosludge and plastic waste conversion underscore the importance of holistic, data-driven approaches to optimize waste-to-resource pathways, informing policy and industrial practice in arid contexts.

**Keywords** Waste to products · Biosludge · Plastics · Eco-industrial parks · Modeling and optimization

## 12.1 Introduction

Arid regions face unique challenges due to their harsh environments, characterized by water scarcity, limited natural resources, and extreme weather conditions. These challenges are emphasized by growing populations and increased urbanization, which place additional strain on already scarce resources. In such regions, traditional approaches to resource management often fall short, necessitating innovative strategies that can ensure long-term sustainability. Resource resilience—the

D. M. Al-Mohannadi (✉)
College of Science and Engineering, Hamad Bin Khalifa University, Education City, Doha, Qatar
e-mail: dalmohannadi@hbku.edu.qa

Chemical Engineering, Texas A&M University at Qatar, Education City, Doha, Qatar

 
A. Angelini et al. (eds.), *Security Enhancement for Climate Change Impacting Urban Resources - SECCURe*, NATO Science for Peace and Security Series C: Environmental Security, https://doi.org/10.1007/978-94-024-2345-7_12

ability to efficiently manage, recover, and utilize resources—is critical for mitigating these challenges and fostering stability. Sustainable Development Goal (SDG) partially covers this 12, which aims to ensure sustainable consumption and production patterns. Achieving this goal requires fostering circular economy models, sustainable production practices, and resource usage efficiency.

The Arabian Peninsula is considered one of the world's most arid zones. It's home to the Gulf Cooperation Council (GCC), a regional, intergovernmental, political, and economic union that includes Bahrain, Kuwait, Oman, Qatar, Saudi Arabia, and the United Arab Emirates (UAE). All GCC nations fall well below the 500 $m^3$ per capita threshold of 'absolute water scarcity [1]. To meet needs, they historically relied on groundwater aquifers, but those are largely non-renewable and are being rapidly depleted by overuse [2]. This chronic water stress, coupled with growing populations and agriculture, has made water security a central challenge for the GCC region. To overcome their lack of freshwater, GCC states have heavily turned to seawater desalination. The gulf region is now a global leader in desalination capacity—accounting for roughly 60% of the world's desalination capacity and about 40% of all desalinated water produced [3]. Over 400 desalination plants are operating across GCC coastlines [3]. However, desalination is highly energy-intensive, creating economic and environmental tradeoffs. A recent analysis noted that in 2016, desalination made up only about 3% of the Middle East's water supply but consumed 5% of the region's total energy usage [4].

Agriculture in the peninsula faces not only water scarcity but also poor soil quality, necessitating heavy use of fertilizers. Much of the terrain is sandy, with low organic content and minimal natural fertility. Thin topsoil, high salinity, and desertification mean that crops cannot thrive without artificial nutrients. The GCC has developed a major fertilizer industry to meet local needs and export abroad.

At the same time, the GCC countries are blessed with massive hydrocarbon resources, which have fueled economic growth. Saudi Arabia alone possesses around 17% of global proven crude oil reserves [5] and is the world's largest oil exporter. Qatar is one of the top holders of natural gas reserves. This has made the Gulf a leading producer of plastics, chemicals, and fuels, but also a major generator of waste.

Recognizing these challenges, GCC countries have the potential to transform waste from oil, gas, and petrochemical processes into useful inputs. A recent report by PwC estimated that raising recycling and recovery rates in the GCC (including energy recovery from waste) could cut primary energy consumption by around 4% [6] Waste-to-energy conversion allows non-recyclable plastics, industrial sludges, and other combustible wastes to generate electricity or heat. Most industrial processes and hydrocarbon conversion in the GCC are set up in industrial clusters, such as Ras Laffan industrial city in Qatar, Al Jubail in Saudi Arabia, Al Ruwais industrial city in the United Arab Emirates, etc. This opens an opportunity for integration.

Process integration and eco-industrial parks play a pivotal role in addressing these challenges by creating systems that minimize waste and optimize resource use. Eco-industrial parks (EIPs) are clusters of manufacturing and services that collaborate to improve environmental, economic, and social performance by sharing resources such as materials, energy, and infrastructure [7]. By fostering synergies among processes, EIPs promote resource efficiency, reduce waste, and enhance sustainability. These approaches enable industries to work collaboratively, transforming waste streams into valuable inputs for other processes. This not only conserves resources but also enhances economic and environmental resilience. In the energy sector, waste-to-energy technologies with energy recovery reduce reliance on fossil fuels, lowering strain on fresh water and fossil fuels. Resource resilience is strengthened through reusing, upgrading, and industrial symbiosis, where waste from one process becomes an input for another, minimizing raw material extraction. Environmentally, reducing landfill waste lowers methane emissions, while circular solutions such as biochar and direct use of biosludge can improve soil quality and agricultural resilience. The next sections look at potential waste streams and will evaluate their role through industrial symbiosis.

## 12.2 Defining Waste Streams and Their Potential

Waste streams refer to the by-products and residual materials generated from various processes, including municipal, industrial, and agricultural activities. Common waste streams in arid regions include biosludge from wastewater treatment, treated wastewater, plastic waste, and even emissions. Traditionally seen as liabilities, these streams hold immense potential for resource recovery.

Through innovative technologies and integrated systems, these waste streams can be transformed into energy, raw materials, and other value-added products. For instance, biosludge can be converted into biochar for soil enhancement, while treated wastewater can support agriculture and cooling systems. Such solutions are particularly relevant for arid regions, where resource scarcity demands efficient and sustainable utilization of all available resources.

This chapter explores the transformative potential of waste streams in building resilience through process integration and the development of eco-industrial parks in arid regions. By examining various waste streams and their applications, the chapter highlights practical tools and methods that can help with resource efficiency. Specifically in the context of arid regions such as those in the GCC, common waste streams include biosludge from wastewater treatment, treated wastewater and plastic waste.

### *12.2.1 Biosludge as a Resource: From Waste to Value*

Treated wastewater plays a crucial role in supporting agricultural activities, particularly in water-scarce regions. As a consistent and nutrient-rich water source, wastewater can improve soil quality, boost crop yields, and enhance food security. Qadir et al. [8] discussed the challenges of wastewater irrigation in developing countries, emphasizing the need for effective policies and infrastructure to ensure safe reuse. Pedrero et al. [9] reviewed wastewater irrigation practices in Spain and Greece, demonstrating that properly treated wastewater can sustain crop yields while reducing dependence on freshwater sources. Paranychianakis et al. [10] examined the necessity for uniform wastewater reuse criteria within the European Union. Toze [11] explored both the benefits and risks of wastewater reuse in agriculture, noting that wastewater can alleviate water scarcity and provide essential nutrients. In the GCC, Aleisa and Al Zubari [12] mentioned that reusing treated wastewater can reduce the pressure on desalinated water production and depleting groundwater withdrawal, and as such reduce associated harmful environmental impacts in the GCC. Treated wastewater presents a valuable opportunity for hydrogen production, offering a sustainable pathway to clean energy. Through electrolysis, wastewater can be treated and converted into hydrogen. This process not only provides a renewable energy source but can address water shortage issues and elevate the need for extra desalination by using already treated wastewater. Ibrahim et al. [13] looked into wastewater for hydrogen production. Regardless of any application of wastewater utilization, it will produce a semi-solid waste stream, namely bio sludge.

Biosludge is a by-product of wastewater treatment processes, generated in significant volumes as a result of biological treatment methods [14]. In arid regions, the management of biosludge poses unique challenges due to the lack of disposal options and limited land availability for landfilling. Improper disposal can lead to environmental contamination, including the release of greenhouse gases, and water pollution.

Traditional biosolids management methods—landfilling, land application, and incineration—are becoming less viable due to land scarcity and public health concerns. Although incineration is cost-effective, it contributes to greenhouse gas emissions, necessitating improved emission control. Alternative thermal and biological conversion technologies, such as anaerobic digestion (AD), pyrolysis, gasification, and hydrothermal liquefaction, offer sustainable solutions [15]. These processes generate energy and valuable by-products like biochar, compost, and ashes for reuse in construction and land reclamation. Each method varies in efficiency, operating conditions, and outputs, with AD producing methane-rich biogas, pyrolysis generating syngas and biochar, gasification yielding syngas at high temperatures, and hydrothermal liquefaction converting wet sludge into bio-crude oil without drying [16].

Biosludge can be converted to Biochar through pyrolysis. Biochar, a carbon-rich material, has significant potential to improve soil quality in arid regions by enhancing water retention, nutrient availability, and microbial activity. The production of biochar from biosludge not only sequesters carbon, mitigating greenhouse gas emissions but also provides a sustainable solution for managing this waste stream. Biochar can be integrated into soil restoration projects, particularly in degraded and desertified landscapes [17].

One of the significant benefits of biochar is its ability to reduce the mobility of heavy metals and organic pollutants in soil. Various modification methods, such as acid, alkali, and thermal treatments, have been developed to enhance the adsorption capacity of biochar, further improving its effectiveness in pollutant removal [17].

Transforming biosludge into energy and biochar provides numerous environmental and economic benefits by diverting biosludge from landfills, and the environmental impact of waste disposal is minimized.

### *12.2.2 Plastic Waste to Value*

Other source streams such as Municipal solid waste (MSW) are a growing concern in arid regions due to increasing urbanization and population growth. The dependency on landfills for waste disposal creates environmental hazards such as leachate contamination, methane emissions, and inefficient land use. These challenges are exacerbated by the arid climate, which limits the natural decomposition of organic waste and complicates landfill management [18]. One key material that can be recovered is plastic waste.

Plastic waste is a growing environmental challenge in arid regions, where high temperatures and a lack of proper disposal infrastructure exacerbate pollution. The slow degradation of plastics leads to long-term environmental contamination, harming ecosystems and human health [19].

Waste plastics can be converted into energy through various technologies similar to biomass and biosludge such as pyrolysis, gasification, and incineration. Pyrolysis breaks down plastics into synthetic fuels, which can be used for power generation or industrial applications. Gasification transforms plastic waste into syngas, a versatile fuel that can be used for electricity production. Incineration with advanced emission control systems offers another route to recover energy while minimizing environmental impact.

Only about 10% of plastic waste is recycled globally [20], the remaining plastics either accumulate in landfills or pollute the surrounding ecosystems. Waste-to-energy (WtE) technologies provide an alternative to mitigate these issues, reducing landfill dependency and limiting environmental damage. **Waste-to-Energy Technologies for Solid Waste** [21].

WTE technologies provide a practical solution to the challenges posed by solid waste including biosludge and plastic waste. These technologies convert waste into usable energy, such as electricity, heat, or fuels, through processes like incineration, gasification, and pyrolysis. Incineration involves burning waste at high temperatures to produce steam that drives turbines for electricity generation, while pyrolysis and gasification break down waste in low-oxygen environments to create syngas, a versatile energy source [21].

The adoption of WTE technologies in arid regions presents multiple advantages, particularly in addressing energy and waste management challenges. WTE plants contribute to energy security by providing a stable and renewable source of energy, reducing reliance on fossil fuels. These technologies play a crucial role in waste reduction by diverting waste from landfills. WTE processes promote resource efficiency by generating valuable by-products such as ash and syngas, which can be repurposed in materials and industrial applications, supporting the transition toward a circular economy.

Integrating WTE plants within eco-industrial parks enables industries to benefit from localized energy production, reducing reliance on external energy sources. Advanced emission control technologies ensure that these processes minimize environmental impact, making WTE a viable and sustainable option for solid waste management in arid regions.

## 12.3 Process Integration for Resource Efficiency and Waste to Value Upgrading

Process integration has been increasingly utilized to enhance resource efficiency across industries by optimizing energy, water, and material flows. Recent reviews highlight the role of Process Systems Engineering (PSE) in advancing circular economy goals, where optimization and life-cycle assessment methodologies help transform waste into valuable feedstock, improving overall resource efficiency [22]. Energy integration techniques such as heat recovery, waste-heat utilization, and energy storage have been shown to reduce emissions and enhance industrial sustainability [23]. Integrated water-energy networks, using mathematical programming and pinch analysis, minimize utility consumption in industrial systems [24]. Process integration techniques applied from individual plants to regional eco-industrial parks, demonstrating economic and environmental benefits across diverse industries, including petroleum refining, petrochemicals, food processing, and wastewater treatment [23]. A systematic approach for the design and integration of eco-industrial parks was done by Lovelady [25], with a focus placed on the efficient management of water resources. For waste, a comparative analysis of energy and carbon footprints of biosolids management strategies in water resource recovery facilities was conducted [26], with technical and economic optimization performed [27]. This all shows that process integration techniques can be used to ensure the minimization of freshwater usage and wastewater generation. These findings emphasize the importance of systematic integration approaches in achieving cleaner and more sustainable industrial production.

Techniques like pinch analysis, heat exchanger network design, and process simulations are employed to optimize energy flows, identify heat recovery opportunities, and enhance system performance in process integration to waste to energy [28–30]. Recently, the resource integration method [28] has shown great potential in synthesizing eco-industrial parks that are capable of looking into resource exchanges, including water, energy, material, and waste between processes.

### *12.3.1 Resource Integration Method for Synthesizing Eco-Industrial Parks*

Ahmed et al. [29] framework provides a systematic method to design integrated processing clusters (e.g. networks of industrial plants) that share and exchange resources (materials and energy) to improve sustainability. As showcased in Fig. 12.1, it represents the cluster using 'resource lines' that track resources (raw materials, intermediates, energy, emissions, waste) across multiple processes. Each process (plant) is modeled as an input–output module that consumes or produces resources via these common resource lines, allowing all possible interactions between plants to be identified. The framework is flexible and can be applied to various multi-process configurations, such as eco-industrial parks or resource-sharing clusters. Its primary aim is to achieve sustainable and profitable operations. A mixed-integer linear programming (MILP) model is used to configure the cluster, maximizing economic profit while meeting environmental goals (e.g., ensuring a required $CO_2$ conversion or emission reduction or zero waste or others). The framework's scope is the optimal synthesis of an integrated network of processes that share resources to enhance efficiency, profitability, and sustainability in industrial applications.

The framework's objective function maximizes the total annual profit of the processing cluster. By maximizing profit, the MILP solver finds the optimal selection

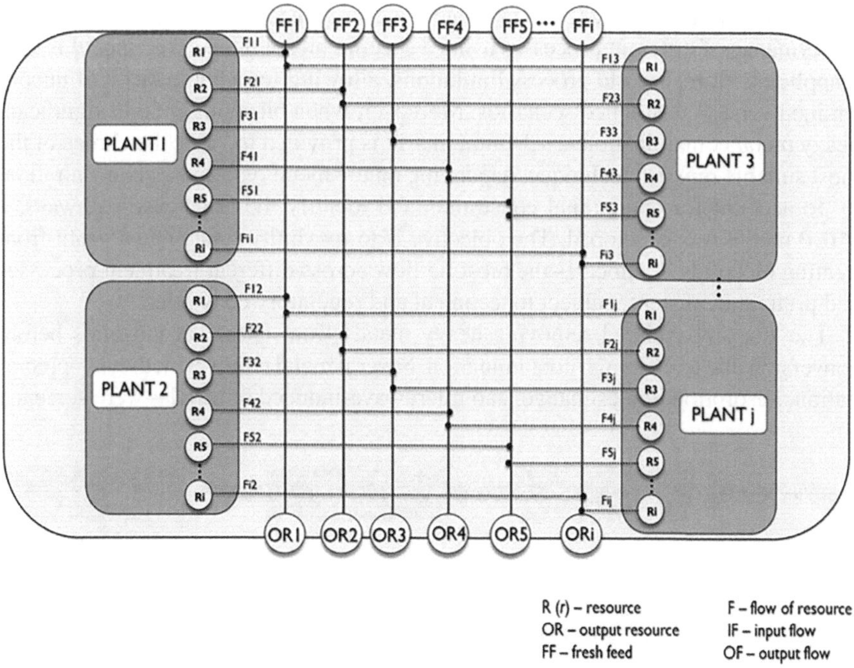


**Fig. 12.1** Resource integration framework adopted from Ahmed et al. [29]

and configuration of processes to ensure financial viability while achieving sustainability targets. This generic model was utilized to analyze waste conversion in two applications, valorization of bioslduge [30] and conversion of plastics [31], which are highlighted in the next sections.

### 12.3.2 *Application 1: Sustainable Valorization of Oil and Gas Industry Biosolids*

In particular, the oil and gas industry—a cornerstone of the GCC economies—generates substantial quantities of biosolids through wastewater treatment processes. The study, originally presented in Elfaki et al., [30] investigated the reuse potential of these waste streams, using data and processes specific to the GCC. The findings demonstrated promising pathways for converting oil and gas biosolids into valuable resources such as biochar, biogas, and construction materials, contributing to both environmental sustainability and industrial resilience in the region. Determining the most effective way to handle an industrial biosolids waste stream is a complex challenge, given the need to accommodate variables such as flow rate, solids content, and contaminant composition. To assist with this, an optimization model is used to select the best treatment approach. A high-level overview of potential treatment routes is presented in Fig. 12.2. To illustrate the model's capabilities, a techno-economic analysis is initially conducted to evaluate the energy, economic, and environmental performance of different processes using a specific biosolid feed. The model is then reapplied with real-world process limitations, allowing for a comparison of unconstrained versus constrained scenarios. Moreover, when biosolids contain significant heavy metal contamination, a selection matrix is provided to guide the choice of the most suitable removal technique, depending on the metal type and its concentration.

To account for operational constraints and identify the best reuse pathways, a MILP model was developed. The objective is to maximize total annual profit from treating biosolids. It allocates the biosolid flow across different treatment processes and pretreatment steps, subject to technical and regulatory constraints.

The study considered removing heavy metals from industrial biosolids before conversion due to their high contamination. Several metal removal methods—precipitation, adsorption, ion exchange, and microwave-induced removal—were screened

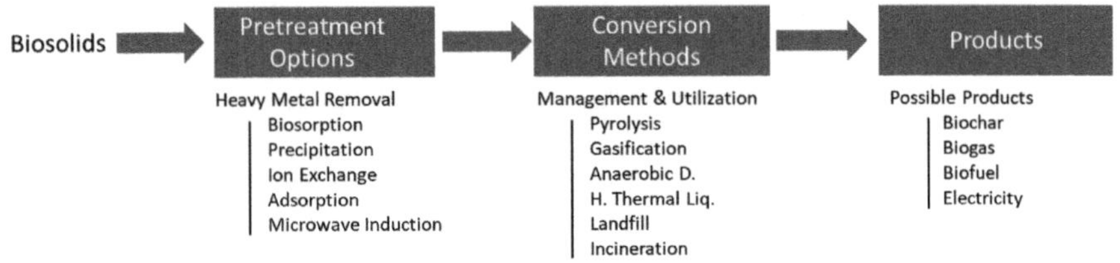


**Fig. 12.2** Biosludge treatment and utilization options adopted from El-Faki et al. [30]

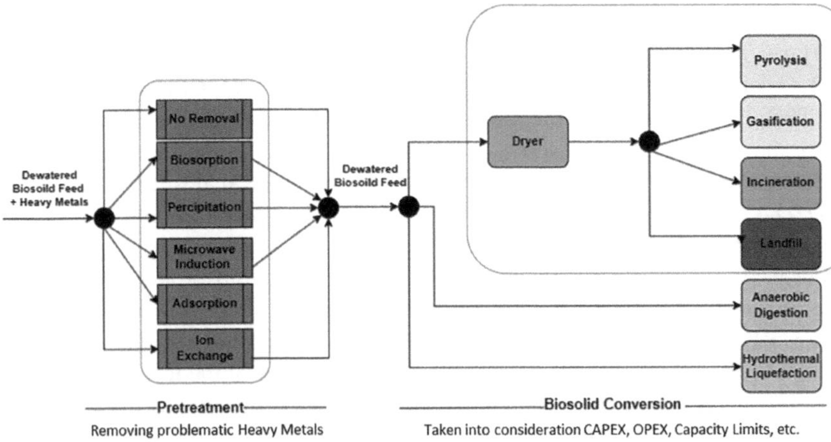


**Fig. 12.3** Treatment system example adopted from El-Faki et al. [30]

and compared for efficacy and cost. This ensures that any subsequent conversion meets regulatory limits on metals, shown in Fig. 12.3.

The case study demonstrates the model using an industrial biosolid waste stream from a gas-to-liquid (GTL) production plant. This stream contains significant moisture and heavy metals, representative of oil and gas industry biosolids.

First, a baseline analysis was performed for various biosolid conversion technologies using a consistent feed of 100 TPD (dry biosolids, negligible heavy metals) using Techno-Economic Analysis (TEA). The method calculated each technology's capital cost, operating cost, revenue from products, and emissions. It provided comparative economic and environmental performance of options like anaerobic digestion (AD), pyrolysis, gasification, hydrothermal liquefaction (HTL), incineration, and landfilling. Pyrolysis had a high annualized CAPEX (~ USD 2.6 million/year) but also the highest revenue (~ USD 4.9 million/year from bio-oil, biochar, and syngas), yielding an annual profit of around USD 1.0 million. Gasification required ~ 15% lower CAPEX (USD 2.2 M/yr) and produced ~ USD 3.8 M/yr revenue (biogas), but its high operating costs left only ~ USD 0.4 M/yr profit. Anaerobic digestion had the lowest costs (<USD 1 M/yr CAPEX and TAC) and decent revenue (~ USD 3.5 M/yr from biogas), giving the highest profit of ~ USD 2.7 M/yr.

Second, three scenarios were analyzed to cover different feed conditions:

1. Scenario 1—Low Moisture, No Heavy Metals: 20% moisture content, negligible heavy metal contamination. The feed rate varied from 50 to 800 tons per day (TPD) to see how the optimal solution shifts with scale.
2. Scenario 2—High Moisture, No Heavy Metals: 60% moisture content, negligible heavy metals. The inlet flow rate again ranged from 50–800 TPD.
3. Scenario 3—Heavy Metals Present: "Dry" biosolids (exceptionally low moisture) contaminated with Zn and Cr (VI) at levels exceeding regulatory limits. Flow rates from 50 to 100 TPD were considered.

It was found that for 20% moisture biosolids (Scenario 1), anaerobic digestion (AD) was most profitable at small scales, yielding ~ USD 13.3 M per year at 300 TPD. Gasification took over at higher feed rates (~ 400 TPD, USD 3 M/yr), and at 800 TPD, landfill was necessary due to capacity constraints (USD—1.3 M loss).

For the 60% moisture biosolids (Scenario 2), drying became necessary for most methods. Pyrolysis with a dryer was optimal at moderate scales (50–350 TPD, ~ USD 0.84 M/yr profit) while Gasification + drying dominated larger scales (400 + TPD, USD 9.5 M/yr profit).

For heavy metal-contaminated biosolids (Scenario 3), biosorption was the best Zn removal method (USD 492,000/yr), while chemical precipitation was optimal for Cr (VI) removal (USD 200,000/yr). The multi-method integration significantly increased profit. At 800 TPD, the model allocated portions of the feed to AD, Pyrolysis, HTL, and Gasification, achieving ~ USD 25 M/yr profit, compared to the single-method landfill scenario which incurred a loss. Environmental analysis confirmed AD as the most sustainable option (~ 2400 tons $CO_2$/yr emissions), while pyrolysis was the most polluting (~ 26,000 tons $CO_2$-eq/yr).

The findings reveal that the model strategically allocates the feed to different technologies by prioritizing the highest-capacity option while ensuring compliance with moisture and heavy metal content constraints. This strategy maximizes profitability and maintains effective heavy metal removal. By optimizing industrial biosolids treatment pathways, the study advances sustainable resource recovery in line with circular economic objectives. The resulting model thus supports informed decision-making in biosolids management and industrial waste treatment.

### 12.3.3 Application 2: Plastic Waste Conversion

One notable application of waste stream utilization that can be applied in arid regions such as the GCC is the recovery and upgrading of plastic waste, as highlighted in the work of Majzoub et al., which presents a regionally grounded approach to converting plastic waste into high-value products through advanced recycling and circular economy strategies [31]. Many paths exist to convert plastic waste into value-added products or to convert the waste to energy as illustrated in Fig. 12.4. For this application, an innovative screening model for plastic waste recycling not only evaluates diverse recycling pathways but also identifies the most suitable products to generate. By integrating upgrading technologies for chemical recycling products, it offers a holistic, sustainable, and circular approach applicable to both closed-loop and open-loop scenarios. However, such upgrading processes often involve trade-offs—particularly in water and energy use. For instance, certain thermal or chemical recycling techniques, including gasification, require significant quantities of water for cooling or processing, which can exacerbate existing water scarcity in arid regions. This tension between technological circularity and resource constraints underscores the importance of context-specific assessments when adopting waste valorization solutions in the GCC. The model's framework supports balanced decision-making by considering multiple perspectives—beyond simply economic, environmental, or circularity factors—ultimately advancing a comprehensive strategy for sustainable plastic waste management.

The objective function maximizes the net profit from converting plastic waste (annual income from products minus production costs) while incorporating a circular economy performance metric into decision-making. Key constraints include material and energy balances (ensuring mass conservation and energy use limits), capacity limits for processes, economic constraints, and a Degree of Circularity Metric (DCM) requirement. The DCM is a composite index that aggregates 10 sustainability indicators, including material efficiency, energy efficiency, water usage, solid waste generation, global warming potential, economic indicator, co-product utilization, recyclability, product quality, and technology readiness. Each indicator is normalized on a 0–1 scale and weighted equally in the base case. The model maximizes profit while ensuring high circularity. Key assumptions include a steady annual waste input, fixed product prices, and technical performance parameters derived from literature.

An example was solved that evaluated handling 100,000 tons per year of mixed plastic waste. Three alternative pathways were analyzed:

1. Gasification to Methanol (GTM)—Partial oxidation gasifies the waste into syngas, which is converted to methanol.
2. Pyrolysis + Refining (PR)—Thermal pyrolysis breaks the waste into pyrolysis oil and gas, which are further processed into olefins and fuels.
3. Incineration for Energy Recovery (IER)—Waste is combusted for electricity.

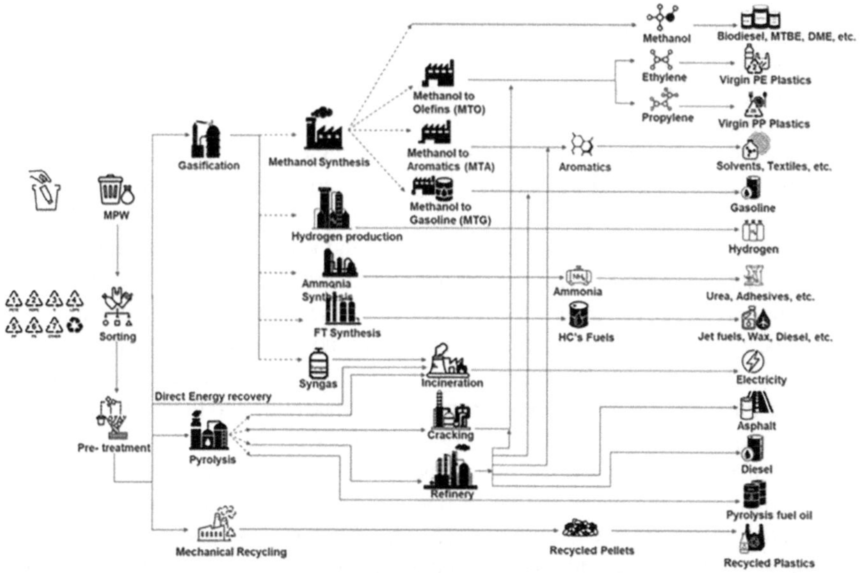


**Fig. 12.4** Screening model for plastic waste utilization adopted from Majzoub et al. [31]

Each route's economic, environmental, and energy performance is assessed, with process data drawn from literature and simulation models. Optimization results indicate that chemical recycling routes (GTM and PR) significantly outperform incineration in both circularity and profitability. Incineration yields an annual net profit of about $11 million with a circularity metric (DCM) of 0.48. Gasification-to-methanol generates around $26 million per year with a DCM of 0.67, nearly 40% higher circularity. Pyrolysis refining performs best economically, with about $33 million per year profit and a DCM of 0.61 (27% above the base case). Both GTM and PR routes offer high material recovery and recyclability scores, while incineration primarily converts waste to energy with minimal material recovery.

The study demonstrates that circular economy considerations can enhance both sustainability and economic returns. Chemical recycling technologies (such as pyrolysis and gasification) outperform incineration in recirculating material value [32]. The Degree of Circularity Metric (DCM) allows policymakers to evaluate trade-offs between sustainability criteria, such as energy efficiency, emissions, and product quality. Incorporating such metrics in policy design can optimize waste-to-resource strategies while maintaining economic viability. The study concludes that a hybrid approach, combining pyrolysis and gasification, can further improve circularity, raising the DCM to approximately 0.70. This suggests that investments in advanced recycling technologies are key to transitioning toward a circular economy for plastics.

## 12.4 Conclusion

The integration of waste streams into sustainable resource recovery systems presents significant potential for enhancing resilience in arid and water-scarce regions such as those in the Gulf Cooperation Council (GCC). By converting solid and liquid waste—ranging from biosludge and treated wastewater to plastic and industrial by-products—into valuable resources, these processes can mitigate environmental impacts, support agricultural productivity, and reduce reliance on finite natural resources.

Waste stream recovery can contribute to food and water security by improving soil fertility, increasing crop yields, and providing alternative water sources for irrigation. However, there are trade-offs to consider. These include high water and energy demands, which may place additional stress on already limited resources in arid environments. Therefore, implementing circular solutions must be context-sensitive and carefully balanced against local environmental constraints and the need for systems analysis tools to help understand the broader impact.

This chapter emphasized the importance of modeling framework and industrial symbiosis, where the waste output of one process serves as the input for another. Such integration not only maximizes resource efficiency but also strengthens local production systems and reduces dependency on imports—key priorities for economic diversification in the GCC.

Moving forward, further research and investment in innovative waste management technologies will be crucial. Policies that support circular economy models and eco-industrial parks should be prioritized to maximize resource efficiency and resilience. A comprehensive approach to waste stream utilization, supported by technology and policy interventions, can transform resource-constrained arid regions into sustainable and self-reliant systems. Through collaboration and innovation, these regions can build long-term resilience while mitigating environmental challenges.

## References


1. Abdelraouf M (2024) Water issues in the GCC countries: status, challenges, and solutions, Gulf Research Center, www.grc.net/documents/6714eeb26d491WaterIssuesintheGCCCountries2.pdf. Accessed on 17 Mar 2025
2. Fanack Water (2023) GCC's water security in the era of climate change
3. Chibani A (2023) The costs and benefits of water desalination in the Gulf. Arab Center DC. https://arabcenterdc.org/resource/the-costs-and-benefits-of-water-desalination-in-the-gulf/
4. IEA (2019) Desalinated water affects the energy equation in the Middle East, https://www.iea.org/commentaries/desalinated-water-affects-the-energy-equation-in-the-middle-east. Accessed on 16th Mar 2025
5. Organization of the Petroleum Exporting Countries (OPEC), Saudi Arabia, https://www.opec.org/opec_web/en/about_us/169.htm. Accessed 17 Mar 2025
6. PwC Strategy& (2020) The $6 billion GCC recycling opportunity, https://www.strategyand.pwc.com/m1/en/articles/2020/the-six-dollar-billion-gcc-recycling-opportunity.html. Accessed on 17th Mar 2025
7. United Nations Industrial Development Organization (n.d.) Eco-industrial parks. United Nations Industrial Development Organization. Retrieved 18 Mar 2025
8. Qadir M, Wichelns D, Raschid-Sally L, McCornick P, Drechsel P, Bahri A, Minhas PS (2010) The challenges of wastewater irrigation in developing countries. Agric Water Manag 97(4):561–568
9. Pedrero F, Kalavrouziotis I, Alarcón JJ, Koukoulakis P, Asano T (2010) Use of treated municipal wastewater in irrigated agriculture—review of some practices in Spain and Greece. Agric Water Manag 97(9):1233–1241
10. Paranychianakis NV, Salgot M, Snyder SA, Angelakis AN (2015) Water reused in EU states: necessity for uniform criteria to mitigate human and environmental risks. Crit Rev Environ Sci Technol 45(13):1409–1468
11. Toze S (2006) Reuse of effluent water—benefits and risks. Agric Water Manag 80(1–3):147–159
12. Aleisa E, Al-Zubari W (2017) Wastewater reuse in the countries of the Gulf Cooperation Council (GCC): the lost opportunity. Environ Monit Assess 189(553)
13. Ibrahim Y, Al-Mohannadi DM (2023) Optimization of low-carbon hydrogen supply chain networks in industrial clusters. Int J Hydrogen Energy 48(14):5332–5345
14. Gupta A, Uddameri V, Majumder A, Nimbhorkar S (2023) Wastewater engineering: issues, trends and solutions. CRC Press, Boca Raton
15. Shahzad HMA, Asim Z, Khan SJ, Almonari F, Mahmoud KA, Mustafa MRU (2024) Thermochemical and biochemical conversion of agricultural waste for bioenergy production: an updated review. Discov Environ 2(134)
16. Gao L, Dong Z, Xu Y, Zhao L, Xing X, Han Z, Jin M, Li X (2025) Advancements in biochar research methods for soil pollution remediation: development and applications. ACS Omega
17. Abdel-Shafy H, Mansourm M (2018) Solid waste issue: sources, composition, disposal, recycling, and valorization. Egypt J Pet 27(4):1275–1290

18. Fayshal MA (2024) Current practices of plastic waste management, environmental impacts, and potential alternatives for reducing pollution and improving management. Heliyon 10(23):e40838
19. Zhang F, Zhao Y, Wang D, Yan M, Zhang J, Zhang P, Chen L (2021) Current technologies for plastic waste treatment: a review. J Clean Prod 282:124523
20. Alao MA, Popoola OM, Ayodele TR (2022) Waste-to-energy nexus: an overview of technologies and implementation for sustainable development. Clean Energy Syst 3:100034
21. Avraamidou S, Baratsas SG, Tian Y, Pistikopoulos EN (2020) Circular economy—a challenge and an opportunity for process systems engineering. Comput Chem Eng 133:106629
22. Klemeš JJ, Varbanov PS, Wan Alwi SR, Manan ZA, Fan YV, Chin HH (n.d.) Sustainable process integration and intensification: saving energy, water and resources. De Gruyter
23. Ahmetović E, Ibrić N, Kravanja Z, Grossmann IE (2015) Water and energy integration: a comprehensive literature review of non-isothermal water network synthesis. Comput Chem Eng 82:144–171
24. Stehlík P, Jegla Z (2022) Applying process integration to thermal processing of waste. In Klemeš JJ, Varbanov PS, Pierucci S (eds) Handbook of process integration (PI): minimisation of energy and water use, waste and emissions, 2nd ed. Elsevier, pp 933–966
25. Lovelady EM, El-Halwagi MM (2009) Design and integration of eco-industrial parks for managing water resources. Environ Prog Sustain Energy 28:265–272
26. Zhao G, Garrido-Baserba M, Reifsnyder S, Xu JC, Rosso D (2019) Comparative energy and carbon footprint analysis of biosolids management strategies in water resource recovery facilities. Sci Total Environ 665:762–773
27. Wzorek M, Tańczuk M (2015) Production of biosolid fuels from municipal sewage sludge: technical and eco-nomic optimisation. Waste Manag Res 33:704–714
28. Ahmed RO, Shehab SS, Al-Mohannadi DM, Linke P (2020) Synthesis of renewable utility networks in water–energy–food nexus with economic and environmental objectives. Comput Chem Eng 135:106690
29. Ahmed RO, Al-Mohannadi DM, Linke P (2021) Multi-objective resource integration for sustainable industrial clusters. Comput Chem Eng 112:247–262
30. Elfaki H, Al-Mohannadi DM, Solim S, Al-Sayed A (2024) Sustainable valorization of oil and gas industry biosolids. Sustainability 16(22):9738
31. Majzoub WN, Al-Rawashdeh M, Al-Mohannadi DM (2024) Toward building circularity in sustainable plastic waste conversion. ACS Sustain Chem Eng 12(23):8642–8661
32. Khan MS, Mubeen I, Caimeng Y, Zhu G, Khalid A, Yan M (2022) Waste to energy incineration technology: recent development under climate change scenarios. Waste Manage Res 40(12):1708–1729

# Chapter 13
# Sustainable Wastewater Utilization for Combined Energy Production and Plastic Pollution Mitigation via Biohydrogen and Polyhydroxyalkanoates

**Georgia Antonopoulou and Ioanna Ntaikou**

**Abstract** In this chapter, an integrated approach to the combined treatment and exploitation of wastewater is presented, focusing on the concurrent production of biohydrogen (BioH2) and polyhydroxyalkanoates (PHAs) through microbiological processes. The principles and microbiology of $BioH_2$ production via dark fermentation are examined in detail, with particular emphasis on the key reactions responsible for generating volatile fatty acids (VFAs), which play a crucial role in controlling the efficiency of $BioH_2$ production and serve as essential precursors for PHA biosynthesis. The fundamentals of PHA biosynthesis by both pure and mixed microbial cultures are further explored, along with a discussion of the current advancements and limitations in industrial PHA production. Special attention is given to the advantages of utilizing acidogenic effluents as substrates for PHA production, addressing the current state of research, identifying existing challenges, and exploring the opportunities offered by this integrated strategy. The presented approach aligns with EU objectives for a circular economy, sustainable energy, and eco-friendly plastic production, providing a promising pathway for waste valorization and environmental protection.

**Keywords** Wastewater valorization · Biohydrogen · Polyhydroxyalkanoates · Circular economy · Sustainable energy · Plastic pollution mitigation

---

G. Antonopoulou
Department of Sustainable Agriculture, University of Patras, 30100 Agrinio, Greece
e-mail: geogant@upatras.gr

G. Antonopoulou · I. Ntaikou (✉)
Institute of Chemical Engineering Sciences, Stadiou St., 26504 Platani, Patras, Greece
e-mail: intaikou@upatras.gr

I. Ntaikou
Department of Civil Engineering, University of Patras, University Campus, 26500 Patra, Greece



A. Angelini et al. (eds.), *Security Enhancement for Climate Change Impacting Urban Resources - SECCURe*, NATO Science for Peace and Security Series C: Environmental Security, https://doi.org/10.1007/978-94-024-2345-7_13

## 13.1 Harnessing Wastewater as a Valuable Resource Within the Circular Economy Framework

### *13.1.1 The Dual Benefits of Simultaneous Wastewater Treatment and Resource Recovery*

For decades, economic development has been based on a linear model consisting of the steps: take/produce—process—consume—discard. As a result, Western societies, due to the reckless production and overconsumption of goods, increasingly deplete natural resources. At the same time, they generate enormous quantities of waste, the accumulation of which leads to the severe degradation of the natural environment [1]. However, understanding the consequences of such a production and consumption culture in Europe led the European Commission to adopt in 2015 its first circular economy action plan, including a package of measures aimed at facilitating the transition from a linear to a circular economy [2]. The circular economy is essentially based on the reuse of waste, wastewater and by-products generated at different stages of the production and supply chain—not only through recycling but also by converting them into other useful products. In this way, goods that have reached the end of their lifecycle and are characterized as "waste" materials may no longer be treated as disposable but as an alternative and valuable source of raw materials, reintegrated into the production chain [3].

A particularly compelling case for this approach is wastewater, which has historically been viewed primarily as a pollutant requiring disposal. Indeed, the treatment of wastewater to remove its polluting load and eliminate potentially harmful or infectious components is a prerequisite for safeguarding environmental and public health [4]. This process might necessitate significant investment in advanced infrastructure and entail high operational costs, which are essential for the effective management of wastewater and the prevention of ecological damage.

However, when wastewater is examined through the lens of the circular economy, its potential extends beyond mere environmental protection. Properly treated wastewater emerges as a valuable resource, offering opportunities for the recovery of essential nutrients (such as nitrogen and phosphorus), high-added-value compounds, and the transformation of its constituents into marketable products through chemical, thermal or biological processes [5]. This paradigm shift not only enhances the economic viability of wastewater treatment but also contributes to resource efficiency and sustainability, creating a closed-loop system where waste is continuously repurposed. Furthermore, the provision of reusable water from treated effluents adds another dimension of value, addressing water scarcity and supporting agricultural and industrial activities [6].

Elements like nitrogen, phosphorus and potassium, which are often present in high quantities in wastewater and may cause eutrophication when released into the environment, can be recovered and repurposed as valuable resources [7]. By transforming these nutrients into high-quality soil amendments and biofertilizers, they enhance soil health, improve its structure and microbial activity, and reduce the reliance on synthetic fertilizers, contributing to more sustainable agricultural practices and mitigating environmental degradation.

Moreover, wastewater with a high organic load serves as a valuable reservoir for the recovery and synthesis of a wide range of bio-based products, which can be obtained through innovative biotechnological processes. These processes, including microbial fermentation, enzymatic conversions, and bio-catalysis, enable the efficient transformation of organic waste into high-value chemicals and materials, such as biopolymers, bioactive compounds, platform chemicals, and organic acids [8]. Furthermore, the integration of physical, chemical, and biological treatment methods facilitates the recovery of renewable energy [9, 10] while simultaneously enabling the production of clean, high-quality water. This treated water can be repurposed for diverse applications, including industrial processes, agricultural irrigation, and even potable uses, promoting a circular economy and sustainable resource management.

Based on the above, it can be concluded that although the initial costs associated with wastewater treatment infrastructure and operations are often significant, the long-term benefits derived from the recovery and reutilization of valuable components make a strong economic and environmental case for investing in wastewater management. However, it is important to note that different types of wastewaters—such as municipal effluents or those from industrial sectors like food processing, textiles, or pharmaceuticals—may be better suited to specific treatment processes due to their unique composition and physicochemical characteristics. Consequently, thorough initial screening and careful selection of the most appropriate valorization strategy should be carried out on a case-by-case basis. For example, food industry wastewater, which is typically rich in organic matter and nutrients, may serve as ideal substrate for biohydrogen and biopolymers production [11], while industrial effluents may contain substrates that are better suited for producing other biofuels, or specialized microbial products or even recover specific biochemical such as phenols [12]. As such, customizing treatment and recovery processes to the distinct properties of each wastewater source can significantly improve efficiency and optimize resource recovery within the circular economy framework.

### *13.1.2 Two-Stage Bioconversion of Organic-Rich Wastewater for Renewable Energy and Bioplastics Production*

A scientifically robust approach to wastewater valorization targets streams with high organic loads, particularly those rich in fermentable sugars, through the utilization of distinct sequential bioconversion processes. This approach combines energy recovery via biohydrogen (BioH2) production and the production of microbial bioplastics, specifically polyhydroxyalkanoates (PHAs). Such a strategy not only helps alleviate energy scarcity but also contributes to addressing the global challenge of plastic pollution.

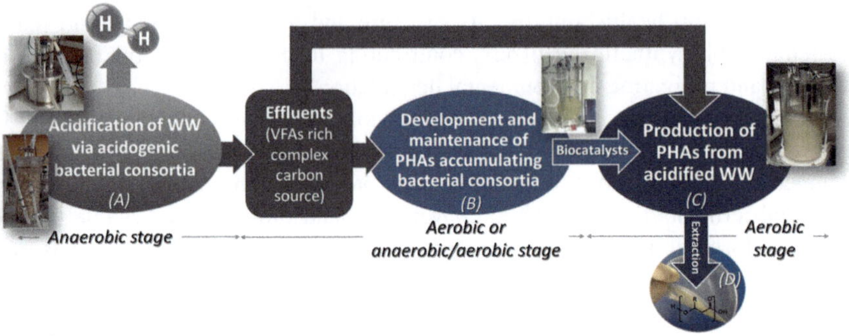


**Fig. 13.1** Schematic illustration of sequential bioconversion process of wastewater for treatment and combined biohydrogen and microbial bioplastics production

Plastic pollution represents a significant environmental threat, degrading both marine and terrestrial ecosystems, creating microplastics, and disturbing ecological balance [11]. Furthermore, the production of conventional fossil-based plastics is linked to the greenhouse effect. In response to the harmful impacts of plastic pollution, the EU adopted a European Strategy for Plastics in January 2018, which is a key element of Europe's transition to a carbon–neutral and circular economy. This strategy includes measures such as banning single-use plastics, ensuring that all plastic packaging is recyclable or reusable, and promoting the gradual substitution of conventional fossil-derived plastics with biodegradable, bio-based alternatives. A critical challenge in the commercialization of sustainable plastics is the high cost of producing biodegradable yet durable materials. This challenge may be addressed by utilizing waste, wastewater, and biomass as renewable feedstocks.

The combined approach of renewable energy and bioplastic production from wastewater typically involves an anaerobic first stage, followed by one or two aerobic stages, as illustrated in Fig. 13.1.

In the first stage (A in Fig. 13.1), dark fermentation (DF) of organic-rich wastewater produces $BioH_2$, a renewable and clean energy carrier with significant potential for integration into energy systems. This process typically utilizes mixed acidogenic anaerobic cultures, which can either be sourced directly from the wastewater (indigenous microflora) or from anaerobic sludge obtained from wastewater treatment plants, whereas the cultures may be applied in either suspended or attached forms but typically in continuous mode [11, 13–15].

The effluent from the DF, which is rich in volatile fatty acids (VFAs) and other carbon-based compounds, serves as an ideal substrate for microbial biosynthesis of PHAs, suitable for various applications, ranging from food packaging and consumer plastic products to advanced biomedical materials [16]. Research has demonstrated that wastewater from industries such as food and beverage processing, which contains high concentrations of fermentable sugars and VFAs is particularly well-suited for this dual exploitation. In the production of PHAs, the effluent from DF not only serves as the primary feedstock for intensive bioplastics production in the final stage

(C in Fig. 13.1), but also for the development and maintenance of the biocatalysts (B in Fig. 13.1) [17]. Although commercial pure carbon sources can also be used for developing robust PHA-producing consortia [18], such an approach might not be sustainable in the long term, since it increases the overall production cost. Therefore, utilizing effluent from the acidification process is preferred, as it aligns with sustainability goals and circular economy principles.

The more detailed fundamental principles and practical considerations of DF for $BioH_2$ production and PHAs production will be examined in the subsequent sections.

As PHAs accumulate intracellularly, they must be recovered from the microbial biomass (D in Fig. 13.1). This is typically achieved through solvent extraction or biomass disruption techniques, which preserve the molecular weight and thermoplastic properties of the PHAs [19]. After extraction, the PHAs undergo a purification process to remove any residual contaminants, ensuring a high-quality product. This recovery and purification process is crucial for optimizing the efficiency of the bioconversion and ensuring that PHAs can be a sustainable alternative to conventional fossil-based plastics.

By integrating these processes into wastewater treatment systems, not only is resource recovery maximized, but this approach also exemplifies the principles of the circular economy, transforming waste into high-value products with broad societal and environmental benefits.

## 13.2 Treatment and Valorization of Wastewaters with Simultaneous Biohydrogen Production

### *13.2.1 Fundamental Characteristics and Principles of Anaerobic Acidogenic Fermentation*

#### Hydrogen

DF produces one of the most interesting molecules obtainable through biorefinery processes, *i.e.*, hydrogen ($H_2$). $H_2$ is the most abundant element in the universe, found in stars, interstellar space, and the troposphere of Earth. At room temperature, $H_2$ is an odorless, colorless, and flammable gas, characterized by light weight and high energy density. $H_2$ bonds with oxygen creating water, the most important element on Earth, while its bond with carbon forms the backbone of a vast collection of organic molecules known as hydrocarbons. Since hydrocarbons are everywhere, $H_2$ is one of the most widely used chemical elements, employed in various industrial applications, such as in chemical industries, metallurgy, and food industries.

Among its other uses, $H_2$ also presents enormous potential as an energy source, due to its high energy content per unit mass, as it has approximately three times the energy density ($kg^{-1}$) compared to conventional gasoline. Furthermore, its use in fuel cells can produce electrical energy of high efficiency and exhaust gases of zero emissions, since when burnt produces only pure water, offering an environmentally friendly alternative to conventional energy sources. For this reason, it is recognized as "the fuel of the future". Among others, $H_2$ can be used in the energy sector, for electrical energy production, in transportation, as fuel for various vehicles, and in industry as the raw material for many chemical reactions. Up to now, its demand has increased by 10% annually, and it is estimated to significantly contribute to the energy sector by 2040 [20].

Since $H_2$ does not naturally occur on Earth in its molecular form, its production relies on compounds that contain it, such as water and hydrocarbons [21]. In this respect, hydrogen is produced through DF from carbohydrates that are contained in wastewater or biomass.

**Anaerobic fermentation**

Anaerobic fermentation is a biological process that converts various organic substrates into energy (in the form of $H_2$) and biochemical products such as carboxylates and ethanol. This process constitutes the initial phase of anaerobic digestion (AD), which continues with the conversion of organic matter into biogas (composed of $CH_4$ and $CO_2$) through a series of subsequent biological reactions carried out by different groups of microorganisms that work together in a syntrophic manner.

Initially, organic matter undergoes hydrolysis, breaking down into monomers and oligomers. These components are then converted into $H_2$, $CO_2$, carboxylates, and ethanol through a phase known as acidogenesis. In the next stage, acetogenesis, bacteria oxidize the carboxylic acids into acetic acid and $H_2$. Homoacetogenic bacteria can convert $H_2$ and $CO_2$ into acetic acid while acetate-oxidizing bacteria can perform the reverse reaction. These reactions are part of the anaerobic fermentation phase. The process culminates in the methanogenesis phase, where hydrogenotrophic bacteria produce $CH_4$ from $H_2$ and $CO_2$ and aceticlastic methanogenic bacteria produce methane from acetic acid, respectively.

Anaerobic fermentation consists of two phases: DF and acetogenic fermentation. Dark fermentation produces $H_2$, carboxylates, and ethanol, while acetogenic fermentation continues to the acetogenesis phase, where acetogens convert carboxylates into acetate and $H_2$, and homoacetogens produce acetate from $H_2$ and $CO_2$. The presence of lactate and ethanol can lead to chain elongation (CE), extending carboxylate chains. These processes typically use mixed microbial cultures (MMC), which support scalability and efficient wastewater conversion.

**Dark fermentation**

During DF carbohydrate-rich substrates undergo microbial conversion, yielding an $H_2$-rich biogas alongside a mixture of fermentative metabolites. The primary metabolic byproducts include carboxylates such as acetate, butyrate, propionate, known as VFAs as well as other organic molecules such as lactate and ethanol [13]. VFAs are short-chain fatty acids typically consisting of $C_2$-$C_5$ atoms, though caproic acid ($C_6$) is sometimes included due to its role in anaerobic fermentation pathways.

These acids have gained significant attention in recent years for their broad applicability in pharmaceuticals, chemicals, polymers, biofuels, and wastewater treatment [22].

The $H_2$ yield can be stoichiometrically correlated with the final fermentative metabolite, which can be produced upon glucose degradation (or its isomers or polymers). It is expressed as mol $H_2$ per mol of consumed sugars and can be calculated from the reactions R1–R3, below. For more complex substrates, its production can be quantified in terms of hydrogen productivity (HP), which represents the fraction of influent substrate electrons allocated to hydrogen gas (both in gaseous and dissolved phases) [23].

$$C_6H_{12}O_6 + 2H_2O \rightarrow 2CH_3COOH + 2CO_2 + 4H_2 \tag{R1}$$

$$C_6H_{12}O_6 \rightarrow CH_3CH_2CH_2COOH + 2CO2 + 2H_2 \tag{R2}$$

$$C_6H_{12}O_6 + 2H_2 \rightarrow 2CH_3CH_2COOH + 2H_2O \tag{R3}$$

As can be seen, acetic and butyric acid production is directly linked to $H_2$ generation, with glucose fermentation to acetic acid yielding the highest theoretical $H_2$ yield of 4 mol $H_2$/mol glucose (HP = 33%) (R1). Butyric acid generation results in 2 mol $H_2$/mol glucose (HP = 17%) (R2), whereas propionic acid generation does not favor $H_2$ (R3). These reactions indicate that shifting metabolism toward acetate and/or butyrate production is essential for achieving a high hydrogen yield. However, in practice, the observed yields are lower (0.5–2.5 mol $H_2$/mol glucose) due to the influence of alternative fermentation routes, such as the lactate and ethanol pathways, which compete with these two fermentation pathways [11, 15] Under mesophilic conditions, hydrogen-producing bacteria (HPB) predominantly belong to the *Clostridium* genus, a group of strictly anaerobic microorganisms capable of fermenting a wide range of carbohydrates through various metabolic pathways, as presented in R4:

$$C_6H_{12}O_6 + H_2O \rightarrow CH_3COOH + CH_3CH_2OH + 2CO_2 + 2H_2 \tag{R4}$$

Clostridium spp. possess the ability to form spores, which makes them highly adaptable and easily enriched in MMC, even after various pre-treatment methods, including heat shock, acid, and alkaline treatments, which are applied in different sources of microbial inoculum [11, 24]. In addition to *Clostridium*, several non-spore-forming, strict anaerobic bacteria also contribute to $H_2$ production, including *Acetanaerobacterium elongatum, Ethanoligenes harbinense*, and *Megasphaera* spp. Notably, *E. harbinense* has been reported to achieve high $H_2$ yields of $3.4 \pm 0.7$ mol $H_2$/mol sucrose consumed (~ 8.9 mmol $H_2$/g COD) even under low pH conditions (3–4.5) [25] In comparison, *Megasphaera* spp. exhibited $H_2$ yields ranging from 0.8 to 2 mmol $H_2$/g COD [26].

Beyond strict anaerobes, certain HPB which are facultative anaerobes, can produce $H_2$ through a metabolic route, with a maximum theoretical $H_2$ yield of up to 2.5 mol /mol glucose (R5). This group of HPB is primarily composed of *Bacillus* species, which typically achieve yields of ~ 2.0 mol $H_2$/mol glucose, as well as members of the *Enterobacteriaceae* family, including *Klebsiella* spp. (2.0 mol $H_2$/mol glucose), *Escherichia-Shigella* spp. (1.44 mol $H_2$/mol glucose), and *Citrobacter* spp. (1.2 mol $H_2$/mol glucose) [27–30].

$$4C_6H_{12}O_6 + 2H_2O \rightarrow 2CH_3COOH + 3CH_3CH_2CH_2COOH + 8CO_2 + 10H_2 \quad (R5)$$

Additionally, acetate or other product formation may proceed via non-hydrogen-producing pathways, or by non–HPB. In general, in mixed fermentation systems, microbial populations can alter their metabolic pathways in response to environmental changes to other non-hydrogen pathways, or under different prevailing conditions, other species, such as non-HPB, can dominate the microbial consortium, producing other products, rather than $H_2$ [10, 13.]

Non-HPB compete with HPB for the substrate. For instance, some members of the *Enterobacteriales* can degrade sugars towards ethanol without $H_2$ production (R6):

$$C_6H_{12}O_6 \rightarrow 2CH_3CH_2OH + 2CO_2 \quad (R6)$$

In addition, the presence of lactic acid bacteria (LAB) such as *Lactobacillus, Sporolactobacillus*, and *Streptococcus* negatively influences DF performance. There are three possible reactions carried out by LAB, such as the homofermentative pathway (R7), which converts sugars exclusively into lactic acid, and heterofermentative pathways (R8, R9), where sugars are converted to other products alongside lactic acid:

$$C_6H_{12}O_6 \rightarrow 2CH_3CHOHCOOH \quad (R7)$$

$$C_6H_{12}O_6 \rightarrow CH_3CHOHCOOH + CH_3CH_2OH + CO_2 \quad (R8)$$

$$2C_6H_{12}O_6 \rightarrow 3CH_3COOH + 2CH_3CHOHCOOH \quad (R9)$$

However, some studies suggest that LAB may enhance $H_2$ production, potentially due to their ability to produce hydrolytic enzymes that break down organic matter, thus increasing substrate availability for HPB [31]. Other studies report that species of the genus *Lactobacillus* release bacteriocins in the fermentation broth, contributing to process instability [32].

**Acetogenic fermentation**

Another group of non-HPB are the $H_2$-consuming homoacetogenic bacteria, which convert molecular $H_2$ and $CO_2$ into acetate (R10). Homoacetogenesis is mainly attributed to high partial pressure of $H_2$ ($pH_2$) in reactors, though its determining factors remain unclear. It is independent of inoculum, substrate, pre-treatment, reactor type, or operation conditions. For this reason, efforts are made to lower $pH_2$ into the reactors to make homoacetogenesis thermodynamically unfeasible. Various bacterial phyla include homoacetogens, with many belonging to *Clostridium* spp. These bacteria can both produce and consume $H_2$ under proper conditions [33].

$$4H_2 + 2CO_2 \rightarrow CH_3COOH + H_2O \quad \text{(R10)}$$

**Chain elongation**

CE reactions have the potential to produce medium-chain fatty acids (MCFAs), which are carboxylic acids consisting of $C_6$-$C_{10}$ atoms. These acids have various industrial applications and increase market demand, and can be used as precursors for PHAs production. CE involves the elongation of carboxylic acid chains, where VFAs typically act as electron acceptors, while lactate and ethanol serve as primary electron donors. When ethanol-rich substrates, are used, CE is more likely to occur, provided that environmental conditions (such as substrate and product concentration, pH, and microbial community) are favorable. The long chains of MCFA allow for their selective removal from the fermentation broth [34]. *Clostridium kluyveri* is the most studied microorganism in chain elongation, with only 16 species known to produce caproic acid until 2020 [35]. Most known chain elongators belong to the Clostridiaceae, Oscillospiraceae, Eubacteriaceae, and Veillonellaceae families [35].

### *13.2.2 Factors Affecting Anaerobic Fermentation*

From R1–R10, it can be concluded that $H_2$ and VFAs production can be achieved by promoting specific metabolic pathways. Different $H_2$ yields and proportions of VFAs depend on parameters like pH, temperature, hydraulic retention time (HRT), substrate composition, co-substrates, and inoculum type. Optimizing the process requires a comprehensive understanding of metabolic pathways and key operating parameters. The role of each parameter in DF is well understood, but their optimum values remain unclear, depending on the individual prevailing conditions.

For instance, pH significantly affects HPB by influencing the activity of hydrogenase, the key enzyme involved in $H_2$ production reactions. Thus, pH influences bacterial growth, enzymatic synthesis and activity, dominant bacterial population, and substrate bioavailability [36]. Whereas the optimal pH range for the typical HPB i.e. *Clostridium spp* is typically 5–7.5 [37] no universal optimal pH exists, especially in MMC. Acidic pH while being effective for methanogens inhibition, its value cannot be too low, as hydrogenase activity drops significantly below pH 4.5. Extremely low pH inhibits hydrolysis and acidogenesis, leading to alternative pathways like lactic acid production. pH also affects carboxylate profiles, with butyrate favored at pH 5–6 and acetate in alkaline conditions [15, 38]. Thus, pH plays a vital role in $H_2$ production and VFAs formation, and final distribution.

Temperature also plays a crucial role in influencing metabolic pathways and overall process efficiency. Temperature affects microbial growth, enzyme activity, and organic matter hydrolysis. Although thermophilic conditions yield higher $H_2$ production due to the suppression of hydrogen-consuming bacteria, mesophilic DF (35 °C) is preferred for its cost-effectiveness, avoiding external heating needs. Regarding VFAs formation, mesophilic conditions are most efficient for acetate and propionate production [39] while butyrate dominates at 55 °C [40].

HRT is crucial for steering process selectivity by washing out undesired microorganisms. Short HRTs (6–12 h) enhance $H_2$ production by removing competitors of HPB [41], i.e. acetoclastic methanogens are eliminated at HRTs of 2.5–10 h. The optimal HRT depends on substrate degradability, as hydrolysis is the rate-limiting step. Simple substrates like sugars require low HRTs (1–12 h), while more complex substrates like starch need longer HRT values (15–17 h) due to slower hydrolysis. Optimal HRTs depend on feedstock type, inoculum, and experimental conditions. A balance between substrate hydrolysis and bacterial selection is needed to maximize $H_2$ yield. Regarding carboxylate accumulation, studies show that increasing HRT can enhance VFA yield up to a threshold, beyond which no significant improvement is observed [42]. HRT also influences specific acid production by selecting microbial populations [15]. However, HRT alone does not determine VFA selectivity, since other operational factors must be considered.

During substrates degradation from MMC, inoculum plays a key role in accelerating the process. Many sources of microbial inoculum are reported for DF, including digestate, wastewater sludge, cow dung, compost, and landfill lixiviates [43]. Different pre-treatment approaches, such as heat-shock, acid, and alkaline treatment are often applied to enrich HPB like *Clostridium* spp., as they can form spores, unlike competing microbes like methanogens [10]. By these methods, VFAs production can be enhanced, and methanogenesis can be suppressed. Interestingly, endogenous bacteria naturally present in substrates can also efficiently produce $H_2$ [13–15, 44]. Additionally, modifying the inoculum can alter metabolic pathways, enhancing specific acid production. Thus, the selective production of one acid can be promoted by using some specific microorganisms, i.e. homoacetogen *Acetobacterium woodii* promoted acetate production [45] while *Propionibacterium acidipropionici* increased propionate levels [46].

Regarding waste biomasses, it is well known that carbohydrate-rich substrates yield the highest $H_2$ output during DF, compared to lipid-and protein-rich biomasses. Substrate selection depends on availability, cost, carbohydrate content, and biodegradability [10, 23]. Glucose, sucrose, or lactose are well-studied sugars since they are easily biodegradable and can be found in waste streams. Agricultural residues such as sunflower straw [47], grass lawn [48], or willow sawdust [49] are also potential feedstocks for $H_2$ generation but need a proper pretreatment method before fermentation. Various wastewaters, such as those from rice, winery, and sugar industries, containing biodegradable organic matter have been explored for $H_2$ in DF [50] demonstrating high hydrogen yields (Table 13.1). One limitation of using industrial effluents as a substrate is their inconsistent availability throughout the year, which affects the continuity of $H_2$ generation [51]. Substrate's characteristics also play a crucial role in VFAs production and composition. It was found that a higher total Kjeldahl nitrogen (TKN) content can enhance propionate yield [52]. Still, an excessively high protein content may inhibit its production [53]. Adjusting the carbon-to-nitrogen (C/N) ratio, such as supplementing carbohydrates to protein-rich substrates and maintaining an appropriate pH, can promote propionate production. Additionally, the co-digestion of different wastes helps balance nutrients, optimize fermentation, and dilute potential toxic compounds [54].

### *13.2.3 Wastewater Treatment by Coupling Fermentation with Other Biological Processes*

In the DF process, the initial organic substrate is converted into bio-$H_2$ and metabolites such as VFAs, lactate, and ethanol. However, only about 20–25% of the total chemical oxygen demand (t-COD) is converted to $H_2$, with the remaining (75–80%) being recovered as fermentative metabolites. From a wastewater treatment point of view, these low conversion yields of organics necessitate a secondary treatment process to maximize COD recovery and enhance the economic viability of the overall process. One option could be the efficient recovery of VFAs from the fermentation broth, through various extraction techniques, the selection of which depends on the intended use and required purity of the final product [61]. On the contrary, biological secondary treatment processes offer a more sustainable alternative, utilizing the effluent's metabolites and nutrients (N, P). These processes can generate products like $H_2$, $CH_4$, lipids, proteins, carbohydrates, and of course PHAs. The choice of secondary process depends on the effluent's composition. For instance, AD can tolerate carboxylates below the inhibitory threshold for methanogenic microorganisms while bioelectro-chemical or photo-fermentative systems are preferred when acetate is produced during DF. However, low-purity VFAs which are commonly produced in mixed acid fermentations using MMC are suitable for applications such as PHAs production, giving a promising alternative in the management of wastewater both from a technical and economic point of view.

**Table 13.1** Wastewaters used as substrates in dark fermentation (DF)

| Wastewater | Operation/reactor | Inoculum | HRT (h) | Conditions | Max $H_2$ production rate | Max $H_2$ production yield | Reference |
|---|---|---|---|---|---|---|---|
| Cheese whey | Continuous stirred tank reactor (CSTR) | Indigenous microbial culture | 24 | Mesophilic | 2.51 L/L/d | 0.9 mol/mol hexose | [55] |
| Cheese whey | CSTR | Indigenous microbial culture | 24 | Mesophilic (alkalinity addition) | 2.9 ± 0.2 L/$L_{reactor}$/d, | 0.78 ± 0.05 mol/mol hexose | [56] |
| Food industry waste extract | Continuous, Anaerobic up-flow column reactor | Indigenous microbial culture | 2–12 | Mesophilic | 2.94 ± 0.37 L /L/d (HRT = 2 h) | 0.14 ± 0.01 mol/mol (HRT = 12 h) | [14] |
| Olive pulp | CSTR | Mixed mesophilic culture | 30, 14.5, 7.5 | Mesophilic | 0.47 L/L/d (HRT = 7.5 h) | 5.9 L/kg TS (HRT = 30 h) | [57] |
| Cassava wastewater | Continuous, Anaerobic fluidized bed reactor | Sludge (from swine wastewater treatment | 8–1 | 28 °C | 8 L H2/L/h (HRT = 1 h) | 1.91 mol/mol (HRT = 2 h) | [58] |
| Beverage wastewater | CSTR | Enriched mixed microflora | 6–1.5 | Mesophilic | 37.5L H2/L/d (HRT = 1.5 h) | 1.62 mol/mol (HRT = 6 h) | [59] |
| Distillery wastewater | Batch | Enterobacter cloacae IIT-BT 08 | – | Mesophilic | 80 mL/L h | 7.38 ± 0.24 mol $H_2$/kg COD reduced | [60] |

## 13.3 Exploitation of Acidogenic Effluents as Substrates to Produce Polyhydroxyalkanoates

### *13.3.1 Fundamental Characteristics and Principles of Polyhydroxyalkanoates Production*

**The nature and properties of PHAs**

PHAs are linear polyesters composed of hydroxyl-fatty acid monomers that can be intracellularly synthesized by various microorganisms, mainly bacteria, and accumulate in small granules within the cytoplasm in particles ranging from 0.2–0.5 μm. Unlike petroleum-derived plastics that need several decades to degrade and be assimilated into the environmental carbon cycle, PHAs can be completely bio-degraded by a variety of microorganisms, like any other natural polymer. Their general structure is formed from repeating units of hydroxyacids, with variations in chain length and composition determining the specific type of PHAs produced (Fig. 13.2). The monomers of hydroxy acids are often in the form of 3-hydroxyalkanoates (3-HAs), but 4-HAs and less commonly 5-HAs or 6-HAs (Table 13.2) with saturated and unsaturated chains containing or not aliphatic and aromatic groups may occur in nature [62]. PHAs are broadly classified into short-chain-length PHAs (scl-PHAs) consisting of 3–5 carbon atoms per monomer (e.g., polyhydroxybutyrate, PHB), medium-chain-length PHAs (mcl-PHAs) containing 6–14 carbon atoms per monomer (e.g., polyhydroxyhexanoate, PHO) and long-chain-length PHAs (lcl-PHAs) which include > 14 carbon atoms in the monomeric chain, although less frequently found in nature [62].

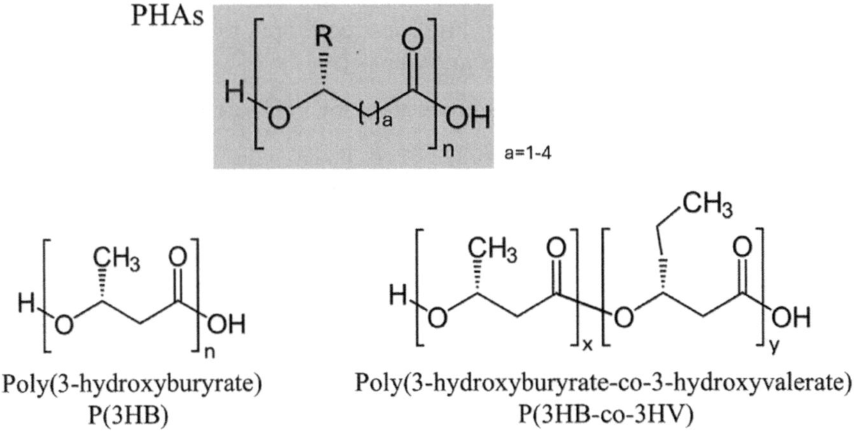


**Fig. 13.2** General structure of polyhydroxyalkanoates (PHAs) molecules and key types

**Table 13.2** Types of polyhydroxyalkanoates (PHAs) based on the side chain (R) and the number of carbon atoms between the hydroxyl and carboxylic carbon

| C atoms between–OH and carboxylic C (a) | Side aliphatic chain (R) | PHAs type | |
|---|---|---|---|
| 1 | H | Poly(-3-hydroxypropionate) | P(3HP) |
| | $CH_3$ | Poly(-3-hydroxybutyrate) | P(3HB) |
| | $C_2H_5$ | Poly(-3-hydroxyvalerate) | P(3HV) |
| | $C_3H_7$ | Poly(-3-hydroxyhexanoate) | P(3HHx) |
| | $C_5H_9$ | Poly(-3-hydroxyoctanoate) | P(3HO) |
| 2 | H | Poly(-4-hydroxybutyrate) | P(4HB) |
| 4 | H | Poly(-6-hydroxyhexanoate) | P(6HHx) |

PHAs exhibit mechanical and thermal properties comparable to conventional petrochemical plastics, such as polyethylene and polypropylene, making them excellent candidates for their replacement. For instance, PHB has similar tensile strength and stiffness to polypropylene, while mcl-PHAs demonstrate flexibility and elasticity akin to low-density polyethylene. This similarity in performance, combined with their biodegradability, biocompatibility, and biobased production, enhances the potential for PHAs to reduce dependence on fossil fuel-derived plastics [63].

Being a versatile family of different homopolymers and co-polymers, their properties, especially the mechanical and thermodynamic ones that are primarily influenced by the type and length of the hydroxyalkanoate monomer, may vary considerably [64]. Moreover, different PHA types may be recovered from MMC in the same batch as blends, further altering the apparent properties [18]. Key properties of different PHAs include high crystallinity, brittleness, and thermal instability for PHB, higher flexibility, lower melting points, and elastomeric behavior for mcl-PHAs and improved ductility and reduced brittleness compared to PHB for copolymers consisting of different types of monomeric units [64].

**Microbial pathways and carbon sources used for PHAs production**

Microbial PHA biosynthesis generally occurs under unbalanced nutrient supply conditions, such as limitation of a main nutrient (nitrogen, phosphorus), with excess carbon availability. Three primary metabolic pathways of PHA biosynthesis have been identified. Via the first one, scl-PHAs occur from sugars, short-chain fatty acids, simple alcohols (i.e., ethanol and glycerol) or amino acids, which are initially converted into two acetyl-CoA molecules. The latter are subsequently transformed into two hydroxybutyryl-CoA molecules through the action of β-ketoacyl-CoA thiolase (PhaA) and acetoacetyl-CoA reductase (PhaB). Finally, these intermediates are polymerized into PHB by the P(3HB) polymerase (PhaCscl) via a condensation reaction. Common scl-PHAs produced through this pathway include the homopolymers P(3HB) and P(4HB) [65] whereas the addition of propionic or valeric acids as carbon source can promote the synthesis of the co-polymer P(3HB-co-3HV) [8], as the propionyl-CoA generated in this pathway serves as a precursor to 3-hydroxyvalerate.

The second metabolic pathway is based on the β-oxidation of fatty acids and leads to the production of mcl-PHAs. In this process, fatty acids are first converted into enoyl-CoA, which is then transformed into R-3-hydroxyacyl-CoA by the enzyme R-3-hydroxyacyl-CoA hydratase. These intermediates are ultimately polymerized into mcl-PHAs by mcl-PHA synthase (PhaCmcl). Examples of mcl-PHAs synthesized via this pathway include P(3HHx), P(3HO), and copolymers like P(3HHp-co-3HN) and P(3HHx-co-3HO-co-3HD) [66]. The third pathway for PHA biosynthesis utilizes intermediates from two different sources: R-3-hydroxyacyl-ACP from in situ fatty acid synthesis or R-3-hydroxyacyl-CoA derived from β-oxidation. In this route, R-3-hydroxyacyl-ACP is converted to R-3-hydroxyacyl-CoA by the enzyme PhaG, making it available for polymerization by PhaC [67]. It becomes thus apparent that the targeted production of specific PHA types can be achieved through the selection of appropriate microorganisms, either in pure or MMC i.e. microbial consortia, and the strategic use of specific carbon sources that direct microbial metabolism [68], enabling the development of custom-designed biopolymers for diverse industrial applications.

**Bioreactor configuration for PHAs production from pure cultures and consortia**

PHAs can be produced from both pure and MMC, each requiring distinct bioreactor configurations to optimize yield and productivity. For pure cultures, systems like batch, fed-batch, and continuous stirred-tank reactors (CSTR) are commonly used [69]. Among these, fed-batch systems are often favored due to their ability to control nutrient limitation during fermentation, which enhances PHAs accumulation. These reactors are designed to maintain a sterile environment, thereby ensuring high substrate-to-product conversion rates and consistent polymer quality throughout the production process. They allow for better management of key parameters, such as pH, temperature, and nutrient levels [70], which is crucial for maximizing PHAs production efficiency.

In contrast, for MMC, bioreactors like sequencing batch reactors (SBR) or anaerobic–aerobic systems are commonly used for the development and maintenance of the MMC, which ensures a diverse microbial community that can thrive under varying environmental conditions [71]. Once the microbial community is established, the main production phase for PHAs typically takes place under batch or fed-batch mode, whereas the use of continuous systems has also been reported [72]. These bioreactor configurations exploit the natural selection of PHAs-accumulating organisms, which is promoted under feast-famine conditions, where microorganisms are exposed to alternating periods of nutrient abundance and scarcity. This approach allows for the utilization of low-cost, non-sterile feedstocks such as industrial and agricultural waste, including food waste, which further reduces operational costs. While MMC systems can result in greater variability in polymer composition due to the presence of different microbial species, they offer significant economic and environmental advantages.

Specifically, MMC-based processes have lower operational costs compared to pure culture systems and contribute to sustainability by reducing the need for sterilization and enabling valorization of waste materials as feedstocks. These benefits make MMC systems particularly attractive for large-scale, cost-effective PHA production in the context of circular economy.

### *13.3.2 Status, Challenges and Limitations for Substitution of Petrochemical Plastics by PHAs*

Despite the significant potential of PHAs as biodegradable and biocompatible bioplastics, and the extensive research invested in supporting their sustainable and economically viable production, the full-scale industrialization of PHAs for everyday applications continues to face numerous challenges.

A major hurdle is the high upstream cost, primarily driven by the expense of carbon substrates such as glucose, sucrose, and hydrolyzed starch. Furthermore, the overall fermentation process for microbial PHA biosynthesis, which is relatively long lasting, further hinders industrial scalability, making the process less competitive with the rapid production rates of synthetic plastics [73].

Additionally, the downstream processing stages—including cell disruption, solvent-based polymer extraction, and purification are labor and resource intensive, significantly contributing to overall costs. These stages are technically challenging, often requiring chemical solvents such as chloroform for polymer recovery and purification, which raises environmental, health, and safety concerns. Although alternative, more sustainable extraction methods are under development, they have yet to achieve widespread industrial adoption. The limited market penetration and demand for PHAs are also attributable to their higher cost relative to conventional plastics, which creates economic barriers despite the evident environmental benefits. Moreover, PHAs face strong competition from other established biopolymers such as polylactic acid (PLA) and starch-based plastics, which are produced more efficiently and at lower costs [74].

Addressing these limitations requires a strong emphasis on the exploitation of waste streams as low-cost and sustainable feedstocks, coupled with the use of MMCs as robust and adaptable biocatalysts. The use of MMCs offers advantages in terms of operational flexibility and resilience to variable feedstock composition, reducing the dependency on pure cultures and refined substrates. By integrating waste valorization strategies with efficient bioprocesses, the economic feasibility of PHAs production can be significantly enhanced. Furthermore, innovations in downstream processing technologies and the development of integrated biorefinery approaches—where multiple valuable products are extracted from a single process—could drive environmental sustainability and broader industrial adoption of PHAs.

As for current market trends in PHAs production, the utilization of waste materials as feedstocks has been adopted by some companies, although this approach is not yet widely implemented across the overall industrial sector. Table 13.3 presents some key industrial producers of PHAs, highlighting that while companies such as Biomer (Germany) and PHB Industrial S.A. (Brazil) continue to rely on sugars like sucrose, Newlight Technologies LLC (USA) and RWDC Industries (USA/Singapore) have transitioned to using waste materials, including greenhouse gas emissions and waste oils, as alternative feedstocks. These advancements represent important steps toward reducing production costs. However, the use of microbial consortia, as opposed to specific biocatalysts (pure cultures), has not yet been fully scaled, contributing to upstream costs due to the need for maintaining sterile conditions. Consequently, scalability and process efficiency continue to pose significant challenges.

### *13.3.3 Advantages of Using Acidogenic Effluents from Dark Fermentation as Substrates for PHAs Production*

As discussed in Sect. 13.2.2, acidogenic effluents from DF are enriched with VFAs, such as acetic, propionic, butyric, and other medium-and short-chain fatty acids. These VFAs serve as high-value carbon substrates for both pure and mixed microbial cultures, facilitating microbial metabolism and PHA biosynthesis. The broad spectrum of VFAs, which includes both even- and odd-chain fatty acids, further enhances the potential of acidogenic effluents. This diversity enables microbial communities to engage in a variety of metabolic pathways, promoting the synthesis of PHA copolymers including scl-PHAs (e.g., P(3HB-co-3HV)) [8], scl- and mcl-PHAs (e.g., P(3HB-co-3HHx)), or mcl-PHAs (e.g., P(3HHx-co-3HO-co-3HD)) [66]. Copolymeric PHAs exhibit improved properties, such as enhanced flexibility, thermal stability, higher impact resistance, and superior processing characteristics. These attributes make them highly advantageous for a wide range of industrial applications, extending their utility beyond food packaging and agricultural sectors to include the medical and pharmaceutical industries [28, 64]. As such, the ability to control the ratio of even-to odd-chain VFAs during DF via selected operational conditions or feed characteristics [39, 46, 52] provides a strategic tool for the design of PHAs with desirable properties, thereby improving the versatility and value of the final biopolymer product.

Moreover, during the acidification step of DF, while carbon sources are biotransformed into VFAs, the overall carbon content in the effluent is not significantly reduced (as mentioned in paragraph 2.3 only about 20–25% of the total t-COD is converted to $H_2$ during DF) This retention of carbon is crucial for the subsequent PHAs production step, as the carbon remains available for microbial assimilation. Unlike other processes where a substantial loss of carbon may occur during the conversion of complex organic matter, acidogenic fermentation ensures that the majority of the carbon is retained in the form of VFAs, which are then directly

**Table 13.3** Overview of current key industrial producers of polyhydroxyalkanoates (PHAs)

| Company | Trademark | PHA type | Substrate | Biocatalyst | 1st year of manufacturing | Current capacity | Website |
|---|---|---|---|---|---|---|---|
| Biomer, Germany | Biomer® | PHB | Sucrose | – | 1995 | | https://www.biomer.de |
| Newlight Technologies LLC, USA | AirCarbonTM | PHB | Air and greenhouse gas emissions ($CO_2$, $CH_4$) | Newlight's 9X biocatalyst | 2003 | 3000 metric tons/y | https://newlight.com https://www.aircarbon.com |
| PHB Industrial S.A., Brazil | BIOCYCLE® | PHB, PHBV | Sugar cane sucrose | - | 2004 | – | www.biocycle.com.br |
| TianAn Biologic Materials Co. Ltd., China | ENMATTM | PHB, PHBV | Dextrose derived from corn or cassava | Non-genetically modified *Ralstonia eutropha* | 2006 | 1.3 M metric tons/y | http://en.tianan-enmat.com/ |
| Danimer Scientific, USA | Nodax® PHA | PHBH | Corn, sugar beet and/or vegetable oils | mcl-PHA bacteria | 2007 | 90,000 metric tons/y | https://danimerscientific.com |
| Kaneka Corporation, Japan | AONILEX® and Green Planet™ | PHB, PHHx | Plant oils | – | 2011 | 5000 metric tons/y | https://www.kaneka.co.jp |
| RWDC Industries, USA, Singapore | Solon® | mcl-PHAs | Vegetable oils, wasted oil | – | 2015 | 32,000 metric tons/y | https://www.rwdc-industries.com |
| Bio-On Srl., Italy | Minerv® | PHB, PHBV | Agricultural waste, sugar beet waste | *Cupriavidus necator* | 2018 | 10,000 metric tons/y | https://www.bionity.com |

utilized by microorganisms for PHA accumulation. This retention of carbon maximizes the efficiency of the process, ensuring that a higher proportion of the waste is converted into the desired biopolymer rather than being lost as microbial biomass.

Another key advantage of using acidogenic effluents is the modification of the carbon-to-nitrogen (C/N) ratio during the acidification step. In the dark fermentation process, nitrogen content from organic waste is reduced during its incorporation into microbial biomass [18, 75]. This is crucial for optimizing PHA production, as an imbalanced C/N ratio, particularly when nitrogen is in excess, can limit microbial growth and PHA accumulation. In raw wastewater or unprocessed organic waste, the C/N ratio is often not ideal for efficient PHAs biosynthesis, potentially resulting in nitrogen-rich conditions that hinder the growth of the microbial cultures needed for polymer synthesis. However, during acidification, the reduction in nitrogen levels improves the overall C/N ratio creating a more favorable environment for increased yield of PHAs [11, 65] supporting the efficient conversion of organic waste into valuable biopolymers.

Consequently, acidogenic effluents provide a highly concentrated and carbon-rich substrate for PHA synthesis, optimizing both the yield and cost-effectiveness of the process.

### *13.3.4 Current Insights and Research Advancements in Integrated BioH$_2$ and PHA Production from Wastewater*

The integrated production of BioH$_2$ and PHAs from wastewater presents a compelling approach that, as discussed above, offers dual benefits: mitigating plastic pollution and reducing reliance on fossil fuels. This process aligns with the EU's sustainability goals, particularly its targets for reducing plastic waste and transitioning to a circular economy. By utilizing wastewater as a feedstock, this integrated system not only addresses the growing environmental challenges posed by plastic pollution but also contributes to the reduction of greenhouse gas emissions and the dependence on non-renewable resources.

However, the large-scale implementation of this technology remains distant, as there are several key challenges that still need to be addressed, as discussed in the previous sections. These challenges include the optimization of both the BioH$_2$ production process and the PHA synthesis process, as well as the integration of these processes for efficient and cost-effective operations. Indeed, when integrating BioH$_2$ production with PHA biosynthesis can amplify challenges, and the synergies between processes need to be carefully balanced to achieve optimal performance. This integration often requires a more complex system design, as the output of one process must be effectively fed into the other, which can increase both capital and operational costs. Furthermore, the integration of these processes requires a careful evaluation of the overall system's efficiency, considering factors such as the flow of resources, energy requirements, and the scalability of the integrated system. While technology holds significant promise from an environmental and sustainability perspective, the economic challenges must be resolved to make it commercially viable.

As noted in the previous sections, ongoing research is addressing these issues, but achieving a cost-effective, large-scale integrated system requires continued optimization of both the individual $BioH_2$ and PHAs processes, as well as their integration, to meet the demands of industrial applications and align with the goals of the circular economy. Ongoing research and statistical evaluations are showing promising results, and further advancements are continuously being reported. Some of these findings are summarized in Table 13.4, reflecting the potential of this integrated approach to drive both environmental and industrial sustainability forward. As shown, for the combined $H_2$ and PHAs production process, wastewaters rich in sugars—such as those from confectionery production, expired fruit juices, dairy processing, and distilleries—have been tested as substrates. These wastewaters serve as excellent fermentation feedstocks due to their high organic content and easily biodegradable carbohydrates [75]. Such substrates support microbial metabolism, leading to the production of diverse VFA mixtures, which are key intermediates for both $H_2$ and PHA biosynthesis, as previously analyzed. It can also be noted that continuous reactor configurations, particularly CSTRs, are preferably employed for the acidification processes, whereas batch, sequential batch, and fed-batch operational modes are applied for PHAs accumulation, particularly with MMC derived from activated sludge or wastewater-adapted consortia. Hydrogen production was particularly notable in crude glycerol and 3-phase olive mill wastewater, reaching $107.3 \pm 0.7$ L/kg WW and 142 L/kg COD, respectively. The data gathered in Table 13.4 highlights the potential of different wastewater sources for bioprocessing, with confectionery wastewater achieving the highest PHAs accumulation ($84.0 \pm 0.9\%$ w/w), while cheese whey and deproteinized dairy wastes also showed significant PHAs yields.

Overall, the data of Table 13.4 highlight the importance of selecting suitable wastewater streams and reactor configurations to optimize biohydrogen and biopolymer production, demonstrating the potential of sugar-rich wastewaters as valuable resources in the circular bioeconomy. Furthermore, the successful application of a two-stage process for energy recovery and bioplastic synthesis presents a promising strategy to enhance energy efficiency and reduce plastic pollution in line with sustainability goals.

**Table 13.4** Process parameters and yields of integrated biohydrogen ($BioH_2$, $H_2$) and polyhydroxyalkanoates (PHAs) production using different wastewater

| Wastewater | Acidification process | $H_2$ max yield | Composition of effluent | PHA Biocatalyst | Reactor operation mode | Max PHA content (%, w/w) | PHA type | PHA yield | Reference |
|---|---|---|---|---|---|---|---|---|---|
| Confectionary wastewater | Continuous, attached microbial consortium in ceramic matrix | $12.70 \pm 0.35$ $m^3$ $H_2/m^3$ WW | Acetate, propionate, butyrate, valerate, caproate, lactate | Anaerobic sludge | Batch | $84.0 \pm 0.9$ | PHBHV | $21.46 \pm 0.13$ kg PHAs/$m^3$ WW | [11] |
| Expired fruit juices | Continuous, CSTR | - | Acetate, propionate, butyrate, valerate, caproate, ethanol, lactate | Indigenous microflora of the waste | Batch | $54.6 \pm 4.3$ | PHBHV | $0.23 \pm 0.01$ g PHAs/ g COD | [18] |
| Crude glycerol | Continuous, anaerobic column biofilm reactors packed ceramic beads | $107.3 \pm 0.7$ L $H_2$/kg WW | Acetate, butyrate, hexanoate, 1,3, propanediol | MMC derived from soil | Fed batch | $40.6 \pm 3.2$ | PHBHV, PHBHHx | $0.38 \pm 0.02$ g PHAs/g COD | [65] |
| Expired fruit juices | Continuous, attached microbial consortium in ceramic matrix | $1.94 \pm 0.08$ L $H_2$/L.d | Acetate, propionate, butyrate, valerate, caproate, ethanol, lactate | Indigenous microflora of the waste | Batch | $49.9 \pm 1.1$ | PHBHV | $0.21 \pm 0.02$ g PHAs/g COD | [75] |

(continued)

**Table 13.4** (continued)

| Wastewater | Acidification process | $H_2$ max yield | Composition of effluent | PHA Biocatalyst | Reactor operation mode | Max PHA content (%, w/w) | PHA type | PHA yield | Reference |
|---|---|---|---|---|---|---|---|---|---|
| 3-phase olive mill wastewater | Continuous, CSTR | 1.32 ± 0.12 L $H_2$/L of OMW | Acetate, propionate, butyrate, residual sugars | MMC derived from activated sludge | Batch, Sequential batch | 8.94 | PHBHV | 24 | [76] |
| Distillery spent wash | Periodic-discontinuous batch mode | - | Acetate, propionate, butyrate, valerate | *Bacillus tequilensis* | Batch | 40 | PHBHV | – | [77] |
| 3-phase olive mill wastewater | Continuous, CSTR | 142 L $H_2$/kg COD | Acetate, propionate, butyrate, residual Sugars | MMC derived from activated sludge | Sequential batch | 24.6 ± 0.2 | PHBHO | 7.58 ± 0.06 g PHAs/L OMW | [78] |
| Cheese whey | Continuous, anaerobic column biofilm reactors packed ceramic cubes | – | Acetate, propionate, butyrate, valerate, hexanoate, heptanoate, octanoate, ethanol, lactate | *Cupriavidus necator* | Fed batch | 71 | PHB | 0.60 g PHAs/ g VFAs | [79] |

(continued)

**Table 13.4** (continued)

| Wastewater | Acidification process | $H_2$ max yield | Composition of effluent | PHA Biocatalyst | Reactor operation mode | Max PHA content (%, w/w) | PHA type | PHA yield | Reference |
|---|---|---|---|---|---|---|---|---|---|
| Oil production plant Wastewater | Continuous, anaerobic baffle reactor | 0.622 mL $H_2$/mg polyacrylate | Acetate, butyrate | – | Sequential batch | 54.2 | PHB | – | [80] |
| Deproteinized dairy wastes | Continuous, CSTR | 1.97 ± 0.38 mol $H_2$/mol sugars | Formate, acetate, butyrate, lactate | MMC derived from activated sludge | Sequential batch | 62 ± 4.5 | PHB | 0.36 ± 0.08 mg COD PHA/mg CODcons | [81] |
| Cheese whey | Continuous, up-flow anaerobic sludge blanket | – | Acetate, propionate, butyrate, valerate, ethanol | MMC derived from fermented cheesewhey | Fed batch | 45–50 | PHBHV | – | [82] |
| Cheese whey | Batch | 5.3 L $H_2$/L WW | Acetate, propionate, butyrate, valerate, lactate | MMC derived from activated sludge | Sequential batch, fed batch | 81 | PHBHV | 0.96 ± 0.07 mmol PHAs/ mmol carbon | [83] |

## References

1. Michelini G, Moraes RN, Cunha RN, Costa JM, Ometto AR (2017) From linear to circular economy: PSS conducting the transition. Procedia CIRP 64:2–6
2. EU (2015) https://environment.ec.europa.eu/topics/circular-economy/first-circular-economy-action-plan_en
3. Camilleri MA (2020) European environment policy for the circular economy: implications for business and industry stakeholders. Sustain Dev 28:1804–1812
4. Fach S, Fuchs S (2010) Design and development of decentralized water and wastewater technologies: a combination of safe wastewater disposal and fertilizer production. Water Sci Technol 62(7):1580–1586
5. Diaz-Elsayed N, Rezaei N, Guo T, Mohebbi S, Zhang Q (2019) Wastewater-based resource recovery technologies across scale: a review. Resour Conserv Recycl 145:94–112
6. Chmielewski AG, Sudlitz M, Żubrowska-Sudoł M (2024) Advanced technology for energy, plant nutrients and water recovery at wastewater treatment plants. Energies 17:2749
7. Qadir M, Drechsel P, Jiménez Cisneros B, Kim Y, Pramanik A, Mehta P, Olaniyan O (2020) Global and regional potential of wastewater as a water, nutrient and energy source. Nat Resour Forum 44:40–51
8. Nielsen PH (2017) Microbial biotechnology and circular economy in wastewater treatment. Microb Biotechnol 10(5):1102–1105
9. Hao X, Li J, van Loosdrecht MCM, Jiang H, Liu R (2019) Energy recovery from wastewater: heat over organics. Water Res 161:74–77
10. Ntaikou I, Antonopoulou G, Lyberatos G (2010) Biohydrogen production from biomass and wastes via dark fermentation: a review. Waste Biomass Valor 1:21–39
11. Kora E, Antonopoulou G, Yan Q, Zhang Y, Lyberatos G, Ntaikou I (2024) Investigating the efficiency of a two-stage anaerobic-aerobic process for the treatment of confectionery wastewater with simultaneous production of biohydrogen and polyhydroxyalkanoates. Environ Res 248:118526
12. Grace Pavithra K, Sundar Rajan P, Arun L, Brindhadevi K, Le Hoang Q, Pugazhendhi A (2023) A review on recent advancements in extraction, removal and recovery of phenols from phenolic wastewater: challenges and future outlook. Environ Res 237:117005
13. Antonopoulou G, Gavala HN, Skiadas IV, Angelopoulos K, Lyberatos G (2008) Biofuels generation from sweet sorghum: fermentative hydrogen production and anaerobic digestion of the remaining biomass. Bioresour Technol 99:110–119
14. Alexandropoulou M, Antonopoulou G, Lyberatos G (2016) Food industry waste's exploitation via anaerobic digestion and fermentative hydrogen production in an up-flow column reactor. Waste Biomass Valor 7(4):711–723
15. Alexandropoulou M, Antonopoulou G, Trably E, Carrere H, Lyberatos G (2018) Continuous biohydrogen production from a food industry waste: influence of operational parameters and microbial community analysis. J Clean Prod 74:1054–1063
16. Castilho LR, Mitchell DA, Freire DMG (2009) Production of polyhydroxyalkanoates (PHAs) from waste materials and byproducts by submerged and solid-state fermentation. Bioresour Technol 100:5996–6009
17. Rangel C, Carvalho G, Oehmen A, Frison N, Lourenço ND, Reis MA (2023) Polyhydroxyalkanoates production from ethanol-and lactate-rich fermentate of confectionary industry effluents. Int J Biol Macromol 229:713–723
18. Kora E, Tsaousis P, Andrikopoulos KS, Chasapis C, Voyiatzis G, Ntaikou I, Lyberatos G (2023) Assessment of production process and characterization of poly(3-hydroxybutyrate-co-3-hydroxyvalerate) from acidified discarded fruit juices, via a robust microbial consortium dominated by Zoogloea sp. J Biol Macromol 226:1500–1514
19. Malashin I, Martysyuk D, Tynchenko V, Gantimurov A, Semikolenov A, Nelyub V, Borodulin A (2024) Machine learning-based process optimization in biopolymer manufacturing: a review. Polymers 16:3368

20. Yusaf T, Laimon M, Alrefae W, Kadirgama K, Dhahad HA, Ramasamy D, Kamarulzaman MK, Yousif B (2022) Hydrogen energy demand growth prediction and assessment (2021–2050) using a system thinking and system dynamics approach. Appl Sci 12:2
21. Hassan Q, Algburi S, Sameen AZ, Jaszczur M, Salman HM (2024) Hydrogen as an energy carrier: properties, storage methods, challenges, and future implications. Environ Sys Decis 44(2):327–350
22. Sharmila GV, Tamilarasan K, Kumar DM, Kumar G, Varjani S, Kumar SA, Banu JR (2022) Trends in dark biohydrogen production strategy and linkages with transition towards low carbon economy: an outlook, cost-effectiveness, bottlenecks and future scope. Int J Hydrogen Energy 47(34):15309–15332
23. Antonopoulou I, Ntaikou I, Stamatelatou K, Lyberatos G (2011) Biological and fermentative production of hydrogen. In: Luque R, Campelo J, Clark J (eds) Handbook of biofuels production process and technologies. Woodhead Publishing Series in Energy, vol 13, pp 305–346
24. Luo L, Sriram S, Johnravindar D, Louis Philippe Martin T, Wong JWC, Pradhan N (2022) Effect of inoculum pretreatment on the microbial and metabolic dynamics of food waste dark fermentation. Bioresour Technol 358:127404
25. Mota VT, Ferraz Júnior ADN, Trably E, Zaiat M (2018) Biohydrogen production at pH below 3.0: is it possible? Water Res 128:350–361
26. Etchebehere C, Castelló E, Wenzel J, del Pilar Anzola-Rojas M, Borzac-coni L, Buitrón G, Cabrol L, Carminato VM, Carrillo-Reyes J, Cisneros-Pérez C, Fuentes L, Moreno-Andrade I, Razo-Flores E, Filippi GR, Tapia-Venegas E, Toledo-Alarcón J, Zaiat M (2016) Microbial communities from 20 different hydrogen-producing reactors studied by 454 pyrosequencing. Appl Microbiol Biotechnol 100(7):3371–3384
27. Niu K, Zhang X, Tan WS, Zhu ML (2010) Characteristics of fermentative hydrogen production with *Klebsiella pneumoniae* ECU-15 isolated from anaerobic sewage sludge. Int J Hydrogen Energy 35(1):71–80
28. Kumar P, Patel SKS, Lee JK, Kalia VC (2013) Extending the limits of Bacillus for novel biotechnological applications. Biotechnol Advances 31(8):1543–1561
29. Oh YK, Kim HJ, Park S, Kim MS, Ryu DDY (2008) Metabolic-flux analysis of hydrogen production pathway in *Citrobacter amalonaticus* Y19. Int J Hydrogen Energy 33(5):1471–1482
30. Seppälä JJ, Puhakka JA, Yli-Harja O, Karp MT, Santala V (2011) Fermentative hydrogen production by *Clostridium butyricum* and *Escherichia coli* in pure and co-cultures. Int J Hydrogen Energy 36(17):10701–10708
31. Sikora A, Baszczyk M, Jurkowski M, Zielenkiewicz U (2013) Lactic acid bacteria in hydrogen-producing consortia: on purpose or by coincidence? In: Lactic acid bacteria—R&D for food, health and livestock purposes. InTech
32. Elbeshbishy E, Dhar BR, Nakhla G, Lee HS (2017) A critical review on inhibition of dark biohydrogen fermentation. Ren Sustain Energy Reviews 79:656–668
33. Westerholm M, Dolfing J, Schnürer A (2019) Growth characteristics and thermodynamics of syntrophic acetate oxidizers. Environ Sci Technol 53(9):5512–5520
34. Steinbusch KJJ, Hamelers HVM, Plugge CM, Buisman CJN (2011) Biological formation of caproate and caprylate from acetate: fuel and chemical production from low grade biomass. Energy Environ Sci 4(1):216–224
35. Candry P, Ganigué R (2021) Chain elongators, friends, and foes. Current Opin in Biotechnol 67:99–110
36. Ghimire A, Frunzo L, Pirozzi F, Trably E, Escudie R, Lens PNL, Esposito G (2015) A review on dark fermentative biohydrogen production from organic biomass: process parameters and use of by-products. Appl Energy 144:73–95
37. Antonopoulou G, Gavala HN, Skiadas IV, Lyberatos G (2010) Influence of pH on fermentative hydrogen production from sweet sorghum extract. Int J Hydrogen Energy 35(5):1921–1928
38. Magdalena JA, Pérez-Bernal MF, Bernet N, Trably E (2023) Sequential dark fermentation and microbial electrolysis cells for hydrogen production: volatile fatty acids influence and energy considerations. Bioresour Technol 374:128803

39. Jiang J, Zhang Y, Li K, Wang Q, Gong C, Li M (2013) Volatile fatty acids production from food waste: effects of pH, temperature, and organic loading rate. Bioresour Technol 143:525–530
40. He M, Sun Y, Zou D, Yuan H, Zhu B, Li X, Pang Y (2012) Influence of temperature on hydrolysis acidification of food waste. Procedia Environ Sci 16:85–94
41. Ueno Y, Sasaki D, Fukui H, Haruta S, Ishii M, Igarashi Y (2006) Changes in bacterial community during fermentative hydrogen and acid production from organic waste by thermophilic anaerobic microflora. J Appl Microbiol 101(2):331–343
42. Dinsdale RM, Premier GC, Hawkes FR, Hawkes DL (2000) Two-stage anaerobic co-digestion of waste activated sludge and fruit/vegetable waste using inclined tubular digesters. Bioresour Technol 72(2):159–168
43. Guo XM, Trably E, Latrille E, Carrre H, Steyer JP (2010) Hydrogen production from agricultural waste by dark fermentation: a review. Int J Hydrogen Energy 35(19):10660–10673
44. Antonopoulou G, Gavala HN, Skiadas IV, Lyberatos G (2011) Effect of substrate concentration on fermentative hydrogen production of sweet sorghum extract. Int J Hydrogen Energy 36(8):4843–4851
45. Yan BH, Selvam A, Xu SY, Wong JW (2014) A novel way to utilize hydrogen and carbon dioxide in acidogenic reactor through homoacetogenesis. Bioresour Technol 159:249–257
46. Chen Y, Li X, Zheng X, Wang D (2013) Enhancement of propionic acid fraction in volatile fatty acids produced from sludge fermentation by the use of food waste and *Propionibacterium acidipropionici*. Water Res 47(2):615–622
47. Antonopoulou G, Vayenas D, Lyberatos G (2016) Ethanol and hydrogen production from sunflower straw: the effect of pretreatment on the whole slurry fermentation. Biochem Eng J 116:65–74
48. Antonopoulou G (2020) Designing efficient processes for sustainable bioethanol and biohydrogen production from grass lawn waste. Molecules 25(12):2889
49. Alexandropoulou M, Antonopoulou G, Ntaikou I, Lyberatos G (2023) The impact of alkaline/hydrogen peroxide pretreatment on hydrogen and methane production from biomasses of different origin: the case of willow sawdust and date palm fibers. Sustain Chem Pharm 32:100971
50. Preethi TM, Usman M, Banu RJ, Gunasekaran M, Kumar G (2019) Biohydrogen production from industrial wastewater: an overview. Bioresour Technol Reports 7:100287
51. Arimi MM, Knodel J, Kiprop A, Namango SS, Zhang Y, Geißen S-U (2015) Strategies for improvement of biohydrogen production from organic-rich wastewater: a review. Biomass Bioenergy 75:101–118
52. Min KS, Khan AR, Kwon MK, Jung YJ, Yun Z, Kiso Y (2005) Acidogenic fermentation of blended food-waste in combination with primary sludge for the production of volatile fatty acids. J Chem Technol Biotechnol 80(8):909–915
53. Feng L, Chen Y, Zheng X (2009) Enhancement of waste activated sludge protein conversion and volatile fatty acids accumulation during waste activated sludge anaerobic fermentation by carbohydrate substrate addition: the effect of pH. Environ Sci Technol 43(12):4373–4380
54. Zhu H, Parker W, Basnar R, Proracki A, Falletta P, Beland M, Seto P (2008) Biohydrogen production by anaerobic co-digestion of municipal food waste and sewage sludges. Int J Hydrogen Energy 33(14):3651–3659
55. Antonopoulou G, Stamatelatou K, Venetsaneas N, Kornaros M, Lyberatos G (2008) Biohydrogen and methane production from cheese whey in a two-stage anaerobic process. Ind Eng Chem Res 47:5227–5233
56. Venetsaneas N, Antonopoulou G, Stamatelatou K, Kornaros K, Lyberatos G (2009) Using cheese whey for hydrogen and methane generation in a two-stage continuous process with alternative pH controlling approaches. Bioresour Technol 100:3713–3717
57. Koutrouli E, Kalfas H, Gavala HN, Skiadas IV, Stamatelatou K, Lyberatos G (2009) Hydrogen and methane production through two-stage mesophilic anaerobic digestion of olive pulp. Bioresour Technol 100:3718–3723
58. Amorim NCS, Alves I, Martins JS, Amorim ELC (2014) Biohydrogen production from cassava wastewater in an anaerobic fluidized bed reactor. Braz J Chem Eng 31(3):603–612

59. Sivagurunathan P, Lin C-Y (2016) Enhanced biohydrogen production from beverage wastewater: process performance during various hydraulic retention times and their microbial insights. RSC Adv 6:4160–4169
60. Mishra P, Das D (2014) Biohydrogen production from *Enterobacter cloacae* IIT-BT 08 using distillery effluent. Int J Hydrogen Energy 39(14):7496–7507
61. Asunis F, Cappai G, Carucci A, De Gioannis G, Dessì P, Muntoni A, Polettini A, Pomi R, Rossi A, Spiga D, Trois C (2022) Dark fermentative volatile fatty acids production from food waste: a review of the potential central role in waste biorefineries. Waste Manag. Res. 40:1571–1593
62. Steinbüchel A, Valentin HE (1995) Diversity of bacterial polyhydroxyalkanoic. FEMS Microbiol Lett 128:219–228
63. Kourmentza C, Plácido J, Venetsaneas N, Burniol-Figols A, Varrone C, Gavala HN, Reis MAM (2017) Recent advances and challenges towards sustainable polyhydroxyalkanoate (PHA) production. Bioeng 4(2):55
64. Laycock B, Halley P, Pratt S, Werker A, Lant P (2014) The chemomechanical properties of microbial polyhydroxyalkanoates. Prog Polym Sci 39(2):397–442
65. Ntaikou I, Koumelis I, Kamilari M, Iatridi Z, Tsitsilianis C, Lyberatos G (2019) Effect of nitrogen limitation on polyhydroxyalkanoates production efficiency, properties and microbial dynamics using a soil derived mixed continuous culture. Int J Biobased Plastics 1(1):31–47
66. Silva JB, Pereira JR, Marreiros BC, Reis MAM, Freitas F (2021) Microbial production of medium-chain length polyhydroxyalkanoates. Process Biochem 102:393–407
67. Choi SY, Cho IJ, Lee Y, Kim YJ, Kim KJ, Lee SY (2020) Microbial polyhydroxyalkanoates and nonnatural polyesters. Adv Mater 32(35):e1907138
68. Zhou W, Bergsma S, Colpa DI, Euverink GJW, Krooneman J (2023) Polyhydroxyalkanoates (PHAs) synthesis and degradation by microbes and applications towards a circular economy. J Environ Manag 341:118033
69. Raza ZA, Tariq MR, Majeed MI, Banat IM (2019) Recent developments in bioreactor scale production of bacterial polyhydroxyalkanoates. Bioprocess Biosyst Eng 42:901–919
70. Bolmanis E, Dubencovs K, Suleiko A, Vanags J (2023) Model predictive control—a stand out among competitors for fed-batch fermentation improvement. Fermentation 9:206
71. Reis MAM, Serafim LS, Lemos PC, Ramos AM, Aguiar FR, Van Loosdrecht MCM (2003) Production of polyhydroxyalkanoates by mixed microbial cultures. Bioprocess Biosyst Eng 25:377–385
72. Chen Z, Huang L, Wen Q, Guo Z (2015) Efficient polyhydroxyalkanoate (PHA) accumulation by a new continuous feeding mode in three-stage mixed microbial culture (MMC) PHA production process. J Biotechnol 209:68–75
73. Leong YK, Show PL, Lan JCW, Loh H-S, Lam HL, Ling TC (2017) Economic and environmental analysis of PHAs production process. Clean Techn Environ Policy 19:1941–1953
74. Moshood TD, Nawanir G, Mahmud F, Mohamad F, Ahmad MH, AbdulGhani A (2022) A literature review on sustainability of bio-based and biodegradable plastics: challenges and opportunities. Energy Eng 119(4):1611–1647
75. Kora E, Patrinou V, Ntaikou I, Antonopoulou G, Terkelekopoulou A, Lyberatos G (2023) Dark fermentation of out-of-date fruit juices for biohydrogen production and biotechnological exploitation of effluents towards bioplastics and microbial lipids. Biochem Eng J 195:108901
76. Ntaikou I, Kourmentza C, Koutrouli E, Stamatelatou K, Zampraka A, Kornaros M, Lyberatos G (2009) Exploitation of olive oil mill wastewater for combined bio-hydrogen and biopolymers production. Bioresour Technol 100(15):3724–3730
77. Amulya K, Reddy MV, Mohan SV (2014) Acidogenic spent wash valorization through polyhydroxyalkanoate (PHA) synthesis coupled with fermentative biohydrogen production. Bioresour Technol 158:336–342
78. Ntaikou I, Valencia Peroni C, Kourmentza C, Ilieva VI, Morelli A, Chiellini E, Lyberatos G (2014) Microbial bio-based plastics from olive-mill wastewater: generation and properties of poly-hydroxyalkanoates from mixed cultures in a two-stage pilot scale system. J Biotechnol 18:138–147

79. Domingos JMB, Puccio S, Martinez GA, Amaral N, Reis MAM, Bandini S, Fava F, Bertin L (2018) Cheese whey integrated valorisation: production, concentration and exploitation of carboxylic acids for the production of polyhydroxyalkanoates by a fed-batch culture Chem. Eng J 336:47–53
80. Zhao L, Han D, Yin Z, Bao M, Lu J (2019) Biohydrogen and polyhydroxyalkanoate production from original hydrolyzed polyacrylamide-containing wastewater. Bioresour Technol 287:121404
81. Colombo B, Villegas Calvo M, Pepè Sciarria T, Scaglia B, Savio Kizito S, D'Imporzano G, Adani F (2019) Biohydrogen and polyhydroxyalkanoates (PHA) as products of a two-steps bioprocess from deproteinized dairy wastes. Waste Manag 95:22–31
82. Lagoa-Costa B, Kennes C, Veiga MC (2022) Influence of feedstock mix ratio on microbial dynamics during acidogenic fermentation for polyhydroxyalkanoates production. J Environ Manag 303:114–132
83. Asunis F, Carucci A, De Gioannis G, Farru G, Muntoni A, Polettini A, Pomi R, Rossi A, Spiga D (2022) Combined biohydrogen and polyhydroxyalkanoates production from sheep cheese whey by a mixed microbial culture. J Environ Manag 322:116149

# Chapter 14
# New Sustainable Technologies for Sewage Sludge Valorization: Energy and Resources from Wastewater Treatment in the Sign of the Circular Economy

**Luigi di Bitonto, Antonella Angelini, Valeria D'Ambrosio, Vito Locaputo, and Carlo Pastore**

**Abstract** Urban wastewater treatment generates approximately 10 million tons of urban sewage sludge annually across Europe. This figure is expected to increase significantly, potentially reaching 13 million tons (91/271/EEC) over the next decade. As a result, managing and treating urban sludge in an economically, environmentally and socially acceptable way is one of the major challenges facing modern society. In recent decades, alternative methods of sewage sludge management have been explored based on the principle of resource recovery and the minimization of waste to be disposed of. A promising approach to solving this problem is to preserve the chemical complexity of the organic components in urban sludge through sustainable processes that selectively recover categories of compounds and convert them into marketable products, in line with the principles of the Circular Economy. Innovative scenarios have been recently developed, focusing on cellulose and lipids, outlining recovery and valorization pathways towards the production of biofuels, biolubricants, and biosurfactants of the new generation. If cellulose and grease in sewage were efficiently separated and valorised, over 2000 ktons of biofuels could be produced on a European scale (approximately 14% of the present continental demand). These

L. di Bitonto · A. Angelini · V. D'Ambrosio · V. Locaputo · C. Pastore (✉)
CNR-Istituto Di Ricerca Sulle Acque, Bari, Italy
e-mail: carlo.pastore@cnr.it

L. di Bitonto
e-mail: luigi.dibitonto@cnr.it

A. Angelini
e-mail: antonella.angelini@cnr.it

V. D'Ambrosio
e-mail: valeriadambrosio@cnr.it

V. Locaputo
e-mail: vito.locaputo@cnr.it

 
A. Angelini et al. (eds.), *Security Enhancement for Climate Change Impacting Urban Resources - SECCURe*, NATO Science for Peace and Security Series C: Environmental Security, https://doi.org/10.1007/978-94-024-2345-7_14

approaches also offer significant environmental benefits related to the exploitation of resources that are typically lost, thereby fully addressing economic sustainability. Over 3850 ktons of $CO_2$ per year would be prevented from being emitted in urban wastewater treatment, which would otherwise be converted into high-value chemicals. These figures demonstrate the promising perspective of this innovative approach towards more sustainable resource management.

**Keywords** Circular economy · Urban sewage sludge valorization · Resource recovery

## 14.1 Urban Wastewater Treatment: Present and Future Perspectives

### *14.1.1 Urban Wastewater Treatment*

The first step towards reducing human impact on the world's natural equilibrium is the containment of potential alterations connected with anthropic activities. A paradigmatic example of this is the recovery of water streams contaminated by human activity. Water treatment is considered one of the key technologies for promoting well-being. Since the early nineteenth century, urban wastewater has been treated before being released into aquatic systems. The most important parameters to monitor when considering the recovery of a water stream are suspended and/or dissolved carbon, nitrogen, and phosphorus. The most common approach involves eliminating contaminants containing these elements through biological processes that convert organic pollutants into carbon dioxide and molecular nitrogen. In the current configuration of wastewater treatment plants (WWTP), incoming species are biodegraded, typically with a focus on oxidation, which increases carbon dioxide emissions. In detail, wastewater treatment is recognized as a high energy-demanding process, which accounts for 2–3% of the overall energy consumption in municipalities, with most of the effort being spent on aeration, which is necessary for the operations of biological oxidation [1]. In addition, there is growing attention towards the effectiveness of the degradation of certain micro- and nanopollutants, (e.g., pharmaceuticals, endocrine disruptors, antibiotics, emerging contaminants in general) since biological mediated degradation faces limitations, with resulting contaminations of global aquatic resources. To mitigate the risk of widespread dispersion of these specific pollutants, the new European Wastewater Treatment Directive [2] defines new frontiers of wastewater treatment, stating the necessity of including "quaternary treatments" in modern WWTPs, increasing the cost and energy consumption to remove emerging pollutants effectively. However, to address the energy issue, a further requirement will be introduced by the same European directive, which addresses the decarbonization goals for 2050. An unprecedented target of energy neutrality has been set for the WWTP of the future: by 2045, all treatment plants must meet their energy needs with renewable energy sources, produced both on-site and off-site by or on behalf of the plant's operators.

This second decarbonization objective could be achieved by avoiding carbon dioxide production, possibly promoting the recovery of carbon and nitrogen dispersed in sewage in a more complex form rather than relying solely on biological oxidation. Nowadays, most organic compounds are typically separated in WWTP in sewage sludge, which need to be "demolished" biologically through aerobic or anaerobic processes. Anaerobic digestion, which generates biogas and can be used as a value-added product in the energy sector, would be considered the most suitable solution, but it is hardly applied for little WWTPs. A large number of WWTPs treat sewage sludge aerobically, increasing the amount of carbon dioxide emitted, either to ensure agitation and aeration (with high energy input) or to break down the organic load aerobically. This means demolition principles based on biological processes cannot be the best practice for addressing the European directive. Especially considering that the energy content of municipal sewage incoming WWTP is, in principle, over ten times the energy consumed in conventional depuration, and almost half of this is related to suspended pollutants [3], in principle easily removable through economic, physical screening/decantation.

An alternative approach would involve designing and implementing cost-effective solutions for the recovery and valorization of organic compounds present in sewage while preserving their initial complexity. In recent years, several studies have been focused on the detailed characterization and recovery of specific compounds for direct valorization. Complex carbohydrates, proteins and grease are the main constituents of the organic fraction. In this work, a detailed description of these technologies will be provided in separate sections, focusing on complex carbohydrates and grease. Then, benefits related to applying these new technologies will be calculated in terms of potential new products and carbon dioxide emissions saved.

## 14.2 A New Concept of Wastewater Treatment in the Sign of Resource Recovery

Cellulose and lipids are two organic categories that can be efficiently and easily recovered and subsequently valorized through low-energy demanding technologies from urban sewage. Cellulose, mainly coming from toilet paper, is present in municipal sewage in the form of fibers (from 150 to 300 mg/L, depending on the geographical area and the WWTP dimension) and possibly separable through a sieving procedure [4]. On the other hand, grease is a species immiscible with water (up to 100 mg/L [3]) and could be potentially separated through spontaneous stratification.

### 14.2.1 Cellulose Recovery from Urban Wastewater Treatment

The dynamic sieving of urban sewage is an innovative procedure that could be easily integrated into the usual pre-treatments units of WWTPs. This technology improves the purification process and produces marketable cellulosic sludge with on-site post-processing of the recovered product until to obtain a cellulose content of 44–48 wt% of the TS. Specifically, besides stages for the sand removal and mechanical coarse screen, cellulose could be intercepted using appropriate machines capable of sieving the fibers from raw wastewater.

The operational performances of this technology and the relevant chemical nature of the cellulosic sludge were investigated and measured on a pilot unit installed in a small WWTP, namely Vernole (6500 Person Equivalent, PE, Lecce, Italy). In this WWTP, 1–1.3 $m^3$ of sewage were collected and treated per day using a sieving pilot unit capable of intercepting all the incoming wastewaters, generating around one ton of wet cellulosic sludge every week.

The chemical characterization and initial processing steps were carried out on samples directly collected from this plant without any further treatment. Residual humidity was determined through thermal dehydration (105 °C for 24 h or 40 °C for 72 h), which obtained an estimation of the total solids (TS) content (% w/w). After drying, the sludge was homogenized and ground (Fig. 14.1) in order to carry out the subsequent characterization steps [5]. Besides the determination of the main macro-constituents (lipids, easily hydrolysable carbohydrates (EHC), cellulose, lignin, proteins and ashes), particular attention was paid to the identification of potential contaminants, including aromatic and polycyclic aromatic hydrocarbons (PAH), polychlorinated biphenyls (PCBs) and C10-C40 mineral oils. On the basis of 30 sampling, analysis of the data revealed a TS content with an average value of 37.1 ± 5.5% w/w. There were no significant changes in the composition of the sludge, with the cellulosic component remaining predominant, accounting for 44–48% by weight of the TS. The metal content analysis showed calcium as the predominant element in the sludge (> 12,000 mg/kg TS), but values related to As, Pb, Cd, Hg, Zn, Cr, Be, Se and Cu were all ignorable or under typical limit values for defining dangerous sludge.

After simple drying, the structural fiber material obtained in this way can be used in the production of various biocomposites. Figure 14.2 shows a general scheme of valorization:.this cellulosic sludge can be used as additives for asphalt composition [4] and insulating materials [6] without further transformation processes. Additionally, cellulosic sludge could be converted by a carbonization process into biochar, which has interesting properties in decontaminating polluted sites [7]. Alternatively, several pilot-scale studies have been conducted that promote the anaerobic fermentation of this residue to produce Volatile Fatty Acids (VFAs), which could find application for the production of biosolvents, biofuels [8], and bioplastics [9]. Moreover, several innovative approaches could be implemented by passing through the hydrolysis of cellulose into a glucose-enriched broth, which could provide a platform for new products. Such a strategy has not yet been systematically explored.

**Fig. 14.1** Dried and ground cellulosic sludge

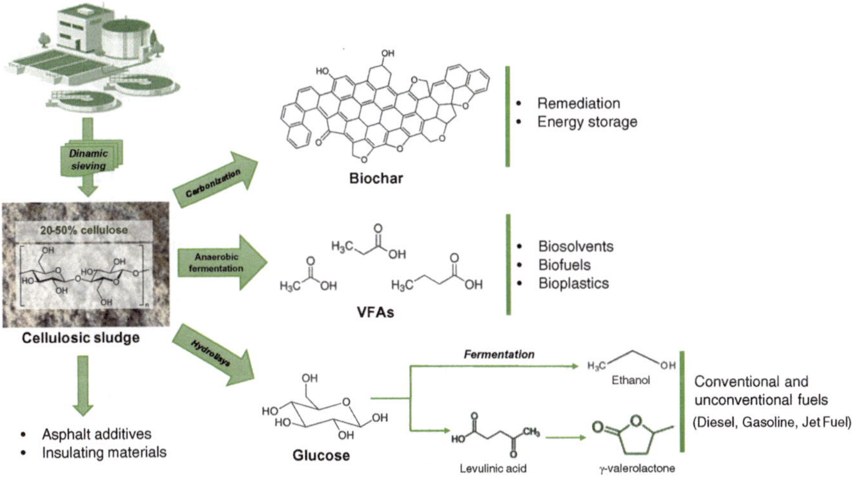


**Fig. 14.2** Scheme of valorization of cellulosic sludge

The valorization procedure of this cellulosic sludge passed through an initial enzymatic demolition of long-chained polymer into glucose, which is usually promoted by commercial *Thricoderma Reesei Cellulases* (Cellic Ctec 2, Novozyme). In contrast to lignocellulosic biomasses, the cellulosic sludge should consist of cellulose and be digested without any further pre-treatment but the study of direct hydrolysis through a commercial cellulase was never explored..

The hydrolysis was carried out in water with a dry sludge loading of 10% by weight (wt%) and an enzyme dosage of about 24 filter paper units (FPU)/g of substrate. The process involved 46.3 g of dried cellulosic sludge, 12.5 g of the abovementioned commercial formulation of the enzyme, and 448.5 g of water, the reaction being was carried out at 50 °C for 72 h. The resulting suspension was then centrifuged, separating the supernatant from the residual solid.

The supernatant was analyzed for glucose content in order to evaluate the efficiency of the enzymatic hydrolysis of cellulose. In detail, the glucose concentration in the suspension was determined to be 43.86 g/L, which corresponds to 41.6 wt% relative to the TS of the sludge. This results in an enzymatic hydrolysis efficiency of 86.5%, confirmingthe high effectiveness of the process, which achieves significant cellulose breakdown under mild conditions and without any further additional pre-treatment of the cellulosic sludge collected from the pilot plant unit.

The effective obtainment of a glucose-rich broth, combined with the potential recovery of cellulase enzymes by size exclusion filtration, which allows enzymes to be reused on fresh cellulosic sludge [10], could open up a variety of more specific applications for cellulosic sludge: from the production of conventional fuels, namely ethanol by yeast-driven fermentation using consolidated processes [11], to new unconventional fuels and materials derived from levulinic acid produced by thermo-catalytical processes [12] (Fig. 14.2).

### *14.2.2 Grease Recovery from Urban Wastewater Treatment*

Urban sewage, which incomes in a WWTP, is an aqueous flow that results from normal human activities and contains water from sink washing of dishes, other cleaning operations and human dejections. In most of these operations, greases and oils are present in different forms of saponifiable fractions, namely glycerides, acids and soaps. Although the quantification of lipids derived from cleaning/washing activities is not of immediate determination, the contribution related to dejections is almost well-defined. Typically, the composition of faeces was well investigated and indicates a value of 4.1 g/cap/day [13], which could increase up to over 6 g/cap/day, representing 8–16% of dry weight. According to these figures, which could be significantly affected by alimentary use correlated with geographical contexts, over 130 ktons of grease could be potentially recovered from human feces only in Italy, representing not ignorable amount with the present waste cooking oil already recovered. In fact, in Italy, over 1500 ktons of vegetal oil are annually consumed, with an annual consumption of 25 kg per-capita. Of these, about 260 ktons are annually turned into waste cooking oil [14], but only 90–130 ktons [15] were regularly harvested through differentiated collection, most of which (over 90%) feed biofuel and biolubricants production. The remaining 130 ktons are supposed to be wasted away improperly, and most presumably, they could be recovered in raw urban wastewater. According to this preliminary data, the recovery of grease from WWTP could play a crucial role in the generation of a new valuable grease feedstock.

Various methods have been proposed [16, 17] to recover grease from sewage sludge. The direct liquid–liquid extraction from wet sewage sludge using immiscible organic solvents seems to be the most effective approach for lipid recovery. Hexane has been explored in studies simulating industrial-scale processes applied to thickened [18] and dewatered [19] primary sludge, yielding reasonable results. However, this process produces residual sludge contaminated with the organic solvent, needing further removal in a separated unit operation, complicating its disposal [20]. Besides hexane, bio-based solvents have also been researched, with ethyl esters of volatile fatty acids [21, 22] being particularly effective for lipid extraction from sewage sludge. These solvents offer the advantage of enabling rapid biodegradability of extraction residues in anaerobic conditions, producing biogas. More recently, ethyl hexanoate was found to be capable of recovering grease from sewage as calcium soaps or free fatty acids, depending on the use of mineral acid [23]. However, the economic benefits of this method are hindered by challenges related to the chemical composition of the sludge and solvent recovery [21]. More recently, solvent-free extraction technologies have been developed to meet the stringent environmental standards of WWTPs [24], although these methods are only suitable for certain types of sludge [25].

Recently, a solvent-free procedure was investigated using a pilot plant capable to process 2 $m^3/h$ of different sewage sludge. This pilot plant was installed in a representative, relatively small WWTP located in Putignano (40,000 PE, Italy). In detail, three different feedstocks were processed, namely floating grease (recovered from preliminary treatment), primary sludge (obtained from primary sedimentation) and mixed sewage sludge, which was a mixture of primary and secondary sludge, as used for the anaerobic digestion.

Figure 3b reports the scheme of the unit operations performed during each processing cycle, which can be described as follow:

Step 1: Pumping of primary or mixed sludge or sewage scum directly from the dedicated unit of the WWTP (primary or mixed sedimentation tank or degreasing unit) to a horizontal centrifuge.

Step 2: First centrifugation at ambient temperature: centrifuged sludge was accumulated in a thermostatic reactor, whereas the aqueous centrate was returned to the head of the WWTP.

Step 3: Once the specified quantity (about 1000 kg, namely 1 ton) of centrifuged sludge has been accumulated in the reactor, HCl conc. 30–33% was added in variable amounts ranging from 20–40 kg per ton of accumulated sludge.

Step 4: After the reagent is added and the mixture is stirred for a variable duration between 1 and 3 h, the reacted mass is heated to temperatures between 60 and 80 °C.

Step 5: The heated reacted sludge is then pumped to a further centrifugation stage, resulting in centrifuged sludge on one side ("Final Sludge" in Fig. 3b) and a fat/water mixture extracted by the centrifugation process (final centrate), which is sent to a thermostated static decanter, where grease floats on the top in a separated layer.

The performance evaluation was made possible by collecting and analyzing sludge at the different steps mentioned above. Chemical analyses conducted on the various

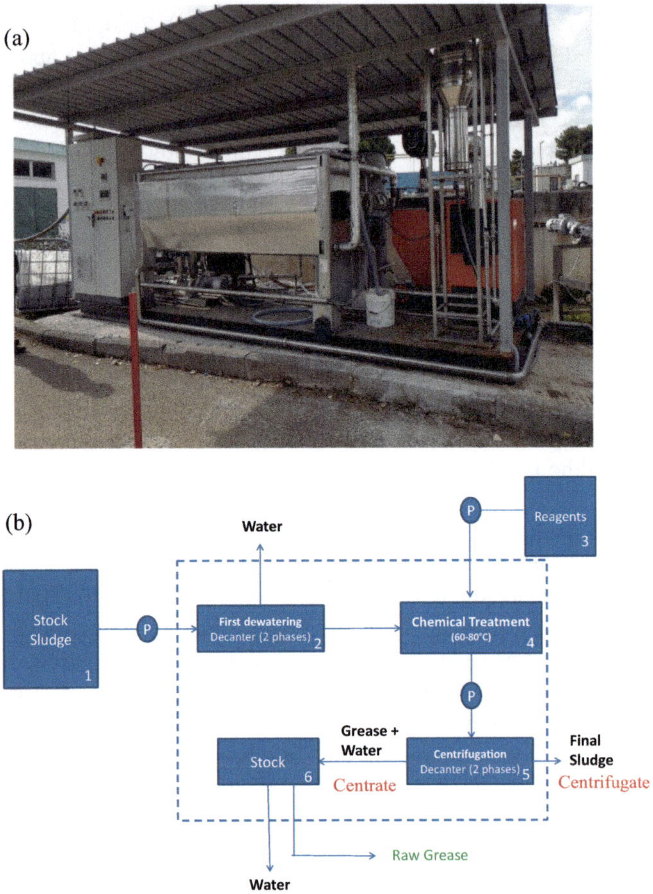


**Fig. 14.3** **a** Picture of pilot plant installed in Putignano, Italy; **b** schematic representation of the grease recovery procedure

samples included pH, acidity, alkalinity, total solids, ash and metal profile, lipid content, fatty acid profile and cellulose content. Table 14.1 presents the results of tests conducted on primary and mixed sludge (predominantly secondary) and sludge from the grease removal section.

Batches 1–3 show the chemical compositions of primary sludge after centrifugation, demonstrating that the sludge can be increasingly concentrated and processed more effectively (higher TS). To achieve this, centrifugal pumping flows of 2 m3/h were used for tests 1 and 2 and 1 m3/h for test 3. In test 1, 11 screw cycles were applied, while tests 2 and 3 used 16 cycles.

The overall lipid content in the primary sludge consistently ranged from 17.1 to 21.8% relative to TS, with cellulose (9.3–11.7%) and lignin (31.4–32.4%) being the other primary components.

**Table 14.1** Representative chemical composition of sludge used in different batch-tests

| Batch | Sludge | TS g/L | Lipids mg/gTS | Ashes mg/gTS | Cellulose mg/gTS | Lignin mg/gTS |
|---|---|---|---|---|---|---|
| 1 | Primary | 78 | 218 | 146 | 93 | 318 |
| 2 | Primary | 90 | 171 | 124 | 115 | 314 |
| 3 | Primary | 149 | 205 | 135 | 117 | 324 |
| 4 | Mixed | 57 | 68 | 154 | 23 | 373 |
| 5 | Sewage scum | 59 | 324 | 265 | 386 | 157 |
| 6 | Sewage scum | 60 | 324 | 256 | 405 | 251 |

Mixed sludges were cold-centrifuged under conditions optimized for primary sludge, achieving a TS content greater than 5–6%, but with a significantly lower lipid content (around 6%).

Table 14.2 presents the key data related to the hot centrifugation of the different batches.

Table 14.2 shows the mass repartition in the Centrate and Centrifugate of about one ton of initial dewatered sludge, as reported in Table 14.1, with the relevant TS values. The grease yield recovery was calculated considering the lipid content in the overall centrate; values marked in brackets specifically refer to the recovered foams that stratified on top of the slurry after spontaneous decantation. In general, such a phase still contained a high moisture content (>50%) and had a lipid/fatty acid content relative to TS of over 60–70%. These recovery values always correspond to the initial fat content present in the starting sludge.

The separated foams were then purified either by further hot centrifugation (80 °C, 4000 rpm for 5 min) on a laboratory scale or by solvent extraction, generating a cleaner and pure grease phase (fatty acids content over 75%).

It is worth noting that higher recovery yields were achieved from sewage scum, with significant recovery yield values (>40%) in terms of separated fatty acids. These

**Table 14.2** Recovery yields of lipids related to the different representative batches while processing different accumulated sludge, as reported in Table 14.1

| Batch | Sludge | Centrate (TS) kg (g/L) | Centrifugate (TS) kg (g/L) | Grease Yield w/w % |
|---|---|---|---|---|
| 1 | Primary | 791 (59) | 208 (150) | 78.0 (9.6[a]) |
| 2 | Primary | 840 (68) | 158 (159) | 74.3 (5.8[a]) |
| 3 | Primary | 750 (104) | 245 (280) | 78.5 (7.3[a]) |
| 4 | Mixed | 910 (48) | 86 (150) | 93.3 (2[a]) |
| 5 | Sewage scum | 945 (45) | 50 (298) | 90 (40.3[a]) |
| 6 | Sewage scum | 960 (49) | 37 (310) | 89 (43.5[a]) |

The amounts of Centrates and Centrifugates (as indicated in Fig. 3b per ton of reacted sludge were reported in kg, whereas the respective TS contents were reported in brackets as g/L.
[a]Grease recovered in the stratified layer on the top of a static sedimentation

results highlighted a critical issue related to the lipid recovery process via a spontaneous decantation alone, most likely due to "natural surfactants" in the sludge. These surfactants stabilized the fatty acids in an aqueous emulsion, with the fatty acids dispersed in the solution at concentrations far exceeding their solubility in water. For instance, the centrate samples contained palmitic acid in the aqueous medium at concentrations greater than 1% (i.e., 10,000 ppm), which is at least three orders of magnitude higher than the solubility of the same acid, which is typically found in the low ppm range (mg/L) in aqueous systems, as reported in the literature [26]. These natural surfactants have primarily been identified as ligno-humic acids and their derivatives, which will be the subject of further study in future research. However, the use of hydrogen peroxide in the reactive stage and the separation of grease accelerated by a vertical separator after the second horizontal centrifugation allowed significant improvement in the grease yield recovery. In this way, over 70% of the initial grease could be efficiently separated in a highly pure stream, in which free fatty acids accounted for over 85% of the final product [27].

As reported in Fig. 14.4, this raw sewage grease could be used and reacted in different ways to produce different final products. Biofuels, in the form of biodiesel [28] could be obtained as the product of direct esterification of grease with methanol or ethanol. Drop-in fuel [29] could be produced through the hydrodeoxygenation of grease. Biolubricants [30] and biosurfactants [31] could also be obtained through the direct esterification of grease and 2-ethylhexanol and glycerol, respectively, significantly implementing the Circular Economy scenario with innovative processes by replacing the use of fossil-derived analogous products.

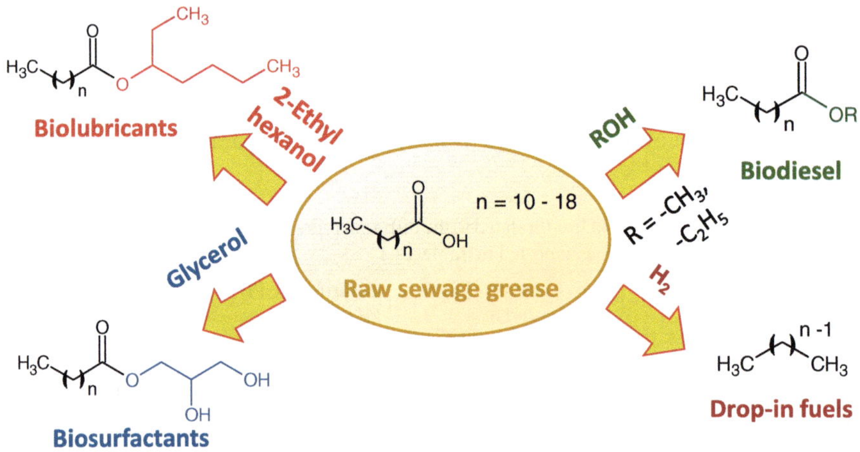


**Fig. 14.4** Possible valorization routes of raw sewage grease

## 14.3 Benefits and Limits of Resource Recovery from Wastewater

The recovery of cellulose and grease from urban WWTPs would be estimated according to the figures reported above. As for cellulose with the use of dynamic sieving, up to 2 kg could be recovered per person annually in Italy as cellulosic sludge, where an overall annual consumption of about 6 kg of toilet paper was considered per person. This recovery yield ignores the amount of cellulosic material already intercepted in the coarse screening. Regarding grease, over 1.5 kg per person could be annually recovered using the solventless procedure described. On a national scale, over 120 and 100 ktons of cellulose and grease could be annually recovered in Italy, respectively. These new raw materials can be valorized to obtain innovative biofuels, materials and commodities in general, capable of replacing present fossil-based production and impacting renewable production. In principle, from this cellulosic sludge, over 60 ktons of bioethanol or about 80 ktons of levulinic acid could be theoretically obtained, with values that would rise to 745 and 1000 ktons in a European context, respectively. Moreover, addressing the cellulosic sludge conversion towards ethyl levulinate, over 800 ktons of this potential bioderived additive to fuel could be produced. Analogously, the continental amount of grease potentially recoverable from WWTPs could guarantee the production of over 1200 ktons of biodiesel from a presently ignored feedstock, only considering grease coming from human dejections while ignoring the amount improperly wasted away of vegetable oils. According to these figures, the possible contribution of WWTPs for the production of new biofuels could be assessed to about 2000 ktons, which would represent 14% of the present continental production of biodiesel, comparable with the present production from waste-derived feedstock [32].

Benefits related to the recovery of these raw materials are not only limited to the generation and exploitation of new feedstock. In fact, these materials would be subtracted from biological oxidation, preventing carbon dioxide emission related to such a process. In detail, the $CO_2$ emission pertinent to the degradation of 1 kton of cellulose could be accounted to be 1.46 ktons of $CO_2$, whereas 2.75 ktons of $CO_2$ would be prevented per ktons of sewage grease separated. According to the abovementioned projections on a continental scale, almost 1100 and 2750 ktons of $CO_2$ would be annually avoided by the present treatment of cellulose and grease, respectively. These figures exclude from the estimation the amount of carbon dioxide which would be saved from avoiding the use of correspondent resources, as well as the emission equivalent related to the operation treatments. Conversely, the $CO_2$ emission equivalent associated with operations required to recover such feedstocks would help define the correct balance [33]. In any case, these approaches produce other benefits in terms of final sludge to be wasted, resulting in a proximate to zero waste discharge process, whose impact in terms of GHG emissions saved would sign a total inversion of trends associated with the urban wastewater treatment.

Technological upgrading of conventional WWTPs represents the first step towards an epochal inversion of the concept of depuration of wastewater. However, it is worth noting that such an innovative approach does not require a complete distortion of the present WWTP configuration, but a simple implementation of preliminary operations would be enough to address and apply this new concept. The innovation of depuration lines would bring other important impacts that could be of interest to society. The transition of WWTP into green and sustainable providers of feedstock would open new job opportunities for medium–high qualified operators on a wide diffuse scale, which includes all territories, even those depressed areas where industrial activities are scarce. In addition, the horizontal necessity of qualified operators would enrich WWTPs with those competencies required to address the transition to a more inclusive and safe society as a sign of a holistic well-being concept.

The effective application of new technologies is subjected to some borderline activities that should be run by specific agencies, decision-makers, and control authorities who have to monitor fundamental aspects with their own competencies. Technical feasibility and economic sustainability are only preliminary steps that need to be done to promote innovation. Circular economy and reuse of resources instead of a linear economic model necessarily pass through a step in which waste becomes a resource. This step is subjected to specific permitting procedures, namely the "End of Waste", which is always a goal to be obtained of absolute value. But especially for the valorization of residual biomasses, besides specific technical aspects which could be defined using measurable quantifications, there are also some perceiving that are out of the most common chemical parameters. With particular regards of sewage, there is a widespread social skepticism which could unconsciously damper any possible applications. Furthermore, the introduction of a new mentality typical for producers of new materials cannot promptly replace the consolidated philosophy of WWTPs, where the most important resource to be recovered still remains water. The importance of social acceptance still remains one of the most important steps to do, with education, to transform a piece of stool into a possible mine of resources.

## 14.4 Conclusions

Wastewater treatment should be seen as a means of water purification and a crucial opportunity to recover resources. By targeting the cellulosic and lipid fractions of wastewater treatment by-products, tailored processes can improve their utilization by avoiding their degradation through biological oxidation and $CO_2$ generation and feeding them into energy and chemical production. The utilization of the cellulosic fraction for the production of ethanol, levulinic acid or biofuels as well as the recovery of lipids for biofuels, biolubricants and biosurfactants production, underlines their strategic value as secondary resources. The recognition of by-products from wastewater as a significant feedstock for the bioeconomy is imperative. A paradigm shift towards their systematic recovery and exploitation is essential for promoting sustainable development and closing resource loops in a circular economy view.

**Acknowledgements** This work was financially supported by European Union—FESR "PON Ricerca e Innovazione 2014–2020. Progetto: Energie per l'Ambiente TARANTO—Cod. ARS01_00637". The authors would like to thank Dr. Enrico Scelsi, Domenico Bellifemine, Domenica Livorti, Andrea Decembrino and Giuseppe Sgaramella for their precious help in the experimental work.

## References

1. Longo S, d'Antoni BM, Bongards M, Chaparro A, Cronrath A, Fatone F, Lema JM, Mauricio-Iglesias M, Soares A, Hospido A (2016) Monitoring and diagnosis of energy consumption in wastewater treatment plants. A state of the art and proposals for improvement. Appl Energy 179: 1251–1268
2. European Parliament and Council: Directive (EU) 2024/3019 of the European Parliament and of the Council of 27 November 2024 concerning urban wastewater treatment (recast). http://data.europa.eu/eli/dir/2024/3019/oj
3. McCarty PL, Bae J, Kim J (2011) Domestic wastewater treatment as a net energy producer-can this be achieved? Environ Sci Technol 45:7100–7106
4. Ruiken CJ, Breuer G, Klaversma E, Santiago T, van Loosdrecht MCM (2013) Sieving wastewater—cellulose recovery, economic and energy evaluation. Water Res 47:43–48
5. di Bitonto L, Antonopoulou G, Braguglia C, Campanale C, Gallipoli A, Lyberatos G, Ntaikou I, Pastore C (2018) Lewis-BrØnsted acid catalysed ethanolysis of the organic fraction of municipal solid waste for efficient production of biofuels. Bioresour Technol 266:297–305
6. Palmieri S, Cipolletta G, Pastore C, Giosuè C, Akyol Ç, Eusebi AL, Frison N, Tittarelli F, Fatone F (2019) Pilot scale cellulose recovery from sewage sludge and reuse in building and construction material. Waste Manag 100:208–218
7. Salmerón I, Núñez-Tafalla P, Venditti S, Hansen J (2024) Biochar obtained from recovered cellulose and its mixture with conventional sources: assessment of its potential for the removal of pollutants in water. Sci Total Environ 954:176357
8. di Bitonto L, Menegatti S, Pastore C (2019) Process intensification for the production of the ethyl esters of volatile fatty acids using aluminium chloride hexahydrate as a catalyst. J Clean Prod 239:118122
9. Conca V, da Ros C, Valentino F, Eusebi AL, Frison N, Fatone F (2020) Long-term validation of polyhydroxyalkanoates production potential from the sidestream of municipal wastewater treatment plant at pilot scale. Chem Eng J 390:124627
10. Qi B, Luo J, Chen G, Chen X, Wan Y (2012) Application of ultrafiltration and nanofiltration for recycling cellulase and concentrating glucose from enzymatic hydrolyzate of steam exploded wheat straw. Bioresour Technol 104:466–472
11. Cardona Alzate CA, Sánchez Toro OJ (2006) Energy consumption analysis of integrated flowsheets for production of fuel ethanol from lignocellulosic biomass. Energy 31:2447–2459
12. Angelini A, Scelsi E, Ancona V, Aimola G, Pastore C (2022) Performic acid pre-treatment of poplar biomasses grown on a contaminated area for enhanced enzymatic digestibility: a viable route to obtain fine-products and recovery of contaminants. J Clean Prod 369:133346
13. Rose C, Parker A, Jefferson B, Cartmell E (2015) The characterization of feces and urine: a review of the literature to inform advanced treatment technology. Crit Rev Environ Sci Technol 45:1827–1879
14. (2023) Società chimica italiana: Utilizzi di oli vegetali usati come materie prime da parte dell'industria chimica. In: La Chimica e l'industria, No. 2, pp 7–9
15. EPR Comunicazione: CONOE a Ecomondo 2023: sostenibilità ambientale e nuove generazioni. https://eprcomunicazione.it/agenzia/conoe-ad-ecomondo-2023-promuovere-la-sostenibilita-ambientale-e-coinvolgere-le-nuove-generazioni/

16. Zarandi M, Torres C, Mateo JM, Jiménez L (2023) Multicriteria analysis of sewage sludge-based biodiesel production. J Environ Manage 348:119269
17. Siddiqui MI, Rameez H, Farooqi IH, Basheer F (2023) Recent advancement in commercial and other sustainable techniques for energy and material recovery from sewage sludge. Water 15:948
18. Olkiewicz M, Torres CM, Jiménez L, Font J, Bengoa C (2016) Scale-up and economic analysis of biodiesel production from municipal primary sewage sludge. Bioresour Technol 214:122–131
19. di Bitonto L, Scelsi E, Locaputo V, Mustafa A, Pastore C (2023) Enhancing biodiesel production from urban sewage sludge: a novel industrial configuration and optimization model. Sustain Energy Technol Assess 60:103567
20. Olkiewicz M, Caporgno MP, Fortuny A, Stüber F, Fabregat A, Font J, Bengoa C (2014) Direct liquid-liquid extraction of lipid from municipal sewage sludge for biodiesel production. Fuel Process Technol 128:331–338
21. D'Ambrosio V, di Bitonto L, Angelini A, Gallipoli A, Braguglia CM, Pastore C (2021) Lipid extraction from sewage sludge using green biosolvent for sustainable biodiesel production. J Clean Prod 329:129643
22. Villalobos-Delgado FJ, di Bitonto L, Reynel-Ávila HE, Mendoza-Castillo DI, Bonilla-Petriciolet A, Pastore C (2021) Efficient and sustainable recovery of lipids from sewage sludge using ethyl esters of volatile fatty acids as sustainable extracting solvent. Fuel 295:120630
23. D'Ambrosio V, Kanda LRS, Corazza ML, Pastore C (2025) Enhancing green lipid extraction from sewage sludge using ethyl hexanoate: a step towards an effective lipid valorization. J Environ Chem Eng 13:115585
24. Scelsi E, Pastore C (2023) Urban sewage sludge valorization to biodiesel production: solvent-free lipid recovery through adsorption on used 3 Ply safety face masks. Environ Technol Innov 30:103072
25. di Bitonto L, Lopez A, Mascolo G, Mininni G, Pastore C (2016) Efficient solvent-less separation of lipids from municipal wet sewage scum and their sustainable conversion into biodiesel. Renew Energy 90:55–61
26. Zhang J, Shi Y, Cao H, Wu Y, Yang M (2020) Conversion of palmitic acid to jet fuel components over Mo/H-ZSM-22 bi-functional catalysts with high carbon reservation. Appl Catal A Gen 608:117847
27. di Bitonto L, Locaputo V, Pastore C (2024) Solventless recovery of lipids from urban sewage sludge: How to sustainably turn a waste into a valuable source of oleochemicals. Chem Eng J 494:152991
28. di Bitonto L, Todisco S, Gallo V, Pastore C (2020) Urban sewage scum and primary sludge as profitable sources of biodiesel and biolubricants of new generation. Bioresour Technol Rep 9:100382
29. D'Ambrosio V, La Parola V, Francesca Liotta L, Roberto E, Carraro G, Savio L, Comparelli R, Lucia Curri M, Pastore C (2024) Catalytic hydrogenation of waste-derived lipids: A route to producing sustainable drop-in biofuels by using Re/$TiO_2$ catalysts. Chem Eng J 499:156648
30. Mustafa A, Faisal S, Ahmed IA, Munir M, Cipolatti EP, Manoel EA, Pastore C, di Bitonto L, Hanelt D, Nitbani FO, El-Bahy ZM, Inayat A, Abdellatief TMM, Tonova K, Bokhari A, Abomohra A (2023) Has the time finally come for green oleochemicals and biodiesel production using large-scale enzyme technologies? Current status and new developments. Biotechnol Adv 69:108275
31. Mustafa A, Fathy S, Kutlu O, Niikura F, Inayat A, Mustafa M, Abdellatief TMM, Bokhari A, Samuel OD, Pastore C, di Bitonto L, Tawfik MA, Munir M, Mohsen R (2023) Cleaner and sustainable synthesis of high-quality monoglycerides by use of enzyme technologies: techno-economic and environmental study for monolaurin. Clean Technol Environ Policy 25:3263–3283
32. European Biodiesel Board (EBB): Statistical report 2023. https://ebb-eu.org/wp-content/uploads/2024/03/EBB_Statistical_Report2023-Final.pdf

33. Notarnicola B, Tassielli G, Renzulli PA, Capua RD, Astuto F, Riela S, Nacci A, Casiello M, Testa ML, Liotta LF, Pastore C (2023) Science of the total environment life cycle assessment of a system for the extraction and transformation of waste water treatment sludge (WWTS)-derived lipids into biodiesel. Sci Total Environ 883:163637

# Part IV
# Process Design, Intensification and Control for a Safe and Secure Grow

# Chapter 15
# Methodology and Technology Development Toward Circular Industrial Clusters: Application to the Energy-Water-Hydrogen Nexus

**Assaad Zoughaib, Thibaut Wissocq, and Rolando Argandona**

**Abstract** Water usage in industry constitutes the largest share of water usage in industrialized countries. This water usage overpass agriculture by far. A large part of water is consumed for process cooling but a non negligible part is used within the process as steam or pure water. More recently, green hydrogen expected production will contribute to highly increase the demand. Waste water can be barely reused in these processes which lead to poor recovery ratios even with a comprehensive mass integration effort. Recently, methodologies coupling mass and energy integration allowed showing that coupling waste heat with waste water allows upgrading waste water into distilled water thanks to thermal membrane distillation. This novel technology has proven it ability to handle water with different concentration in minerals making it preferable to reverse osmosis when it comes to highly concentrated waste water. This paper focuses on water upgrading opportunities in the industrial sectors thanks to the waste heat valorization. Waste heat quality and availability is screened together with the need for pure water. Several matching are identified showing that the development at industrial scale of thermal membrane distillation technologies allows important energy and water savings.

**Keywords** Thermal water distillation · Energy and mass integration · Water energy nexus

A. Zoughaib (✉) · T. Wissocq · R. Argandona
Mines Paris, PSL University, Paris, France
e-mail: Assaad.zoughaib@minesparis.psl.eu



A. Angelini et al. (eds.), *Security Enhancement for Climate Change Impacting Urban Resources - SECCURe*, NATO Science for Peace and Security Series C: Environmental Security, https://doi.org/10.1007/978-94-024-2345-7_15

# 15.1 Introduction

## *15.1.1 Role and Challenges of Hydrogen in Industry Decarbonation*

The fight against climate change has intensified, with hydrogen (H2) emerging as a key player in decarbonizing the energy sector [1]. It's versatile for various sectors, including those hard to electrify like transport. However, H2 demand projections vary widely, and its production methods significantly impact its environmental benefits.

H2 is classified by production methods: grey (steam methane reforming), blue (with carbon capture), turquoise (methane pyrolysis), and green (water electrolysis using renewable energy). Only green H2 is considered clean, but it currently makes up just 4% of total H2 production due to higher costs.

Boosting green H2 production through water electrolysis using renewable energy sources (RES) could decarbonize the energy sector and address the climate crisis. Green H2 can store excess RES energy and offers benefits like compactness and independence from electrical infrastructure. Despite current cost challenges, green H2 production is expected to grow significantly by 2050 as technology advances and RES electricity prices decrease. Several authors [2–5] have taken different base scenarios to make their projections, stating that it will be required from 50 Mt/y of $H_2$ [5] to 2.3 Gt/y by 2050 [4]. Despite the lack of homogeneity in the projections, a considerable amount of $H_2$ would be required, and therefore it would be necessary to find a reliable source from which to obtain it.

The production of green H2 involves RES and water electrolyzers, which require near-ultrapure water. This will create an additional burden on the global water resource that is already in stress due to an increasing demand (Fig. 15.1) but also a decrease in the resource as a consequence of the climate change [7]

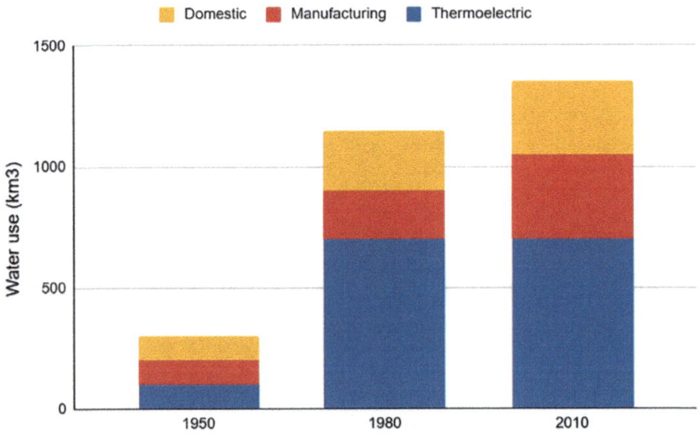


**Fig. 15.1** Global water demand by sector data extracted from [6]

### *15.1.2 Energy and Mass Integration Methodologies for a More Circular Industrial Hubs*

The concept of industrial ecology has led to the development of eco-industrial parks (EIPs) and synergies between industries. These aim to recover untapped resources like waste heat or wastewater and facilitate the exchange of energies and resources among industries, thereby reducing their resource consumption and environmental impact. Various methods and mathematical optimization models have been employed to design such systems, but they often focus on a single resource type, such as heat, water, or hydrogen.

Integrating both mass and energy considerations simultaneously can lead to more efficient resource management, especially when conversion systems like heat pumps, electrolyzers, or water treatment systems are involved. Some studies have developed simultaneous approaches for this purpose. For instance, a multi-objective approach in [8] addresses water-energy networks, aiming to minimize water consumption and achieve economic objectives for companies, with direct heat integration.

Both sequential and simultaneous approaches for heat and mass design problems have been explored at industrial process and territory scales [9]. These problems are often non-linear and non-convex, making them challenging to solve. A linear formulation for mass and heat recovery networking, including heat exchanger networks, is presented in [10]. Additionally, dedicated approaches for EIP design have been developed to minimize total costs, including carbon taxes, by creating energy and material synergies between industries [11–14].

However, those previous works do not address all the issues related to EIP design. Resource availability may not align with demand, and industries may have different temporal dynamics, necessitating storage solutions to enhance resource recovery. Economic divergences can also occur, as each actor prioritizes their own interests. To address these challenges, [15] proposes a multi-period MILP (mixed-integer linear problem) model for designing heat and mass recovery networks in EIPs. The model identifies recovery paths for recycling waste resources and includes conversion system design to upgrade co-products into raw materials. This methodology is presented more in detail in the next section and its application to a case study will show the importance of the thermal membrane distillation technology paving the way to larger circularity opportunity for heat and water.

The last part of the paper will focus on the performance of a thermal membrane distillation prototype showing case for industrial waste water recycling.

## 15.2 Methodology for Mass and Heat Valorization Networks Design

### *15.2.1 Overview of the Model's Aspects*

This section presents previous works [15] by considering industrial resource demands and availabilities over time periods $t \in T$. An industrial cluster comprises various objects representing resource demands or availabilities:

- **Heat Streams**: Defined by temperatures $T_{inh}$ and $T_{outh}$ and power $Q_{h,t}$. They can represent heat demand (cold streams) or waste heat availability (hot streams).
- **Thermal Utilities**: Similar to heat streams but with variable power $Q_{hu,t}$ and associated operating costs $C_h$. They satisfy heating or cooling needs.
- **Process Sources**: Defined by mass flow $L_{j,t}$ and pollutant composition $y_{j,k,t}$ representing available resources like water effluents.
- **Process Sinks**: Defined by mass flow $G_{i,t}$ and acceptable pollutant composition range $z_{mini,k,t}$, $z_{maxi,k,t}$, representing resource demands.
- **Variable Sources**: Optimized mass flow rates $L_{vj}$, to satisfy process sinks, representing inputs like storage or resource purchases.
- **Variable Sinks**: Optimized mass flow rates $G_{vi}$, for discarding unusable resources, representing outputs like storage or disposal.

More complex systems can be defined based on these components:

- **Heat Networks**: Modeled by utility pairs representing hot and cold sides, with temperatures Thot and Tcold. The network layout is determined by optimization.
- **Thermal Storages**: Modeled as stratified tanks with hot and cold zones, characterized by investment costs.
- **Conversion Systems**: Associate variable sources, sinks, and thermal utilities to valorize resources, modeled as grey-boxes with efficiency coefficients and investment/operating costs.

To manage computational complexity, a non-uniform time representation is used, reducing the number of time steps from 8760 (hourly precision over a year) to 96; more precisely, each season is represented by a typical day. Evry typical day is repeated until the next season. The optimization variables include network layout, thermal storage volume, conversion systems, and heat/mass flow dispatching over time.

### *15.2.2 Model Formulation*

The main equations cover mass conservation, transhipment, network modeling, conversion systems, thermal storage, and the objective function.

- Mass and Energy Integration

Mass and energy integration equations are based on previous works [12, 14], extended with a temporal dimension. They ensure mass conservation and classic transhipment at the territory scale. Heat exchangers are considered in parallel configurations.

- Network Formulation

Geographical clusters represent industrial sites with thermal streams and source/sink processes. Mass and thermal networks facilitate resource exchanges between industries, characterized by linear investment costs. Mass networks are modeled by variable sources and sinks, while thermal networks are characterized by hot and cold utilities.

- Conversion System Modeling

Conversion systems transform resources using variable sources, sinks, and utilities. They are defined by linear equations coupling mass flows with electrical and thermal power, with associated investment and operating costs.

- Thermal Storage

Thermal storage is modeled as a stratified tank with hot and cold zones. The storage behavior during loading and unloading phases is described, with investment costs depending on total volume.

- Objective Function

The objective is to minimize total actualized costs, including investment (CAPEX) and operating costs (OPEX). CAPEX covers pipe, conversion system, heat exchanger, and storage costs, while OPEX includes resource purchase, disposal, utility, and electricity costs.

### *15.2.3 Solving Methodology*

The solving methodology involves two steps: cooperative and non-cooperative governance.

- Cooperative Scheme

The cooperative approach optimizes the territory's industrial clusters collaboratively, seeking to minimize global costs. Investment and operating costs are shared, and economic benefits are redistributed.

- Non-Cooperative Scheme

The non-cooperative approach simulates negotiations between producers and consumers, introducing a third-party network operator [14]. The methodology ensures that each actor benefits economically, with buying and selling prices adjusted to reach an equilibrium between supply and demand.

This approach guarantees that all participants in the synergies gain economic benefits, addressing individual economic interests.

## 15.3 Case Study

This methodology is applied to a real-world industrial case study involving ten industries with varying heat, cold, and water demands, as well as waste heat availability (see Fig. 15.2). Each industry is modeled by its water demand using source-sink representations and by thermal streams for heat demand (≤ 100 °C), cold demand, or waste heat availability. Thermal utilities and variable sources represent the current solutions, such as gas boilers, chillers (≥ 0 °C), cooling towers, and existing industrial and drinking water networks, each with associated operating costs.

The costs of the utilities is as follows: hot utilities cost 30 €/MWh, air-cooling towers cost 2.5 €/MWh, chillers cost around 23 €/MWh (depending on efficiency),

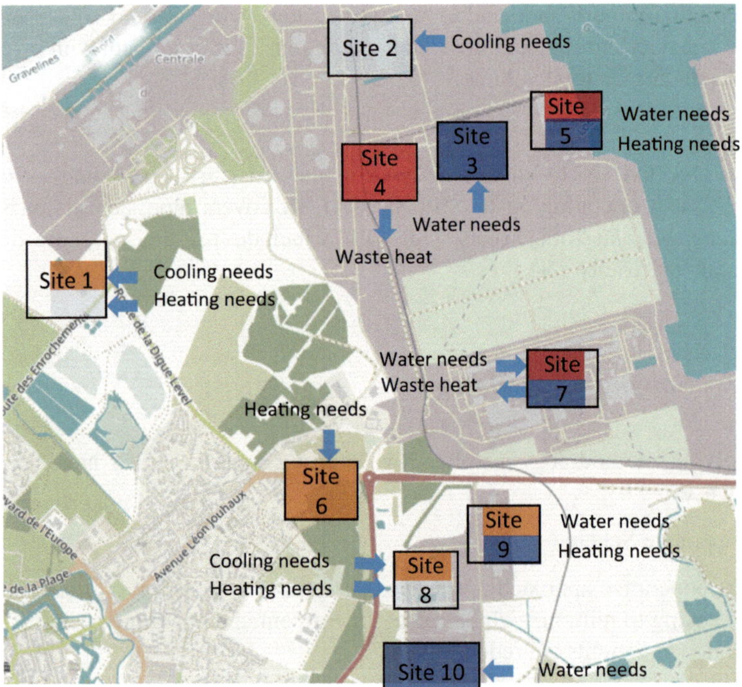


Fig. 15.2 Eco industrial park case study from [15]

drinking water costs 1.5 €/$m^3$, and industrial water costs between 0.4 and 0.6 €/$m^3$ (depending on consumption level).

### 15.3.1 *Thermal Membrane Distillation: A Water Conversion System Valorising Waste Heat*

Membrane distillation (MD) is a non-isothermal membrane process based on generating a vapor phase within a hot solution stream near the surface of a membrane [16]. The vapor can pass through the membrane (the hot side of the system) and condenses on the cold side, where cold distilled water circulates. This process is made possible by using hydrophobic membranes, which only allow vapor and volatile compounds to pass through while retaining the liquid portion of the hot stream (which contains dissolved salts and other non-volatile compounds). The hot feed water evaporates, and the water vapor diffuses through the membrane. The vapor condenses into liquid on the cold side, resulting in a permeate (distilled water) on the cold side of the system. This distinguishes MD from other membrane processes that use absolute pressure difference, concentration gradient, or electrical potential gradient to achieve mass transfer.

The driving force of MD is the partial vapor pressure difference between the feed stream and the permeate, responsible for converting some of the feed liquid molecules into vapor, thus enabling the mass transfer process. This partial vapor pressure difference is related to the temperature gradient across the membrane sides, so MD systems are considered thermally driven systems. MD technology does not require high operating temperatures; it works efficiently at feed liquid temperatures between 50 and 90 °C and develops well from a temperature difference between the hot and cold sides of at least 20 °C.

This feature makes this technology ideal for waste heat valorisation for industrial waste water upgrade into pure water adapted for reuse in the process.

This conversion unit is modelled along the other systems and considered in the case study described previously.

### 15.3.2 *Main Results of the Case Study*

The two steps of the methodology are applied sequentially; first the cooperative scheme that allows to obtain the maximum synergy potential then the non cooperative scheme that allows to analyse the winning synergies based on individual benefits for each industrial.

- Cooperative scheme results

  The initial step of the methodology leads to the following synergies:

- **Cluster 4** supplies heat to clusters 1, 2, 6, 9, and 10 through the heat network. Within clusters 1, 2, and 8, absorption machines convert waste heat to meet cold demands.
- **Cluster 7** utilizes its own waste heat for a local Thermal Membrane Distillation (MD) system, reducing its water consumption.
- **Clusters 4 and 5** use waste heat for their MD systems to meet internal needs, with surplus distilled water sent to clusters 3, 9, and 10 via a distilled water network. This distilled water fully replaces drinking water demand and partially replaces industrial water.

The total investment cost is 4 million euros, including 770,000 euros for absorption machines, 400,000 euros for TMD systems, and 2.8 million euros for heat and water networks. Annual operating expenses are estimated at 338,000 euros, covering MD and absorption machine operations and the purchase of fresh drinking and industrial water. The heat recovery process completely substitutes the need for hot and cold utilities like chillers and gas boilers.

- Non cooperative scheme results

The second step of the methodology involves non-cooperative governance. A third-party network operator is introduced to set buying and selling prices for resources and to coordinate bids and offers between producers and consumers. The following assumptions are made:

- **Absorption Machines**: Operated and purchased by the network operator at a capital cost of 250 euros per kW. Thermal Membrane Distillation (MD) systems are bought and operated by the industries, with surplus water injected into a dedicated water network.
- **Resource Pricing**: The network operator buys heat at 2.5 euros per MWh. The selling prices for heat and water is subject to sensitivity analysis.
- **Depreciation**: The depreciation period is set at 10 years with an actualization rate of 5%.

This setup results in 12 different configurations, depending on the variations in water and heat selling prices.

The solution obtained earlier in the cooperative scheme is not obtained here, demonstrating that competition between network operator and industrial interests exists. For a buying price of water lower than 0.8 Eur/m$^3$, the water network is not profitable and is not invested. MD are only used in self-consumption in these cases and we get (with only heat and cold transactions.

When the selling price of heat is below 7 Eur/MWh, the heat network is not profitable for the network operator and it is not invested. The only network is the water network. The other configurations involve heat, cold and water synergies and highlight competition for waste heat use.

Heat and cold supply are favoured because industrial operating costs are reduced to a greater level than water supply from MD. Two main situations occurred:

- When the heat selling price is too high (> 14 Eur/MWh), industrial with cold demand are not interested in the synergies. Waste heat can thus be used for water production.
- When the heat selling price is too low (< 9 Eur/MWh) to be profitable for the network (pipe and absorption machine investment), waste heat can be used for water production.
- In one configuration waste heat is shared between water, heat and cold supply. It shows that water production from MD is lower because heat and cold are favoured since it allows higher industrial gains.

This case study illustrate the strength of such methodologies to assist engineers in designing the highest possible synergies reaching circular industrial hubs. Water and energy nexus can be used positively for closing the water loop and upgrading waste water into reusable water thanks to waste heat.

The MD technology allowing this loop is therefore a young technology and needs to be pushed into market. The next section shows practical results achieved with an industrial pilot of MD.

## 15.4 AQUAHIVE: A Pilot Scale MD Experimental Results

The previous section showed that MD is a promising technology for waste water updgrading using waste heat. This requires scaling of this technology into industrial scale and a verification of its polutants rejection effectiveness.

MD uses hydrophobic membranes, which allow only the passage of vapor and volatile compounds while retaining the liquid part in the hot stream (which contains dissolved salts and other non-volatile compounds). Moreover, as the solution treated is not in direct contact with the membrane, fouling is reduced compared to other distillation membrane processes.

These characteristics made it possible for the MD to treat different inlet solutions as, in general terms, the quality of their distillate is not affected by the initial operating conditions (temperatures, flows, concentrations) or surface area. However, the presence of volatile ionic species could affect the distillate quality. MD has already proved its effectiveness in the desalination sectors; however, its operation with other solutions has been explored less.

This section aims to present the Aquahive system, an innovative design for the MD process, and its distillate quality performance using several water solutions. The Aquahive system exploits the advantages of the air gap membrane distillation (AGMD) configurations and introduces some innovations to improve their performance. This study evaluates the following inlet solutions: tap water, oil-field produced water, distillery stillage, and chlorinated water.

### 15.4.1 *Industrial Scale AQUAHIVE Pilote Module Description*

The Aquahive module tested had 10 membrane surfaces grouped into five membrane frames (two membrane surfaces on each frame) and six condensers. Table 15.1 summarizes the module's main dimensions.

Several modules are integrated into a container allowing to test its performance in real conditions on industrial sites (Fig. 15.3).

**Table 15.1** Aquahive pilot module dimensions

| Sub-component | Parameter | Value |
|---|---|---|
| Condenser | Quantity | 6 |
| Air gap | Width | 7 mm |
| Hot solution channel (Membrane frame) | Effective height | 2 m |
| | Depth | 0.5 m |
| | Width | 10 cm |

**Fig. 15.3** AQUAHIVE modules integrated on industrial site

### *15.4.2 Distillary Stillage Upgrade*

AQUAHIVE pilot shown above was installed in an alcohol distillery in the north of France. The aim of the demonstration is to show how hot distillary stillage can be upgraded into steam generation grade distilled water.

Distillery stillage is wastewater produced during the manufacture of ethanol and other alcohols. It is typically generated in the distillation column, where valuable alcohol is recovered at the top, and stillage, usually in liquid form, is removed from the bottom. Stillage is abundantly produced as a by-product in alcohol production, with 10 to 15 L generated for every liter of distillate. Its high volume and chemical characteristics necessitate careful management.

Stillage is a low-solids fluid, typically not exceeding 8% by weight, with a chemical composition dependent on the raw material used for alcohol production and the distillation process. Generally, it is acidic with high concentrations of organic compounds and a high chemical oxygen demand (COD). It also contains significant amounts of nitrogen, phosphorus, and potassium.

High COD values can deplete oxygen in water bodies, inhibiting microorganism activity and leading to eutrophication.

Due to its properties, not all treatment technologies can handle stillage effectively. Membrane distillation (MD) is a viable option because its acidic nature and chemical components do not directly affect the system.

The stillage sample was treated at an average hot side temperature of 62 °C, and the distillate was collected over four hours. Samples of the distillate and stillage were analyzed in a specialized laboratory, with results summarized in Table 15.2.

The results indicate that the initial electrical conductivity of the stillage was relatively low, suggesting it was not heavily loaded with dissolved solids. The Aquahive system significantly reduced conductivity to levels comparable to de-ionized water and decreased chloride and sulfate concentrations. The modest reduction in chloride levels can be attributed to its already low initial concentration. For COD, the rejection rate could not be precisely calculated due to instrument limitations, but it achieved a notably high reduction.

**Table 15.2** Distilled water from stillage quality results

| Parameter | Unit | | Stillage | Distillate | Rejection rate (%) |
|---|---|---|---|---|---|
| pH @20 °C | | | 6.59 | 6.09 | |
| Conductivity | μS/cm | | 129.5 | 5.4 | 95.8 |
| Chlorides | mg/L | | 0.14 | 0.07 | 50 |
| Sulfates | mg/L | | 17.49 | 0.33 | > 98 |
| COD | mg/L | | 179 | < 20 | > 89 |

## 15.5 Conclusions

Industrial water usage accounts for the largest share of water consumption in industrialized nations, significantly exceeding that of agriculture. A substantial portion of this water is used for process cooling, while a notable amount is utilized as steam or pure water within industrial processes. The anticipated production of green hydrogen is expected to further increase water demand.

Despite comprehensive mass integration efforts, wastewater reuse in these processes remains limited, resulting in low recovery rates requiring further methodology development and technology demonstration.

In this paper, recent methodologies combining mass and energy integration have been presented. The formulation coupling energy and mass integration allowed to highlight the role of waste heat to upgrade wastewater into distilled water through thermal membrane distillation (MD).

This innovative technology has proven effective in handling water with varying mineral concentrations, making it preferable to reverse osmosis for highly concentrated wastewater.

The pilot scale AQUAHIVE MD shown in this paper explores opportunities for upgrading water in industrial sectors by valorizing waste heat. Its testing in a alcohol distillery in the North of France demonstrated its high pollutant rejection rate for upgrading the stillage into pure water reusable for steam generation in the plant.

The obtained water quality is compatible with pure water that can be used before the electrodeionization (EDI) step in the water treatment plant of an electroliser; paving the way to replace two stages of reverse osmosis. This will result in a valorisation of the stack cooling recovered heat and the reduction of the electrical consumption of the water treatment plant while valorising low grade waters. This perspective application is under study for a near future demonstration.

## References

1. Gao F-Y, Yu P-C, Gao M-R (2022) Seawater electrolysis technologies for green hydrogen production: challenges and opportunities. Curr Opin Chem Eng 36:100827
2. Woods P, Bustamante H, Aguey-Zinsou K-F (2022) The hydrogen economy—where is the water? Energy Nexus 7:100123
3. Simoes SG, Catarino J, Picado A, Lopes TF, di Berardino S, Amorim F et al (2021) Water availability and water usage solutions for electrolysis in hydrogen production. J Clean Prod 315:128124
4. Oliveira AM, Beswick RR, Yan Y (2021) A green hydrogen economy for a renewable energy society. Curr Opin Chem Eng 33:100701
5. Lane B, Reed J, Shaffer B, Samuelsen S (2021) Forecasting renewable hydrogen production technology shares under cost uncertainty. Int J Hydrog Energy 46:27293–27306
6. Floerke M (2013) Domestic and industrial water uses of the past 60 years as a mirror of socio-economic development: a global simulation study. Global Environmental Change
7. Mekonnen MM, Hoekstra AY (2016) Four billion people facing severe water scarcity. Sci Adv 2 :e1500323. https://doi.org/10.1126/sciadv.1500323

8. Ramos MA (2016) Optimisation Bi-Niveau d'Ecoparcs Industriels pour une Gestion Durable des Ressources. Ph.D. Thesis, Institut National Polytechnique de Toulouse
9. Liu L, Song H, Zhang L, Du J (2018) Heat-integrated water allocation network synthesis for industrial parks with sequential and simultaneous design. Comput Chem Eng 108:408–424. ISSN0098-1354
10. Ghazouani S (2016) Modèles linèaires d'optimisation pour la conception simultanée de réseaux de matière et de chaleur d'un ecoparc industriel. PhD thesis, Mines Paris PSL University
11. Kantor I, Robineau J-L, Butun H, Mar´echal F (2020) A mixed-integer linear programming formulation for optimizing multi-scale material and energy integration. Frontiers Energy Res 8:49
12. Chahla GA, Zoughaib A (2019) Agent-based conceptual framework for energy and material synergy patterns in a territory with non-cooperative governance. Comput Chem Eng 131:106596
13. Farhat A (2015) Intégration Energétique multi-site: Applications à la Conception d'Eco-parcs Industriels. PhD thesis, Mines Paris PSL
14. Wissocq T, Ghazouani S, Le Bourdiec S (2019) A methodology for designing thermodynamic energy conversion systems in industrial mass/heat integration problems based on MILP models. Energy 185:121–135. ISSN 03605442
15. Wissocq T, Le Bourdiec S, Assaad Z (2021) Mass and heat valorization network design for eco-industrial parks in non-cooperative scheme. In: ECOS 2021—34th international conference on efficiency, cost, optimization, simulation and environmental impact of energy systems, June 2021, Taormina, Italy (hal-03277675)
16. Argandoña R, Assaad Z, Mustapha R (2023) A comprehensive review of water distillation technologies for green hydrogen production. In: 36th international conference on efficiency, cost, optimization, simulation and environmental impact of energy systems (ECOS), Jun 2023, Las Palmas, Gran Canaria, Spain (hal-04394670)

# Chapter 16
# Biorefinery Process Design for the Integrated Valorization of Biomass

**Mariano Martín and Mahmoud El-Halwagi**

**Abstract** The valorisation of biomass residues is a challenge and an opportunity. The large number of possible added value products and electricity that can be obtained results in a difficult product and process design problem. Integrated processes would allow fully reutilization of the residues, providing, at least, part of the utilities and intermediates are internally produced. Systematic process synthesis techniques are presented to show the development of integrated processes towards circularity. In addition, with residues, building a circular economy occurs at different levels, at process level, reusing the waste to produce the utilities and to recover nutrients that would help produce the next harvest, and along the food chain, so that the different echelons of the chain are integrated towards developing a circular economy. This chapter presents the mathematical optimization approach developed for the simultaneous process and product design and several examples are presented to show the application to the valorisation of biomass such as orange peels, wine, coffee and oil production residues, or synthesizing products from manure developing fully integrated processes. In addition, a larger scale has also been considered such as the integration of the operation of farms and agricultural exploitation defining the animal feed for the optimal integration of the different business and reducing the impact by reducing the need for inorganic fertilizers.

**Keywords** Circular economy · Added value products · Integrated biorefinery · Process design · Multiscale analysis

---

M. Martín (✉)
Department of Chemical Engineering, University of Salamanca, Salamanca, Spain
e-mail: mariano.m3@usal.es

M. El-Halwagi
Gas and Fuels Research Center, Texas A&M Engineering Experiment Station, College Station, USA

Artie McFerrin Department of Chemical Engineering, Texas A&M University, College Station, USA

 
A. Angelini et al. (eds.), *Security Enhancement for Climate Change Impacting Urban Resources - SECCURe*, NATO Science for Peace and Security Series C: Environmental Security, https://doi.org/10.1007/978-94-024-2345-7_16

## 16.1 Introduction

Among the sustainable development goals of the UN, #2 zero hunger, aims at providing the necessary means to feed the ever-growing population. With a limited arable ground expansion, the yield of the different species needs to improve by increasing the efficiency of food production techniques but also the use of food, reducing waste, and its residues. That efficiency can also be linked to other goals such as the use of less energy and clean energy for any of the operations, #7, as well as make the best of the use of water and avoid water pollution, which is in line with the clean water goal, #6, since the residues from food production are somehow responsible for nitrate and phosphate building up in the water bodies. In addition, #11 aims at sustainable cities and communities, while # 12 targets the responsible consumption and production. While these objectives are broad, they are to be achieved though improving all the operations across the different stages of the food value chain.

The *food value chain* (FVC), see Fig. 16.1, consists of all the stakeholders who take part in the production and value-adding activities that are needed to make food products all the way from the growing/feeding, the transportation, the processing, the packing and the distribution for its use. However, once the food has been consumed there are residues from packing and from the food itself. The global food service market is projected to grow from \$2540.05 billion in 2022 to \$5194.60 billion by 2029, at a compound annual growth rate of 10.76% in the forecast period. Circular economy can be created not only at this last point, for the reuse of the residues but also at the stages of food production, food processing and finally after food preparation and consumption. Therefore, as shown by Fig. 16.2, the circularity is a multiscale problem, all the way from the product, a particular chemical, to the units that process the biomass, the processes and the network of echelons that include food production, waste collection, waste processing, food transportation, use and waste management that represents another processing step.

The food production can be at the agricultural sector, fruits, cereals, vegetables, etc. as well as the farms to produce meat from different animals (chicken, pigs, cows, fish, etc.). Over the last years the specialization and the need to increase the production capacity, has separated the farms from the agricultural production. However, both are related. Animals eat forage from the fields, mostly residues from grain production, and generate residues that contain nutrients that can be used to grow the next harvest. This is a natural circular economy that was operating as such for years and now it has to be promoted but also improved. In addition to the big picture, within the food production/processing further circularities can be developed for the better usage of the residues of industry and finally the residues of the food. This chapter aims at presenting the possibilities of creating circular economies at different levels and stages of the value chain that would improve the efficiency of the transformations and the entire system. Process system engineering approaches provide the means of designing such solutions by presenting the so call systems approach that takes a wider view to the problem and identifies the connections and synergies towards more efficient production processes [1].

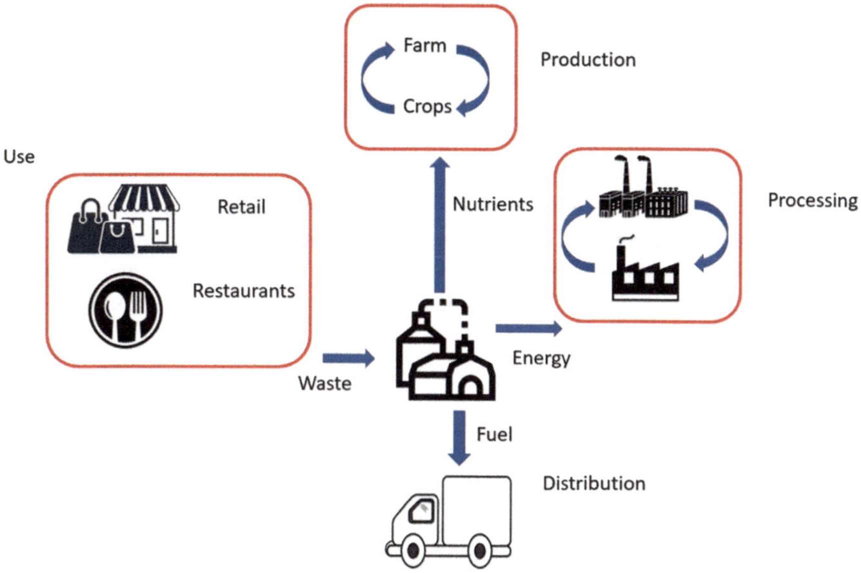


**Fig. 16.1** Scheme of the circular economy for the food industry and value chain

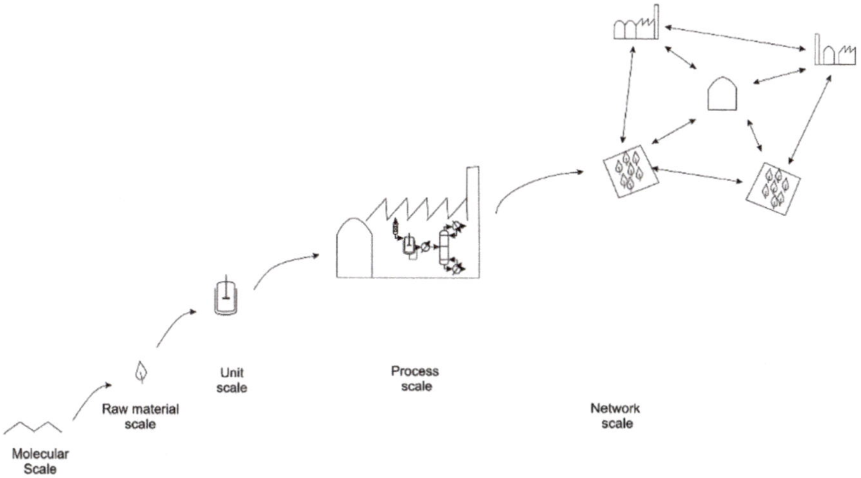


**Fig. 16.2** Multiscale analysis of biomass/waste valorization

## 16.2 Process Design Principles

Building circular economies require the use of systematic approaches because the connections are not straightforward, and the potential alternatives are numerous. For instance, to identify the resources out of the waste streams and the best use that will allow developing circularities at different stages of the value chain. An effective approach to embedding the various design options is to create a superstructure the implicitly contains all the alternatives of interest. Process synthesis (via modelling, optimization and integration) is achieved by optimizing the created superstructure to systematically evaluate the integration opportunities. Formulating one of them boils down to the models of the units and the network involved in the processing of the food residues. This section presents the principles that allow generating such structures. It does not aim at being fully comprehensive, since there are books devoted to each one of the concepts presented, but rather introducing the methods and preparing the reader for the examples that have benefited from these concepts to address the problem at hand.

### 16.2.1 Concepts of Process Synthesis Subsection Sample

Identification of the best alternatives towards the exploitation of the resources requires a systematic approach. Finding and solving trade-offs require the use of techniques that allow their comprehensive analysis and comparison from economic, social, and environmental points of view. The formulation of superstructures of alternatives that allow a systematic comparison and evaluation has become a widely used technique [2, 3]. The starting point corresponds to modelling the units to formulate the process [4] and the network [5]. This section is divided into the challenges in process modelling and the process design methods.

**Process modelling challenges**

There are some classifications for the types of models of the units involved in a processing plant. Lately they have been divided into first principles and empirical. Among the first ones, the level of detail determines whether they are shot cut models, simple ones, consisting of a reduced number of equations, that have the drawback of lack of accuracy for complex systems or detailed models that include rigorous thermodynamics, flow dynamics, mass, and heat transfer. These last kinds of models allow predicting the yield of the operations by they are computationally expensive. In this category the main challenge is the fact that novel units/technologies or complex ones are difficult to be modelled since identifying all phenomena involved is hard and for theoretical models to fit experimental data additional experimental effort are required. Empirical models, including the use of experimental data to fit several parameters and those using machine learning, have the disadvantage that they rely on data that can be only valid for the units used to obtain them, with scale up issues and their validation to different systems requires a large amount of experimental data. However, the current access to large sets of data allow to develop them quicker and update them over time as additional data become available. The model of the units can also be considered as a multi scale problem so that the mechanism behind the operation/reaction are included. The multiscale determines the level of detail as well. For instance, a reactor can be modelled using an input–output approach, using experimental data and the stoichiometry, but we can use the kinetics as a second level of detail or all the way to simulations with computational fluid dynamics (CFD) so that the fluid dynamics of the system can also be considered. Similarly for particle technology units such as biomass dryers, the single particle drying model can be included as more detailed model level compared to a simple mass and energy balance but also the particle dynamics can be evaluated using Discrete element methods (DEM).

**Process design: methods and approaches**

There are two major schools to design processes, heuristics [6] based largely on rules of thumb and experience and the systematic approach using graphical, algebraic, and mathematical optimization techniques [7–10]. However, both approaches can be coupled to use the best features of each of them creating hybrid design methods. These coupled methods use heuristics to narrow down the alternatives based on experimental know-how and formulate a mathematical optimization problem to decide on the portfolio of products, processing technologies and operating conditions [11]. Over this chapter we present this last alternative as it is the most promising to address the design of processes involving the processing of waste towards added value products.

### *16.2.2 Process Integration Towards Sustainability*

Circular economy is a recent term that aims at the reuse and recycle of waste to reduce and avoid the amount of residues so as to save and reduce the use of natural resources [12]. However, transposing this topic into chemical processes, it is possible to argue that it was already around, and it has been put into practice since well before the term appeared in the literature. One example is the Solvay process. Already back in 1860s no intermediates were released since all of them where reused and recycled towards the production of the main product, sodium carbonate. Therefore, the process industry has extensive experience in developing efficient processes. Three principles are presented towards such efficient internal reuse of resources, mass integration via recirculation, energy integration, a different type of integration since no direct contact between streams is necessary, and process integration itself. All of them are presented below.

**Mass integration. Circularity**

Mass integration is "a systematic methodology that provides a fundamental understanding of the global flow of mass within the process and employs this holistic understanding in identifying performance targets and optimizing the generation and routing of species throughout the process" [10]. The first major contribution in the field of systematic approaches for mass integration was the introduction of mass-exchange networks [13, 14] which is a subset of mass integration dealing with the removal of pollutants from a set of rich (waste) streams to a set of lean streams which can be used in a circular manner or as once-through. Since then, numerous mass-integration categories and approaches have been developed to address the broader aspects of the field. A key characteristic of many of these tools is based on tracking chemicals and formulating material conservation problems [15–17]. They have typically been applied to a wide variety of problems especially for the conservation of natural resources and the mitigation of pollution. These approaches include aggregate representations via pinch-based approaches (e.g., [18]), algebraic approaches (e.g., [19, 20]), and generic superstructures [21–23]. Automatic process flowsheeting is becoming a topic of interesting in the interface of machine learning and chemical engineering. Integrated processes have been developed recently for the reuse of waste where the intermediates are produced from the waste and the final product does not require additional chemicals beyond residues or waste stream [24, 25]. Furthermore, multiscale mass integration techniques have been used for the optimal design of eco-industrial parks based on the concept of circular economy and synergistic integration. An important category deals with the design of carbon-hydrogen–oxygen symbiosis network (e.g., [10, 26–29]).

**Energy integration (heat and power)**

Systematic heat integration steams from the Pinch method to identify the minimum utilities costs [8]. Thus, the high costs of energy, thermal and power, have been the driving force for the design of heat exchanger networks (using sequential [30] and simultaneous [31] approaches) as well as coupling heat with power within the same structure [18, 32, 33]. These networks aim at determining the best use of the thermal energy available to reduce the use of utilities as well as, in the case of power, reduce the use of electricity by using the power obtained in the expansion of streams. The operation with gases is particularly interesting for the last case, while the design of biochemical processes benefits specially from thermal energy integration. The major challenges are the low temperature of operation of exothermal reactors used for biochemical reactions, which result in the additional consumption of cooling water as well as the complex prediction of the heat of reaction when biomass is involved, especially due to the microorganism energy consumption to carry out the operations. We have therefore introduced the water-energy nexus for what is work within biorefineries operation, a topic evaluated back in 2010s [34]. This nexus also relates mass exchange networks with energy exchange ones.

### *16.2.3 Process Integration. Integrated Processes Challenges and Opportunities*

While it is possible to evaluate mass and energy separately, the best results as well as the trade-offs can always be found and identified when the entire process is analysed. Several examples in the field of biofuels [35] and hybrid process integrating biomass and fossil resources have made use of these approaches [36] presenting the advantages of the integrated approach. However, identifying synergies by integrating technologies results in large and complex mathematical problems that demand the use and development of algorithms and solution procedures. It is here where formulating the most adequate model makes a difference. In addition, it represents an opportunity for the development of tools and algorithms that would allow us to design more efficient processes that is the primary task of process system engineering.

### 16.2.4 The Concept of Product Design

For years the design of products was left to the marketing departments because of the direct link between the consumer expectations and the features that the product must achieved. However, following the same reasoning as for the integrated process design, when considering process and product simultaneously it is possible to identify non straightforward relationships. In this chapter we only focus on chemical/compounds but the mathematical principles can be extended to other fields. This subsection is divided into formulated products, those that are made of by mixing a number of ingredients and whose properties depend on the composition (i.e. detergents [37] or animal feed [38, 39]), the so-called refinery problem, since biomass consists of a mixture of compounds. This problem is similar to crude-based refineries where a portfolio of products can be produced out of the raw material and finally, due to the link between the ingredients and final products and their geographical availability, the final subsection is devoted to present the facility location / supply chain problem.

**Formulated products: Animal feed**

The problem to address here is also known as blending or pooling problem [37, 40–42]. The final product is obtained from mixing a number of ingredients aiming at achieving the properties desired or expected by the user, ie. A limit of pollutants in the emissions, performance of the blend, as a function of the proportion of the ingredients. Figure 16.3 shows this problem. It is based on the selection of the appropriate ingredient and its amount. The challenge in the formulation comes from the mathematical complexity related to the bilinear terms. These terms appear when computing the composition of the final product, that determines its performance, leading to local solutions and the need to use a global optimization solution approach. In addition, developing models to estimate the performance of the product is a problem in itself, since the relationship between the ingredients and the performance of the product are hard to obtain. Note that economics, environmental and social objectives are to be included in the formulation so that the problem becomes multiobjective. This type of problem is typical in the elaboration of animal feeds to meet certain nutrition requirements. An extension of this problem would include selecting the suppliers of the ingredients and the markets [43, 44] to formulate a supply chain model resulting in a very large nonconvex MINLP that is difficult to solve requiring Taylor-made algorithms.

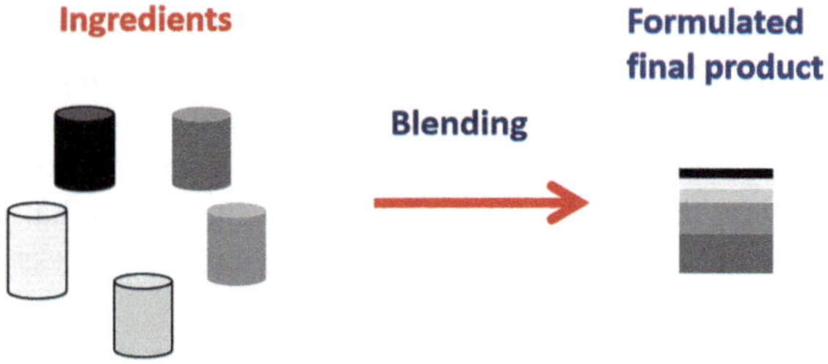


**Fig. 16.3** Blending problem scheme

$$
\begin{aligned}
&obj = \text{Product cost} \\
&\text{st.} \\
&\text{Mass Balances(x)} = 0; \\
&\text{Product performance(x)} = 0; \\
&\text{RawMat Availability(x)} \leq A; \\
&\text{Product Demand(x)} \leq D; \\
&\text{Process constraints(x)} \leq P_i; \\
&\text{Min: Environmental Burden} \\
&y \in \{0, 1\} \\
&x \in R
\end{aligned}
\tag{16.1}
$$

**Refinery problems: Extraction and recovery of chemicals**

Biorefineries are multiproduct facilities that typically process a raw material that is made of different compounds heavily linked, aiming at separating it into its constituents, see Fig. 16.4. Two main processes are to be designed, extraction of species that are as such within the biomass superstructure or breaking down the biomass solid into the components. The first one aims at larger and typically high added value products and milder operating conditions are to be expected [11, 45]. The second is the most common facilities aiming at the production of sugars, bioethanol or syngas-based products [35, 46]. In this second case, the smaller the intermediate, the higher the operating conditions and the larger the number of process units. At some point the intermediates are used to synthesize the final product maintaining, reducing, or increasing the molecular weight of the product compared to that of the intermediate. Therefore, more complex flowsheets are to be expected but more mature technologies can be used. For instance, once syngas is available the typical technologies from the petrochemical industry can be used with no particular difference. The typical formulation for these problems can be that of a superstructure of alternatives, a non-convex MINLP where the binary variables (y) represent the technologies, the equality constrains (h(x,y) = 0) the mass and energy balances as well as the thermodynamics of the system, and the inequality constraints are the bounds of variables and operating conditions (g(x,y) < = 0). If the number of alternatives is limited, decomposition approaches are the most used.

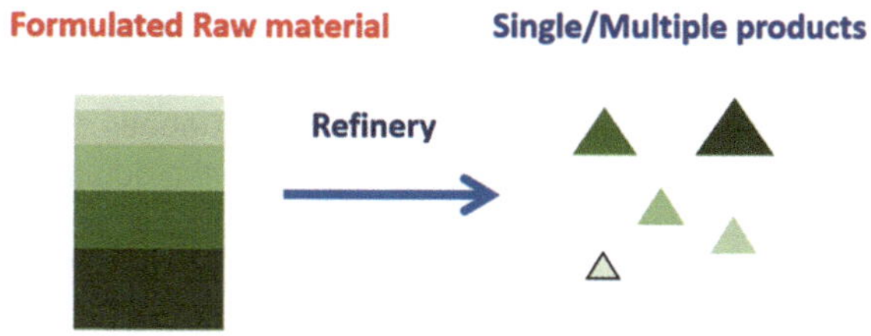


**Fig. 16.4** Refinery problem scheme

$$
\begin{aligned}
&obj = \text{Profit}\\
&\text{st.}\\
&h(x, y) = 0;\\
&g(x, y) \leq 0;\\
&y \in \{0, 1\}\\
&x \in R
\end{aligned}
\tag{16.2}
$$

**Process and product design**

The variability in the composition of the raw material, the biomass, presents an additional problem that has opened a full new line of work, the simultaneous process and product (biomass) design. The transformation of the biomass towards one or several products is linked to the biomass composition. Algae are mostly constituted by lipids, carbohydrates, and protein among other compounds while lignocellulosic biomasses are made of lignin, cellulose, and hemicellulose. Carbohydrates are the raw material for glucose, lipids are the starting point for oils or even alcohols [47], cellulose is also a polymer of mostly glucoses, hemicellulose is a polymer or xylose, maltose, among others [48]. It is possible to use each fraction of the biomass as raw material for a particular final product, i.e. glucose for ethanol, furfural or ibutene production, xylose for ethanol or xylitol production. Beyond these single products it is possible to use a set of intermediates to produce more complex final products. Biodiesel can be obtained from the transesterification of the algae lipids oil and the glucose-based ethanol generating an integrated biorefinery [47] that can be extended by using a fraction of the glucose to produce ibutene and react the glycerol with it towards glycerol ethers increasing the yield to biodiesels substitutes [49, 50]. The mathematical problem for the design of the integrated process, an NLP or MINLP, is extended by considering the raw material composition as a variable.

#### 16.2.4.1 Network Modelling. Facility Location Problems

Once the facility is analyzed at process scale, the study of the exploitation or use of biomass or waste as a resource requires an analysis at a larger scale, the strategic scale. In addition, the collection of the ingredients for a particular product also represents a supply chain kind of study. This strategic scale analysis can be considered as an extended facility location problem [51, 52] that would allow selecting the size and location of the processing facilities as well as the suppliers for the raw materials and the dispatch of the products. The objective in (P3) considers the fixed costs related to the deployment of facilities and the continuous costs due to product shipment. The constraints represent the need to satisfy the demand of each consumer, $d_j$, but only the production capacity of the facilities, $U_i$, can be shipped, $x_{ij}$. It is also possible that budget limitations, emission limitations can be added to the formulation to become multiobjective problems.

$$\min Z = \sum_i f_i y_i + \sum_{i=1}^{n} \sum_{j=1}^{m} c_{ij} x_{ij}$$

s.t.

$$\sum_{i=1}^{n} x_{ij} = d_j; \quad \text{j} = 1,2,\ldots\text{m}$$

$$\sum_{j=1}^{m} x_{ij} - U_i y_i \leq 0; \quad \text{i} = 1,2,\ldots., \text{m}$$

$$U_i : \text{ Maximum capacity}$$

$$y_i = 0, 1; x_{ij} \geq 0 \qquad \text{(P3)}$$

## 16.3 The Food Value Chain

### *16.3.1 The Agricultural Business: Product Design and Circular Economy*

The specialization of the production has created a mismatch between nutrient availability and demand and consumption within livestock and cropping systems that makes these sectors responsible for 24.5% of greenhouse gas emissions. It is possible to reduce the gap by integrating livestock and crop management. Location is a key factor in the sustainable operation of these integrated systems since this variable affects both the economic and environmental dimensions of the design of the farm, see Fig. 16.5.

The idea behind that integration is to simultaneously address a product design problem, the formulation of the feed required by the animals, with a process design one that is devoted to the design of the nutrient recovery process and a network or supply chain design problem to decide on the location of the facilities, and its size so that the system operates with favourable economic and environmental term [38].

Such a large problem requires a two-step solution approach. A pre-screening step allows to discard locations unable to meet environmental constraints. The second step is based on an optimization framework that integrates all three problems, process,

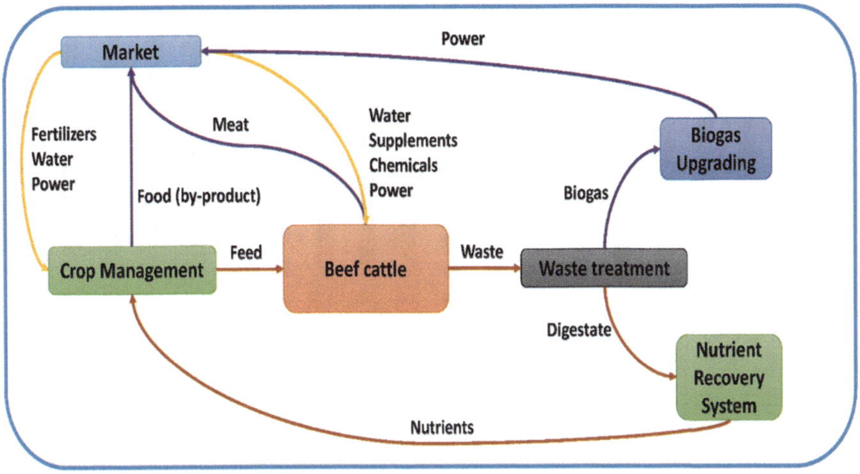


**Fig. 16.5** Circular economy between the agricultural and farm sectors. [38]

product, and supply chain ones. The model formulated integrates empirical models that estimate the nutritional needs of the animals, fertilizer consumption, waste production, as well as the effect of selection of locations and the size of the farm on the objective function. Agricultural exploitation is designed to produce the feed required by the animals reusing the waste from them and for the exploitation itself to produce fertilizers and biogas. The final aim is to implement a circular economy. A multi-objective and multi-period optimization model is formulated.

The results show that the optimal location from economic and environmental points of view is closely related to the size of the farm, finding the best value of 1000 initial animals in the agricultural district of 'Bureba-Ebro'. The profit is 2.78 times higher than the second-best option, 'Campos'. This demonstrates the importance of considering location and farm design simultaneously. For a facility located in Bureba-Ebro, the portfolio of crops necessary to satisfy the nutritional needs of the animals is presented in Fig. 16.6. It can be seen that, except for the first years, the most important crops are Alfalfa and Barley (straw). The variation in the fraction of each of these two crops depended on the number, age, and type of animals predominant each year, forming a total of 5 loops. It is important to point out that the multi-objective scenario represents a 35% reduction of the environmental impact (35%) by selecting crops with a higher yield per hectare (alfalfa and vetch) and using barley grains to animal feed instead of sending them to the market. As a result, the total crop area is reduced, reducing the total nutrient consumption (12%). This solution has its cost, reducing the profit by 4.4% compared to an optimal economic solution. The production cost of meat cost is 8.87 €/kg (1.6 €/kg corresponds to the waste treatment). Another important result is that the most profitable economic activity is crop production (representing between 55 and 65% of total revenues depending on the scenario considered). Therefore, the integrated operation can be seen as a multi-product operation where the main product is the crops with meat being a by-product

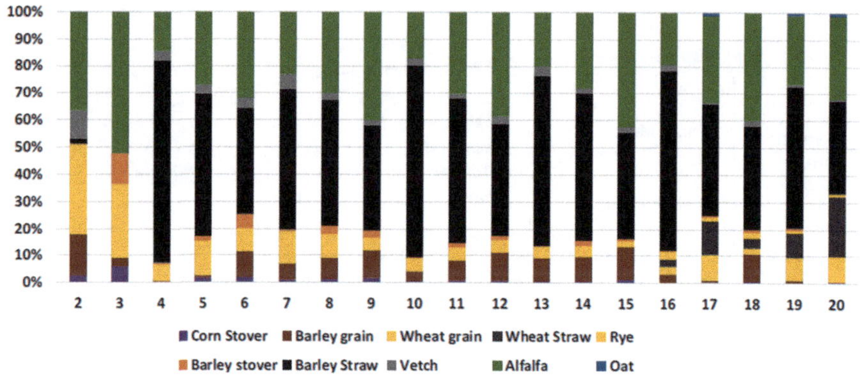


**Fig. 16.6** Feed distribution over time [38]

of the facility. Meat production cost can be reduced to 1.51€/kg by considering the income from crop sales. In addition, power production through biogas does not represent an important source of income and the most important costs are those associated with crops (34.10%), fertilizer (29.11%), and waste treatment (18.13%) in the multi-objective scenario.

## *16.3.2 Waste Valorisation. Product Portfolio and Process Synthesis*

The second type of example is the valorisation of waste to produce added value products as well as others to provide additional credit for the biorefinery. The problem represented in this section consists of the definition of the portfolio of promising products to obtain and, at a second stage, the definition of the technologies and flowsheet as well as the final set of products [11]. This is biomass specific. In certain cases, the portfolio of products can be identified straightforwardly, while in some other cases it is not obvious, and the systematic techniques presented in this chapter are key to unveil the process structure and the portfolio of products. Several of the most common products are studies such as the olive oil industry, the juice industry using the case of the oranges, the coffee, and the wine industries.

**Integrated facility for the processing of oranges** [53]. The process can be divided into five stages. First the fruit is processed to extract the juice using a screw. The juice is to be pasteurized before packing requiring thermal energy. The peels are the raw material for the rest of the process and will allow the process integration and the creation of a circular process. Limonene is extracted from the peels. This chemical is already present as such. Two extractive technologies are considered including the use of hexane or steam explosion. Next the limonene is purified. In the case of using hexane, it is put into contact with the milled peels, centrifuged and the hexane is recovered in a stripping column. The hexane is condensed and recycled.

The disadvantage is the fact that the limonene produced in this way cannot be used within the food industry. The steam explosion pretreatment consists of feeding steam to a batch tank where the milled biomass is put into contact with steam. After a few minutes, the pressure is released and the limonene together with water are evaporated. The rest of the water and the residual biomass are sent to anaerobic digestion. No need for additional heating is required. From the digester onwards, the facility is a power plant. The advantage of removing the limonene is that the digestion progresses better since limonene typically acts as inhibitor. At the digestor, biogas and fertilizer are produced. The biogas contains small quantities of $H_2S$ and $NH_3$ that poison the catalysis and must be removed. After biogas clean-up, two alternatives are considered, either the use of a combined cycle, where first a Brayton cycle is used to produce electricity from the biogas and, in a second stage, the hot flue gas is used to generate steam to feed the steam turbine and generate electricity. This alternative reduces water cooling needs per kW produced. A second alternative consists of the use of the hot flue gas to produce thermal energy to help self-sustain the operation of the facility. To analyse all the alternatives, a process systems approach is used. The units are modelled one by one based on first principles. The flowsheet model has been formulated as an MINLP that is solved as four different NLP's, one per topology configuration and an economic evaluation of the alternatives is performed, see Fig. 16.7.

The use of steam explosion requires 50% lower production costs but 30% higher investment to treat the waste. The use of hexane consumes 4 times more thermal energy, but the processing of the waste requires less digesters. Combined cycles are less profitable. The lower production cost of the steam explosion allows for smaller facilities to become profitable before the ones using hexane limonene recovery. By including the waste processing into the juice industry, the facility does not only remain profitable, due to the high price of limonene, but also the circular economy concept reduces the residue and the external energy consumption providing additional byproducts of added value such as fertilizers.

**Olive oil waste biorefinery** [11]: When the biomass contains a large number of promising chemicals the design of the process is more complex. Therefore, a multi-stage multilayer systematic procedure is used to select the optimal product portfolio. It consists of a hybrid heuristic, metric based, and optimization methodology that

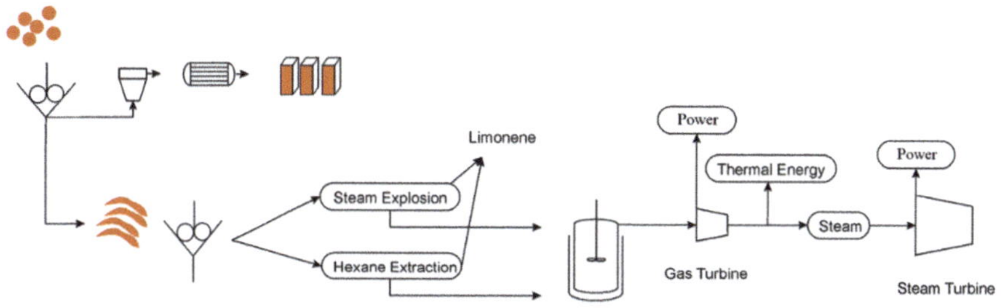


**Fig. 16.7** Superstructure of alternatives for the production of orange juice and its waste [53]

evaluates the economic and environmental performance of added value products from a particular raw material. The first stage preselects the promising products. Next, a superstructure optimization problem is formulated to valorize or transform waste into the optimal set of products. The methodology has been applied to evaluate the best use of the biomass residue from the olive oil industry towards food, chemicals, and energy. The heuristic stage is based on literature review to analyze the feasible products and technologies. Next, simple metrics have been developed and used to preselect products that are promising. Finally, a superstructure optimization approach is used to design the facility that processes leaves, wood chips and olives into final products, see Fig. 16.8 [11]

The results show that the optimal portfolio of products corresponds to the recovery of phenols, hydroxytyrosol and tyrosol from alperujo and oleuropein from olive leaves and branches, based on the economic indicator. After the selection of the products to obtain, a superstructure is formulated and optimized. As a result of the optimization, pre-treatment and membrane technology should be used from alperujo to recover of a mix of phenols while in the case of olive leaves and branches, a dryer followed by a mill, a pressurized liquid extractor, filter under pressure and resin adsorption are used. The investment required to process waste adds up to 110.2 million € for a 100 kt/yr of olives production facility while the profit depends on the level of integration. If the facility is attached to an olive oil production installation, the generated profit ranges between 14.5 MM €/yr (when the waste is purchased at prices of 249 € per ton of alperujo and 6 € per ton of olive leaves and branches)

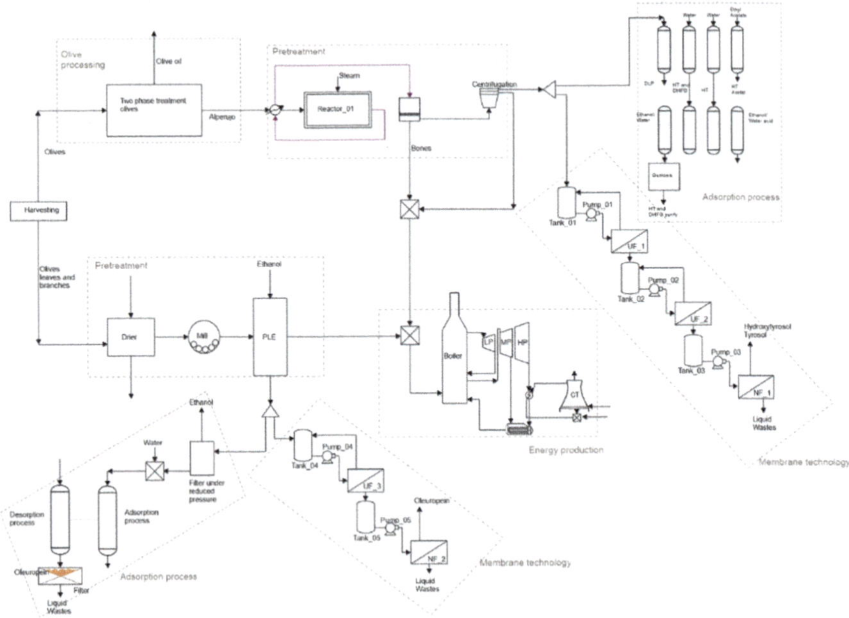


**Fig. 16.8** Superstructure of alternatives for valorising waste from the olive oil industry [11]

and 34.3 MM €/yr, when the waste material is acquired for free. Regarding energy, solid waste is used to generate the utilities employed in the recovery process. A power plant is integrated into the facility whose power is around 12.95 MW, which is sufficient for the recovery process.

**Spent coffee ground biorefinery** [45]. Among the products of the food industry, one stands out above the rest, coffee. Coffee is the second most important consumer product after oil, with a production of 10.16 billion kilograms between 2018 and 2019. Its production is mainly concentrated in countries such as Brazil, Vietnam, Colombia, Indonesia, Ethiopia, India, and Honduras, with Brazil being the largest coffee exporter in the world. In the production of coffee, a large amount of waste is generated, especially the spent coffee grounds (SCG) and the coffee silver skins (CSS). In the coffee industry, 650 kg of SCG per ton of green coffee beans and 2 kg per kilogram of soluble coffee produced are generated. If this type of waste ends up in landfills, it causes a series of environmental problems such as soil contamination, due to the presence of toxic substances such as caffeine or other polyphenols, the production of greenhouse gases such as $CH_4$ and $CO_2$, due to the decomposition of organic matter, and the release of large amounts of $CO_2$ in incineration processes. Alternatively, SCG can be used to produce a wide variety of high-added-value products due to its composition. Due to the decentralized production of the spent coffee grounds and its high-water content, it is difficult to process the residues for restaurants and small coffee shops and the work focused on processing the residues from the production of soluble coffee, and a fraction of the remaining energy of the recovery process is used to drive the main production process. Five products, dry natural extract, natural pigment for the textile industry, biogas, digestate, and electrical energy, and three processes are considered. On the one hand, the spent coffee ground extract shows interesting values of phenolic compounds such as caffeine and chlorogenic acids that show antioxidant and antitumor activity. Furthermore, caffeine is related to the decrease in depression, fatty liver, and other diseases. On the other hand, the use of natural pigments in the textile industry can increase the safety of the dyeing process due to the low toxicity of this pigment and the increased sustainability in terms of chemistry and energy consumption. Most of the pigments and natural extracts are obtained from vegetable or animal sources requiring cultivation or harvesting of the natural environment in which they are produced, causing a negative environmental impact. The natural extracts and pigments obtained from waste do not only reduce the environmental impact of the process in which they are generated but also represent a valorisation of the waste promoting circular economy. The digestate obtained from the SCG can be used as a natural fertilizer and substitute part of the mineral fertilizer used for the production of coffee beans. All of the products that can be obtained from the SCG do not compete for part of the market but rather replace part of the current products with others with a more sustainable origin and favour the circular economy of a process with high environmental impact. Three main processes and two subprocesses derived from Process 1 (Process A1 and Process A2) are considered for the valorisation of the SCG, see Fig. 16.8. The modelling of the processes is carried out using mass and energy balances, phase equilibria, experimental yields, and rules of thumb to describe the yield and performance of each one of the units. Process 1 consists of an extraction and filtration system for the production of a natural coffee extract of high added value. This process generates two residues that are valorized through anaerobic digestion (Process A1) to produce biogas and digestate and a filtration and drying process (Process A2) to produce natural pigment. Process A1 and Process 2 use the same technology, but the difference is the raw material. SCG is used as a raw material for Process 2, while Process A1 uses the residue from the decanter of Process 1. Finally, Process 3 uses the SCG to produce electrical energy using a gasifier and a combined cycle. In the design of the superstructure, the integration of energy and water is considered. The energy required for the anaerobic digestion, filtration, and drying processes is generated within the facility through an auxiliary process. Mathematical optimization techniques are used for the selection of the products and the technologies from SPG. See Fig. 16.9.

It turns out that two processes are economically viable, the production of energy, Process 3, and the production of natural extract and pigment, Process 1. Nevertheless, Process 1 shows a profit 40 times higher than Process 3 due to the high sale price of the natural pigment and of the natural extract, and therefore, this process is chosen for the valorization of the SCG. Between these two products, the income from the sale of natural pigments is 3.9 times higher than the natural extract, which makes natural pigment the most interesting product. The annual profit of Process 1 is 65 MM€/year, while operating costs are 4.59MM €/year. Regarding investment costs, 13.97 MM€ are necessary to start up the new production line based on Process 1. The digesters are the most expensive equipment (48% of the total equipment cost); nevertheless,

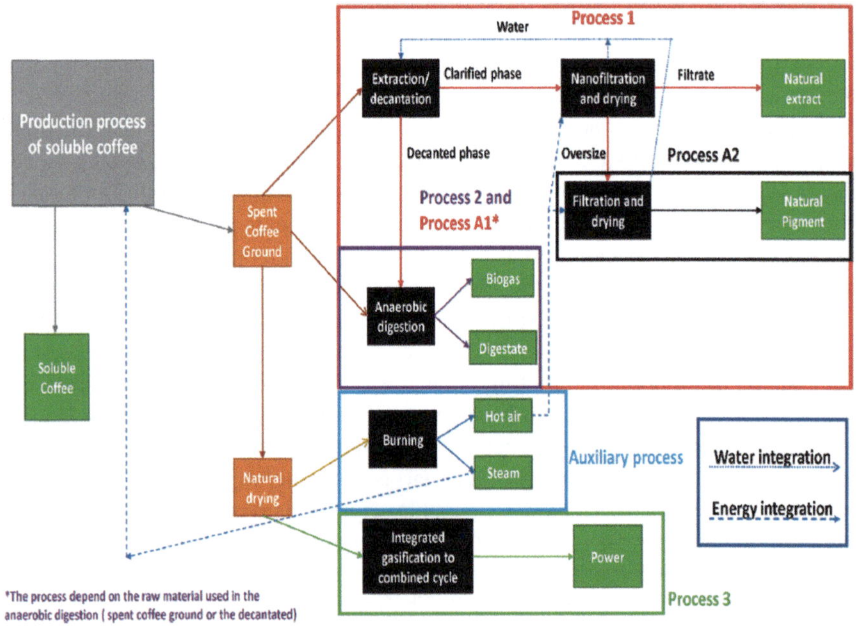


**Fig. 16.9** Alternative processes from spent coffee ground Adapted from [45]

they are necessary for the treatment of the waste produced in the decantation process. The treatment of these wastes was maintained to comply with the treatment of all wastes generated since the benefits of the sale of digestate and biogas (0.77 MM€/year and 0.15 MM€/year, respectively) are negligible compared to other products. Up to 7.5% of the energy required by the production of soluble coffee can be obtained from the waste processing section [45].

**Wine waste valorization process** [54]. Wine industry is one of the most profitable ones. Bottles of the finest wine can reach thousands of euros. During the wine production process, a series of wastes are generated, among which grape pomace, grape stalks, and wine lees. The great variety of high-value-added products that can be obtained from these residues together with other types of products (biofuels, energy) has led to a large number of techno-economic studies for the different treatment processes available. However, to the best knowledge of the authors, there is no study in the literature that considers all these processes simultaneously and analyses the best one considering different objectives and different sizes of the vineyards.

Therefore, up to 8 different processes (extraction-purification system, anaerobic digestion, composting, fermentation, combustion, gasification, pyrolysis, and adsorption) are proposed in this work to obtain up to 8 products (grape seed oil, polyphenols, tannins, tartaric acid, biogas, fertilizer, biochar, and power). Each of these processes is modelled through first principles (material balances, empirical yields, thermodynamic equilibria, among others) and integrated into a superstructure that considers all processes simultaneously and selects the best option from different points of view (economic, environmental, and social). Integration of energy and other resources (such as water or other raw materials) between the different processes is considered. Solutions are stratified to address the different cases of study depending on the size of the winery and its waste treatment budget. The processes are modelled, through material balances, thermodynamic equilibria, empirical correlations and performances. Analysing the largest wineries in California, their production ranges from 2 million cases (9-L boxes) to 53 million cases (Downey, 2022). Therefore, these industries can generate between 18 and 477 million liters of wine per year. This is equivalent to a grape pomace production between 0.1 and 2.5 kg/s. Thus, 0.1 kg/s, 1 kg/s and 10 kg/s are evaluated for each of the alternative processes.

For small (less than 0.1 kg/s) or very large (greater than 10 kg/s) capacities, the production of high value-added products outperforms the other processes in all three impacts and in profitability. For intermediate capacities, combustion and gasification have a low economic impact and a high environmental impact, anaerobic digestion stands out as having the lowest environmental impact, and tannin production is the best-balanced process, with a high economic and social impact. Pyrolysis is the worst process in all three impacts. Therefore, it is very important to establish the available capital to invest, the production capacity of the wine industry and the weight of the different impacts to decide the best process to treat this solid waste. ROR on investment of each process is considered as a metric for the profitability of the processes. The results show that, for the highest production case, that is, 10 kg/s, there is one process that is much more profitable than the rest, the integrated multiproduct system (IMPS). On the contrary, in the case of 0.1 kg/s there are several processes that are not profitable (combustion, pyrolysis and tannin production). In this scenario, the only promising process is the IMPS. Therefore, for both capacity (0.1 kg/s and 10 kg/s), a CAPEX of group 1 is the most recommended. However, in the intermediate capacity (1 kg/s) there are several processes with very similar ROR. On the one hand, IMPS has a ROR identical to tannin extraction, but with a much higher CAPEX (5.15 times, see Fig. 16.4). On the other hand, anaerobic digestion has a ROR very similar to combustion, but with a much lower environmental impact (11 times). In this case, a CAPEX of group 3 could be much more suitable than one of group 1.

### *16.3.3 Integrated Processes: Polymers and Surfactants*

**Integrated plastic production facility** [55]. The use of plastics and polymers has improved society´s lifestyle. Their properties have provided advantages in many fields from the automobile industry to electronics that, together with a lower unit cost and improvements in their performance specifications, results in an increasing use of plastics. Over one third of the total consumption of plastics is employed in packing applications making recycling a need. However, a secondary effect of employing large amounts of this material is the impact of microplastics in water bodies (e.g. oceans, seas) since it compromises the health and integrity of ecosystems. For this reason, a substitution of the conventional polymers by biodegradable ones can provide an immediate positive impact.

In order to achieve this goal, an integrated process is conceptually designed for the production of high value products as biopolymers from renewable and waste sources as sawdust, manure and $CO_2$, obtaining the intermediates required (e.g. methanol, glycerol, starch) for the synthesis of the biopolymer, a polyol (plasticising agent), as well as useful by-products such as biodiesel, promoting a circular economy. The process consists of six sections, see Fig. 16.10, including biogas synthesis from manure wastes, syngas reforming and clean-up, methanol synthesis, algae growing to produce oil and starch, biodiesel synthesis and biopolymer production from the by-product glycerol, sawdust, and the algae starch. Organic waste, manure or sludge, is digested anaerobically for the production of biogas and digestate. Two lines are operated in parallel. The digestate is used as a source of nutrients, phosphorous, nitrogen and phosphorus together with $CO_2$, for algae growth. The source of $CO_2$ can be two-fold, the one captured at the methanol synthesis section, and an additional stream. The algae, when reached the appropriate size, are harvested and the oil and starch are extracted. The biogas obtained from anaerobic digestion is reformed (dry and steam) to obtain syngas. The syngas is to be used for methanol production for the transesterification of the oil. Therefore, the proper $H_2$ to CO ratio is to be achieved. The methanol is used for glycerol production, required at the polymer synthesis section. However, it is the by-product of biodiesel, obtained as major product in this section. This section has produced glycerol that is to be mixed with the lignocellulosic waste, sawdust. Before using it, it needs to be dried. Next, a fraction is used as raw

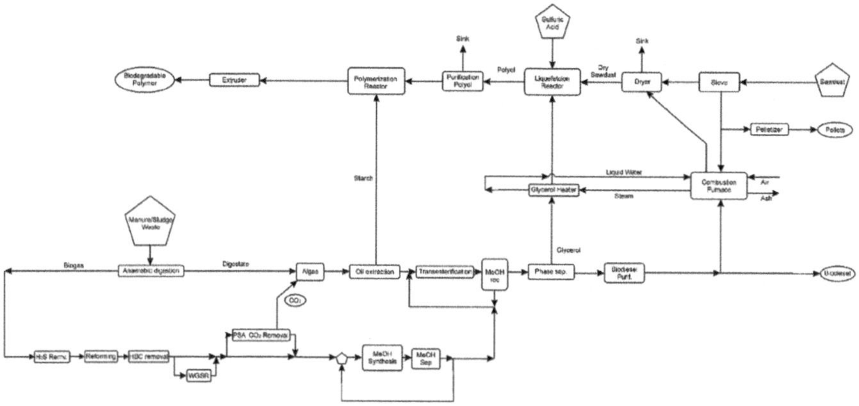


**Fig. 16.10** Integrated facility to produce renewable polymers. Adapted from [55]

material and the rest may be used as energy source for the process and added to the liquefaction reactor together glycerol as solvent and sulfuric acid as catalyst to produce polyol. The polyol obtained is purified by mechanical centrifugation. The biodegradable polymer is obtained by mixing the polyol with the polymeric matrix of starch. Then, the biopolymer is processed in an extruder to obtain pellets, used as plastic-based pieces or to produce agriculture plastic films.

To decide on the optimal integration of the sections, a process model is formulated based on mass and energy balances, including the thermodynamic equilibria involved in the biogas reforming and methanol synthesis together with surrogate models for the transesterification to biodiesel and polyol synthesis. The problem becomes an NLP problem with 5680 equations and 7870 variables solved in GAMS®. Heat integration and a detailed cost estimation are performed next.

Optimal solution suggests full use of the sawdust in the production of biodegrable polymers. In this way, the facility processes 20 kt/yr of sawdust and 4723 kt/yr of waste (50% cattle manure and 50% pig manure). The plant produces 354 kt/yr of biodegradable polymer, 85 Mgal/yr of FAME and 35 kt/yr of methanol, a production size according to algae-based biodiesel and sawdust-based plastics production plant. Once the heat integration is perfomed and the heat exchanger network is designed, 0.0035 kg of steam per kg of biopolymer produced are required where a 27% of the total biogas produced is employed to satisfy the energy requirements of the reformer. Regarding the economic assessment, a total investment cost of 1039 M$ are required with a biopolymer cost of 0.98 $/kg.

**Integrated biosurfactant production facility** [24]. Surfactants are specialty chemicals whose purpose is to wet, emulsify, help solubilize or soften species. They are used in laundry as well as in other industries such as gas recovery from fracture and, so far, they have been typically produced from crude oil. These uses result in surfactants ending at waterbodies. Even with proper wastewater treatment plants, it is helpful is the surfactants are made from renewable resources and they are biodegradable. There are already synthetic paths for sustainable biosurfactants

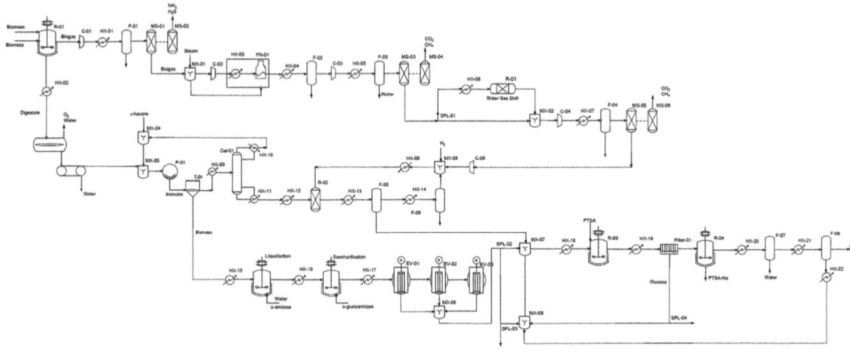


**Fig. 16.11** Integrated facilyt for the production of sustainable surfactants. Adepated from [24]

(Kumar et al. 2022; Reis et al. 2013) but they produce the entire molecule at once. Surfactants consist of two sections, a hydrophilic head group and a hydrophobic alkyl chain. Therefore, it is possible to generate both fractions from biomass to build alkyl polyglucosides, APG's, that can be produced from manure and $CO_2$.

An integrated facility is designed for the production of sustainable surfactants from waste, including $CO_2$ and manure via algae growing and biogas production. From the algae, intermediates such as glucose and lipids are obtained. From biogas, hydrogen is produced. Next, alcohols are produced via lipids hydrogenation so as to synthesize the APG's from them and the glucose, see Fig. 16.11. Following this structure, a circular economy around the production of the APG's is developed. The design problem is formulated as a large NLP for the selection of processing route of the intermediates and final product as well as the operating conditions. Such formulation allows for the selection of the algae growing and its composition for the limited use of additional utilities and chemicals. An economic analysis and process scale down has also been performed.

The algae composition recommended consists of 60% lipids and 27% carbohydrates. The yield of the facility reaches 0.47kg/kg algae (0.08 kg/kg manure). For the production of 252 kt/yr of APG, the facility consumes 17.7 MW of thermal energy and steam and 7.9 MW of electricity, absorbing 788 kt/yr of $CO_2$ after discounting the emissions related to thermal and power consumption. The investment adds up to 196 M€ for a production cost of $0.17€/kg_{APG}$. The large amount of manure to be processed, corresponding to 300 k cows, suggests a scale down study to evaluate the effect on the investment and production costs. Current market price of APG can be achieved processing the waste of over 2.5k animals.

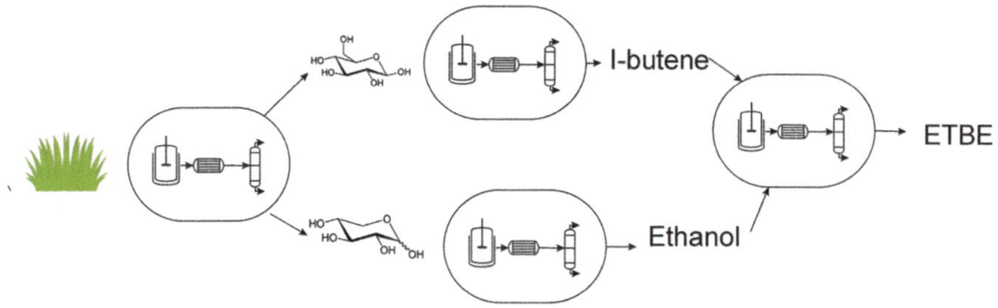


**Fig. 16.12** Conceptual process for ETBE production [48]

### *16.3.4 Process and Product Design*

Lignocellulosic biomass is a complex raw material containing a range of chemicals. The composition of the cellulose, hemicellulose and lignin provides the source for the simultaneous production of a portfolio of products to make the most of the biomass.

The production of renewable ETBE from ethanol and i-butene from switchgrass can be viewed as a process and product design [48]. A superstructure embedding a number of alternatives is proposed. Two technologies are considered for switchgrass pretreatment, dilute acid and ammonia fibre explosion (AFEX), so that the structure of the grass is broken down. The experimental results available only prove that glucose can be fermented to produce i-butene, while xylose is used to produce ethanol as second-generation bioethanol. To reduce the energy consumption in the dehydration of ethanol, a multieffect column with 3 effects is implemented, as part of the heat integration efforts, followed by molecular sieves and a PSA-membrane system is used to recover the i-butene, see Fig. 16.12. The problem is formulated as an MINLP and solved by decomposing it into four NLP's, one per pretreatment. After process design, an economic evaluation is performed to assess the investment and production costs. Biomass composition determines the by-product obtained.

Dilute acid is selected as pretreatment because of the largest yield to sugars and the possibility of adjusting the production of both, i-butene and ethanol for the needs to obtain ETBE. For a facility that processes switchgrass with a production capacity of 90 kt/yr of ETBE, the investment adds up to 160 M€ for a production cost of 0.61 €/kg. By processing the same amount of an engineered biomass, with 40% cellulose, 21.9% hemicellulose instead of the ratio of switchgrass 32/25, it is possible to increase the production capacity by almost 16%, reducing the production cost by that much too. The cost estimation results in an investment of 165 M€ to produce 112 kt/y and a production cost of 0.51 €/kg. It represents a reduction of 16% in the production cost when using the generic biomass presented in the base case.

Beyond the optimization of the biomass as presented in case previous case, it is possible to select the proper biomass for the task. The example consists in the design of a facility for the simultaneous production of xylitol and sorbitol from lignocellulosic biomass, that is the source for xylose, and glucose from the hemicellulosic and cellulosic fractions respectively. Xylitol can be obtained from xylose via fermentation or catalytic. Sorbitol is obtained from glucose, but preferable from fructose, also via fermentation or catalytic hydrogenation. Fructose can be obtained from glucose via isomerization. The fermentation of xylose takes place at mild temperature with conversions of 85–92% while the catalytic synthesis can achieve 100% operating up to 120 °C and 60 bar. The yield from glucose in case of the catalytic path is far lower.

A superstructure of alternatives, see Fig. 16.13, is formulated to process different lignocellulosic biomasses (i.e. corn stover, miscanthus, other crop residues). Various pretreatments, such as dilute acid, AFEX, organosolv. etc. for the fractionation of the biomass are considered. Next, after hydrolysis, the C5 and C6 sugars are processed separately, for which either a catalytic stage or a fermentation one is considered. Glucose is to be isomerized before it can be processed, achieving 56% of isomerization, with a glucose use lower than the catalytic process. Finally, the purification of any of the products is carried by crystallization in a multi-state evaporator system towards energy integration within the process flowsheet. Surrogate models based on experimental data for the different stages of the process including the pre-treatment and the conversion and first principles models for crystallization are developed to formulate the superstructure MINLP optimization model [56]

The optimal results suggest the use of dilute acid and the catalytic synthesis, and a system of three crystallizers is selected. The fermentation-based processes require additional investment due to the need for isomerisation of the glucose and the high residence times in both the isomerisation and the fermentation tanks, that doubles the one when catalytic hydrogenation is considered. In addition, the larger consumption of steam in the case of fermentation also increases the production costs which double the ones using catalytic synthesis. This optimal flowsheet is used to evaluate the performance of different raw materials into the product portfolio as a function of the biomass composition as well as the design of the optimal biomass. The production cost is optimized among 8 different biomasses, agricultural residues such as wheat straw, sugar bagasse or corn stover, energy crops such as switchgrass or hybrid poplar and forest residues (Birch, pone and spruce). Out of them, corn

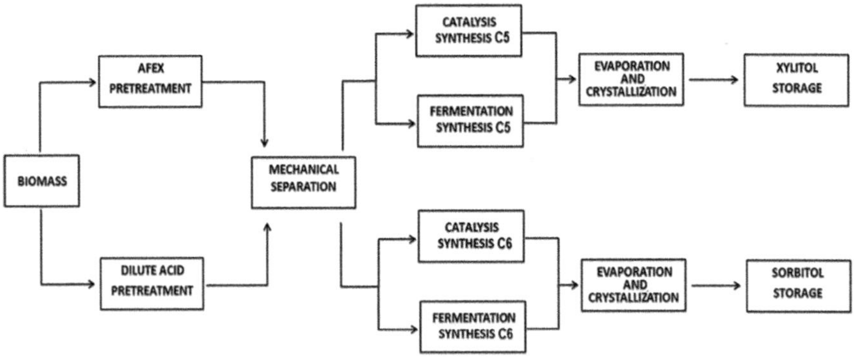


**Fig. 16.13** Structure of the alternative pathways to xylitol and sorbitol. Adapted from [56]

stover is selected with a production cost of 18.7 M€/yr, and investment of 111.7 M€ and a production cost of 0.22€/kg.

However, the market prices for xylitol and sorbitol are not the same. The inverse engineering of biomass was also performed resulting in a composition of 15% water, 20% cellulose, 40% hemicellulose, 15% lignin and 5% ash. The closest biomass corresponds to *Sargassum* (brown algae), that is capable of producing 230.5 kt/yr of xylitol and 116 kt/yr of sorbitol with investment and production costs of 120.5 M€ and of 0.25 €/kg of products, respectively.

## 16.4 Conclusions

The food industry represents an opportunity for waste valorisation and to improve the efficiency of its operation. The residues generated represent a challenge in terms of the amount but also the composition that can even be dangerous for the environment. This chapter presents the methods and approaches that allow the development of a circular economy around the food industry at different levels, from the process itself, reducing the use of additional inputs and maximizing the raw materials, to the integration of different businesses to reduce the need for additional natural resources, ie. Fertilizers. The techniques are based on process system engineering that provides the tools for the systematic analysis of technologies, products, and operating conditions. Several problems are formulated, from process to process and product design but also strategic ones at the level of supply chain. Process level studies show how circularity results in higher investment costs to make use of the residues, and therefore the strategy is first a waste management approach creating additional byproducts that valorise waste (i.e. limonene, phenols) and reduce energy consumption (heat or power integration) in the integrated facility or provide credit. It can be seen how by simultaneously considering various objectives as well as integrating technologies and business more efficient solutions are found since better use of the resources can be achieved. Strategic decision tools provide guidelines for biomass valorization

**Acknowledgements** The authors acknowledge the funding received from the European Union's Horizon 2020 research and innovation programme under the Marie Sklodowska-Curie grant agreement No 778168.

## References

1. FBI (2025) Food service market size, share & industry analysis
2. Yeomans H, Grossmann IE (1999) A systematic modeling framework of superstructure optimization in process synthesis. Comput Chem Eng 23:709–731
3. Mencarelli L, Chen Q, Pagot A, Grossmann IE (2020) A review on superstructure optimization approaches in process system engineering. Comput Chem Eng 136:106808

4. Martín M, Grossmann IE (2012) BIOptPACK: a library of models for optimization of biofuel production processes. Comput Aid Chem Eng 30:16–20 (Elsevier, Oxford. UK)
5. Taifouris M, Martín M (2023) Towards energy security by promoting circular economy: a holistic approach. Appl Energy 333:120544
6. Douglas JM (1988) Conceptual design of chemical processes. McGraw-Hill, New York
7. Biegler L, Grossmann IE, Westerberg A (1997) Systematic methods of Chemical process design. Prentice Hall, New York
8. Smith R (2005) Chemical process: design and integration. Wiley, New York
9. Foo D (2016) Process integration for resource conservation. CRC Press, Boca ratón
10. El-Halwagi MM (2017) Sustainable design through process integration: fundamentals and applications to industrial pollution prevention, resource conservation, and profitability enhancement, 2nd edn. IChemE/Elsevier
11. Guerras L, Sengupta D, Martín M, El-Halwagi M (2021) Multi-layer approach for product portfolio optimization: waste to added value products. Acs Sust Chem Eng 9(18):6410–6426
12. Kirchherr J, Denise R, Marko H (2017) Conceptualizing the circular economy: an analysis of 114 definitions. Resour Conserv Recycl 127:221–232
13. El-Halwagi MM, Manousiouthakis V (1989) Synthesis of mass exchange networks. AIChE J 35(8):1233–1244
14. El-Halwagi MM, Manousiouthakis V (1990) Simultaneous synthesis of mass exchange and regeneration networks. AIChE J 36(8):1209–1219
15. El-Halwagi M (2023) Handbook of process integration (PI): minimisation of energy and water use, waste and emissions, 2nd edn
16. Ponce-Ortega JM, Hortua AC, El-Halwagi MM, Jiménez-Gutiérrez A (2009) A property-based optimization of direct-recycle networks and wastewater treatment processes. AIChE J 55(9):2329–2344
17. Gabriel F, El-Halwagi MM (2005) Simultaneous synthesis of waste interception and material reuse networks: problem reformulation for global optimization. Environ Prog 24(2):171–180
18. El-Halwagi MM, Harell D, Spriggs HD (2009) Targeting cogeneration and waste utilization through process integration. Appl Energy 86(6):880–887
19. Foo DCY, Kazantzi V, El-Halwagi ZMM, Manan ZA (2006) Surplus diagram and cascade analysis techniques for targeting property-based material reuse network. Chem Eng Sci 61:2626–2642
20. Almutlaq A, Kazantzi V, El-Halwagi MM (2005) An algebraic approach to targeting waste discharge and impure-fresh usage via material recycle/reuse networks. J Clean Technol Env Policy 7(4):294–305
21. Nápoles-Rivera F, Rojas-Torres MG, Ponce-Ortega JM, Serna-González M, El-Halwagi MM (2015) Optimal design of macroscopic water networks under parametric uncertainty. J Clean Prod 88:172–184
22. Rubio-Castro E, Ponce Ortega JM, Serna-González M, Jiménez-Gutiérrez A, El-Halwagi MM (2011) A global optimal formulation for water integration in eco-industrial parks considering multiple pollutants. Comput Chem Eng 35(8):1558–1574
23. Ahmetović E, Grossmann IE (2011) Global superstructure optimization for the design of integrated process water networks. AIChE J 57(2):434–457
24. González-Núñez S, Martín M, Amador C (2024) Optimal integrated plant for renewable surfactants production from manure and CO. Biochem Eng J 202:109148
25. Pham V, El-Halwagi MM (2012) Process synthesis and optimization of biorefinery configurations. AIChE J 58(4):1212–1221
26. Topolski K, Lira-Barragán LF, Panu M, Ponce-Ortega JM, El-Halwagi MM (2019) Integrating mass and energy through the anchor-tenant approach for the synthesis of carbon-hydrogen-oxygen symbiosis networks. Ind Eng Chem Res 58(36):16761–16776
27. Al-Fadhli FM, Baaqeel H, El-Halwagi MM (2019) Modular design of carbon-hydrogen-oxygen symbiosis networks over a time horizon with limited natural resources. Chem Eng Process Process Intensification 141:107535

28. Panu M, Topolski K, Abrash S, El-Halwagi MM (2019) $CO_2$ footprint reduction via the optimal design of carbon-hydrogen-oxygen symbiosis networks (CHOSYNs). Chem Eng Sci 203:1–11
29. Noureldin MMB, El-Halwagi MM (2015) Synthesis of C-H-O symbiosis networks. AIChE J 64(4):1242–1262
30. Floudas CA, Ciric AR, Grossmann IE (1986) Automatic synthesis of optimum heat exchanger network configurations. AIChE J 32(2):276–290
31. Yee T, Grossmann IE (1990) Simultaneous optimization models for heat integration—II. Heat exchanger network synthesis. Comput Chem Eng 14(10):1165–1184
32. Fu C, Vikse M, Gundersen T (2018) Work and heat integration: an emerging research area. Energy 158:796–806
33. Al-Azri N, Al-Thubaiti M, El-Halwagi MM (2009) An algorithmic approach to the optimization of process cogeneration. J Clean Tech Env Policy 11(3):329–338
34. Martín M, Ahmetovic E, Grossmann IE (2011) Optimization of water consumption in second generation bio-ethanol plants. Ind Eng Chem Res 50(7):3705–3721
35. Martín M, Grossmann IE (2012) Energy optimization of lignocellulosic bioethanol production via hydrolysis of switchgrass. AIChE J 58(5):1538–1549
36. Baliban RC, Elia JA, Floudas CA (2012) Simultaneous process synthesis, heat, power, and water integration of thermochemical hybrid biomass, coal, and natural gas facilities. Comput Chem Eng 37:297–327
37. Martín M, Martínez M (2013) A methodology for simultaneous process and product design in the consumer products industry: the case study of the laundry business. Chem Eng Res Des 91:795–809
38. Taifouris M, Martín M (2021) Towards a circular economy approach for integrated intensive livestock and cropping systems. Acs Sust Chem Eng 9(40):13471–13479
39. Taifouris M, Martín M (2022) Integrating intensive livestock and cropping systems: sustainable design and location. Agri Syst 203:103517
40. Shih JS, Frey HC (1995) Coal blending optimization under uncertainty. Eur J Oper Res 83, 3(22):452–465
41. Conte E, Gani R, Ng KM (2011) Design of formulated products: a systematic methodology. AIChE J 57:2431–2449
42. Pham V, Laird C, El-Halwagi MM (2009) Convex hull discretization approach to the global optimization of pooling problems. Ind Eng Chem Res 48(4):1973–1979
43. Taifouris M, Martín M, Martinez A, Esquejo N (2020) On the effect of the selection of suppliers on the design of formulated products. Comp Chem Eng 141:106980
44. Taifouris M, Martín M, Martinez A, Esquejo N (2021) Simultaneous optimization of the design of the product, process, and supply chain for formulated product. Comp Chem Eng 152:107384
45. Taifouris M, Corazza M, Martín M (2021) Integrated design of biorefineries based on spent coffee ground. Ind Eng Chem Res 60(1):494–506
46. Martín M, Grossmann IE (2011) Energy optimization of lignocellulosic bioethanol production via gasification. AIChE J 57(12):3408–3428
47. Martín M, Grossmann IE (2013) Optimal engineered algae composition for the integrated simultaneous production of bioethanol and biodiesel. AIChE J 59(8):2872–2883
48. Galán G, Martín M, Grossmann IE (2019) Integrated renewable production of ETBE from Switchgrass. ACS Sust Chem Eng 7(9):8943–8953
49. Martín M, Grossmann IE (2014) Simultaneous dynamic optimization and heat integration for the co-production of diesel substitutes: biodiesel (FAME & FAEE) and glycerol ethers from algae oil. Ind Eng Chem Res 53:11371–11383
50. De la Cruz V, Hernández S, Martín M, Grossmann IE (2014) Integrated synthesis of biodiesel, bioethanol, Ibutene and glycerol ethers from algae. Ind Eng Chem Res 53(37):14397–14407
51. Grossmann IE (2021) Advanced optimization for process systems engineering. Cambridge Series in Chemical Engineering, UK
52. Babinski ML (1965) Integer programming: methods, uses, computations. Manage Sci 12:253–313

53. Criado A, Martín M (2020) Integrated multiproduct facility for the production of chemicals, food and utilities from oranges. Ind Eng Chem Res 59(16):7722–7773
54. Taifouris M, El-Halwagi MM, Martín M (2023) Evaluation of the economic, environmental and social impact of the valorization of grape pomace from the wine industry. ACS Sust Chem Eng 11(37):13718–13728
55. Roldán San Antonio JE, Martin M (2023) Optimal integrated plant for biodegradable polymers production. ACS Sust Chem Eng 11(6):2172–2185
56. Galán G, Martín M, Grossmann IE (2021) Integrated renewable production of sorbitol and Xylitol from Switchgrass. Ind Eng Chem Res 60(15):5558–5573

# Chapter 17
# The Role of Process Integration in the Nexus Energy-Water

**Miguel Castro Oliveira and Henrique A. Matos**

**Abstract** Water and energy are two categories of resources which are fundamental for all human activities in general. In the context of end-use sectors, these are in general present in all sites (for instance, in the case of process industry, all plants possess a determinate number of water-using and energy-using processes). The research on all potential interdependencies possible to be established between water and energy resources, as encompassed by the concept of water-energy nexus, is relevant for the understanding of the improvement of the use of both these resources. A commonly implemented concept for the assessment and performance of the improvement of the use of several categories of resources in a determinate site has been process integration. Similarly to the concept of water-energy nexus, it deals with potential interdependencies between several resources, in this case with all potential interconnections between existing processes established by material and energy stream recirculation. The application of process integration principles to the case of the improvement of water and energy resource use in a simultaneous manner is thus significant of the promotion of the water-energy nexus character of a determinate site. In this chapter, it is performed and exploitation of the role of process integration in the context of the water-energy nexus, by exploiting several existing methodologies, with a particular focus being given to innovative paradigm of Water and Energy Integration Systems (WEIS) and its digital twin-based counterpart, designated as ThermWatt.

**Keywords** Water-energy nexus · Process integration · Water and energy integration systems

M. Castro Oliveira (✉) · H. A. Matos
CERENA - Centro de Recursos Naturais e Ambiente, Department of Chemical Engineering, Instituto Superior Técnico, Lisbon, Portugal
e-mail: miguelmcoliveira@tecnico.ulisboa.pt

H. A. Matos
e-mail: henrimatos@tecnico.ulisboa.pt

M. Castro Oliveira
Research, Development and Innovation, ISQ - Instituto de Soldadura e Qualidade, Porto Salvo, Portugal

 
A. Angelini et al. (eds.), *Security Enhancement for Climate Change Impacting Urban Resources - SECCURe*, NATO Science for Peace and Security Series C: Environmental Security, https://doi.org/10.1007/978-94-024-2345-7_17

## 17.1 Introduction

The natural occurrence of human activities entails the use of significant quantities of external natural resources [1, 2], with two of the most important of these resources being water and energy [3]. While energy is used in diverse forms (such as fuels and electricity) [4], the use of water is not generally significantly differentiated (a water flow may contain several dissolved contaminants in different quantities, although such represents a characterization of a single resource rather than a differentiation within a single category of resource) [5]. In the context of the industrial sector, the operational improvement of production processes passes not only by the increase of production capacity but also by the reduction of the total use of external resources for the same exact end [6].

In recent years, several approaches have been taken to improve water and energy use in end-use sectors. For water efficiency improvement, the installation of wastewater treatment units coupled with water recirculation has been proposed for several case studies as a set of measures which allow the overall decrease of freshwater consumption and also the reduction of contaminant discharges [7]. For energy efficiency improvement, measures such as waste heat recovery [8], thermal energy storage [9–11] and high-efficiency electric motor installation [12] have been proposed. Advanced categories of measures include energy recovery from water and wastewater [13, 14], in which specific discharges of water systems are used to bring benefits at the level of energy use.

In order to more effectively analyse the improvement of water and energy use in end-use sectors from a holistic perspective, the concept of water-energy nexus has been introduced [15]. This concept deals with all potential interdependencies between the use of water and energy resources, whose exploitation must be performed so to bring an overall compromise between economic viability, environmental burden reduction and social stability, so to promote the sustainability character of physical systems [16]. A specific application of the water-energy nexus which has been most recently proposed by the definition of the concept of Water and Energy Integration Systems (WEIS) [17, 18]. These are systems which encompass a set of combustion-based processes and a set of water-using processes and the installation of several types of technologies, which in a holistic perspective bring benefits at the level of the reduction of overall water and energy inputs and gas and solid contaminant discharges.

This chapter presents the role of process integration in the context of the water-energy nexus. This is primarily performed by exploitation of several existing methodologies (implementation in real life and computational modelling). A particular focus is given to the innovative paradigm of Water and Energy Integration Systems (WEIS) and its digital twin-based counterpart, designated as ThermWatt.

## 17.2 Strategic Framework

The conceptualization and further implementation of water efficiency and decarbonisation measures (which include energy efficiency improvement, renewable energy integration and alternative fuel integration) subsist on the specific aims associated with the most recent sustainability policies, most relevantly the European Union strategy for energy system integration [19]. This strategy is based on three pillars.

The 1st pillar of the Strategy (Energy Efficiency and Circular Economy Nexus) subsists on:

- Energy efficiency-first principle (giving priority to energy demand-side solutions concerning energy supply-side ones, in the case that these are more cost-effective);
- Waste heat recovery (WHR) from industrial sites at the centre of intra-plant energy efficiency improvement and the functioning of district heating and cooling networks (which is based on the promotion of WHR practices on a broader perspective and the surpassing of existing barriers);
- Energy recovery from wastewater (mainly through the production of biofuels).

The 2nd pillar of the Strategy (Renewable-based Electrification) subsists on:

- Compensation of growing electricity demand through the promotion of the use of renewable energy resources as primary energy forms;
- Electrification of industrial processes;
- Application of energy storage technologies.

The 3rd pillar of the Strategy (Alternative low-carbon fuels) subsists on:

- Promotion of the use of green hydrogen in sectors with more difficult decarbonisation;
- Promotion of carbon capture and storage (CCS) and carbon capture and use (CCU).

The conceptualization of water recirculation and energy recovery systems (either the simultaneous application of these practices or standalone application of these) overall consists of the application of the principles of water integration and energy integration, which are specific applications of the process integration methodology for the water and energy resources, respectively [20–22]. The study of water and energy integration has the ultimate goal of studying the potential improvement of water and energy efficiency [22]. The study of energy efficiency has been the scope of the majority of international energy and environmental policies, namely the most recent European 2030 climate & energy framework [23], European Green Deal [24], European 2050 long-term strategy [25] and the Paris Agreement [26].

The interdependencies of water and energy resources, which include the understanding of the improvement of water use through the knowledge of the use of energy, have been studied within the scope of the water-energy nexus [27]. The water and energy integration methodologies may be applied in a high number of industrial processes and also a high number of industrial sectors [22]. The concept of water-energy nexus and the methodology of process integration are aligned in the aspect that simultaneous water recirculation and energy recovery systems are possible to be commissioned by the sharing of discharge streams, which contain considerable quantities of water and energy that may be furtherly valorised [28, 29].

## 17.3 Process Integration Methodologies

A commonly proposed and effectively implemented concept oriented to reduction of resource input through the valorisation of material and energy outputs. Process integration may be regarded as an assessment of the set of all processes and systems existing in a plant and all potential interdependencies to optimise the external resources usage. Process integration is essentially used within the field of process engineering and used to improve resource efficiency in process industries [30].

The concept of process integration may be implemented in several end-use sector cases (such as in the context of a process industry plant) by applying several methodologies. The most generic methodologies subsist on the conceptualisation of superstructures, which encompass all the processes existing in a determinate site set to serve as the object of analysis and all potential technologies planned to be implemented [18, 31]. Implementing these most general methodologies may be performed based on virtual reality through the development of computer models, consisting of digital twins of the real-life installations [32]. These models are set to allow the assessment of specific conditions associated with the operation of industrial processes and the systems encompassing the sharing of streams from these processes [27, 33]. The main technical purpose of such a development effort is the assembling of simulation and optimisation models for plants (or rather sections of plants which encompass a multitude of intensive water and energy processes, whose interdependencies may be exploited for the promotion of resource efficiency).

### *17.3.1 General Methodologies*

The most general process integration methodologies may be categorised into:

- **Standard methodologies**, primarily based on the application of the pinch analysis method (in the case of the handling of water and heat-related resources, water pinch analysis and heat pinch analysis);
- **Aggregated methodologies**, which deal with a high number of processes and a high number of set-to-be-implemented technologies of several categories in a determinate site.

In Table 17.1 a characterisation of the more generalist process integration methodologies is presented.

**Table 17.1** Characterization of process integration (PI) methodologies

| Methodology | Characterization | Ref |
|---|---|---|
| *Standard methodologies* | | |
| Pinch Analysis-based Heat Integration | Application of pinch analysis-based heat integration in industry, including:<br>• Assembling of heat exchanger networks (HEN) in a plant;<br>• Application of pinch technology to improve the operation of industrial processes and systems (for instance, ORC and Kalina cycle);<br>• Simultaneous application of pinch and exergy analysis methods;<br>• Implementation of algorithms for the assembling of energy management scenarios | [34–53] |
| Pinch Analysis-based Water Integration | Implementation of strategies for the minimisation of freshwater consumption and reduction of water footprint, through:<br>• The application of water pinch analysis (WPA) and water cascade analysis (WCA) methods;<br>• Study of water management strategies;<br>• Study of the reduction of wastewater discharge | [54–63] |
| *Aggregated methodologies* | | |
| Combined Water and Energy Integration (CWEI) | The object of study is a water network constituted by a set of water-using processes and wastewater treatment units, considering the following aspects:<br>• It is based on the application of steady-state based optimisation methodologies for the assessment of the most favourable levels of water stream recirculation in several points of the water network;<br>• Being based on the use of a resource as both an additional water and heat source, it only considers water streams for recirculation | [64–68] |
| Total Site Integration (TSI) | The object of study is a set of energy-using processes connected by a common utility system, considering the following aspects:<br>• It is based on the analysis of heat recovery between several processes through consumption and generation of utilities, in each the heating and cooling needs of each process are identified for the purpose of utility targeting;<br>• The supply of utilities with the aim to improve energy use may be performed through several measures (utility replacement, recirculated heat streams and renewable energy resources);<br>• For variable energy demand and supplies, it may involve the use of thermal energy storage units;<br>• It is itself an extension of the simpler heat integration methodologies (such as pinch analysis) | [69–73] |

### *17.3.2 Optimisation Methodologies*

The conceptualisation of energy recovery and water recirculation systems is set to be analysed by assessing the optimal point of the operation of the overall system, which consists of the point at which the water and energy consumption are the least possible while respecting all the operational constraints. In light of computational works, such assessment may be performed by the application of optimisation methodologies. In Table 17.2, a synthesis of all the optimisation methods applicable for heat recovery systems and water-energy networks is performed.

In relation to the overall aspects presented in Table 17.2:

- The pinch analysis method is associated to several benefits in relation to mathematical programming (MP), such as being based on a holistic perspective for the application of a systematic approach for the planning of industrial systems, although it majorly suffers from drawbacks at the level of simulation performance (it is a method conceptualized in a steady-state perspective) and at the level of modelling (simple models of the whole systems considering a high number of assumptions on fluid properties and temperature profiles);
- The research on the optimisation of water systems (WS) is, to date, a much more profoundly developed area in comparison to energy recovery systems (ES). The field of the optimisation of WS has been associated with the development and implementation of several proper state-of-the-art models and algorithms, while for ES the applied methods are summarized to the standard optimisation methodologies;
- Nonetheless, it is possible to verify the lack of the existence of dynamic programming (DP) studies for WS. Such lack of studies may be simply attributed to the fact that the operational conditions associated to industrial processes encompassed in WS do not appreciably vary during a determinate operational time given for the analysis, thus not requiring to analyse the minimization of a water and energy cost function for different successive time intervals.

In addition to the affirmations as mentioned above, further aspects must be taken into account concerning the use of pinch analysis and mathematical programming (MP)-based methods as alternatives for the solving of the same optimisation problems:

- The mathematical programming (MP)-based methods are characterized by the opportunity to consider either a steady-state or a transient-state perspective of the operation of real-life systems, as well as a higher degree of accuracy with respect to the modelling of physical phenomena in relation to the counterpart heuristic-based pinch analysis methods. Nonetheless, the methods that subsist in pinch analysis are associated with greater facility in considering a holistic system analysis perspective, owing to the simplification of computational methods. The computational efforts related to the running of pinch analysis-based models are reduced due to modelling simplifications, thus systems with more components

**Table 17.2** Characterization of optimisation methodologies implemented in the context of energy recovery and water systems

| Method | Characterization | Ref |
|---|---|---|
| Energy Recovery Systems | | |
| Linear and Non-linear Programming | It is based on the application of standard optimisation methods in which the governing equations of industrial systems are formulated as equality constraints and the objective functions are set as the summation of capital costs (associated to the system implementation) and operating costs (associated to water and energy use);<br>Under determinate circumstances, the models may be adapted so that non-linear equality constraints are linearized | [74–81] |
| Multi-objective Programming (MOP) | It is based on the attainment of the minimization of several indicators related to energy use and equipment sizing;<br>It is commonly based on bi-objective problems, in which the Pareto set encompasses two of these variables: energy efficiencies, exergy efficiency energy consumption levels, investment costs and equipment-related areas;<br>It is commonly implemented for the optimisation related to ORC encompassing systems;<br>A potential disadvantage is the issue of rank reversal associated to the use of TOPSIS procedures | [82–85] |
| Dynamic Programming (DP) | It is applicable for heat recovery systems with time-dependent variables, such as the ones containing TES units;<br>The objective function is formulated (in a general form) as the minimization of supplied thermal power during a determinate time period (defined as an integral) | [86–90] |
| Graphical Methods | It is based on the application of the pinch analysis methodology for the planning of heat exchanger networks (HEN);<br>The application of graphical methods may be concomitantly use for equipment design purposes, namely in respect to the calculation of heat transfer areas | [91–95] |
| Water Systems<br>(Water Recirculation and Treatment Systems) | | |

(continued)

**Table 17.2** (continued)

| Method | Characterization | Ref |
|---|---|---|
| Linear and Non-linear Programming | It is based on the application of standard linear and non-linear methods;<br>It generally subsists on the planning of water systems superstructures, with the objective function being the summation of water and energy costs (which are generally linear) and heat exchanger-related costs (which are generally non-linear);<br>It has limitations in the finding of optimal solutions, as in general only local optimal solutions are attained | [7, 31, 96–109] |
| Multi-objective Programming (MOP) | It is based on the minimization of pairs of variables associated to the water systems, namely water costs, utilities costs and heat exchanger units;<br>It has associated limitations in terms of the handling of the genetic algorithms with equality constraints (such as mass and enthalpy balance equations that must be necessarily enunciated within the model) | [107, 110] |
| Graphical Methods | The two types of graphical-based approaches are applied for water system conceptualization:<br>• Combined application of water pinch analysis and heat pinch analysis methodologies;<br>• P-graph framework (based on the mapping of all operating units and potential material flows) | [21, 68, 111–114] |

(for instance, in the case of large-scale problems which consider a higher number of unit operations) are more easily modelled with these methods;;

- Concerning water systems-related issues, solving multi-contaminant problems through water pinch analysis-based methods is impossible in some cases [115, 116]. Solving the same problems with MP-based methods is also tricky, but not impossible [115, 116]. In this case, MP-based methods represent a universal manner to solve water system-related problems, although being associated with the same issues in terms of computational efforts in relation to pinch analysis-based methods mentioned above;
- In addition to its limitations related to multi-contaminant problems, water pinch analysis-based methods are also claimed to fall short in relation to MP-based ones on aspects related to scope (MP-based ones allow in principle the consideration of more system design scenarios) [117] and constraints (water pinch analysis considers less categories of constraints) [116]. The combination of all these issues related to water pinch analysis render it as an overall less effective set of methods in relation to mathematical programming methods applied for water networks in respect to global optimisation (the obtention of a global optimum for a determinate problem obtained in the follow-up of the consideration of all aspect to

system constraints, the most accurate physical phenomena modelling and the minimization of freshwater input).

### 17.3.3 Water-Energy Nexus and Process Integration

From the general analysis of the aspects presented in Table 17.2 (as well as the aspects presented in the methodological framework presented in Table 17.1), it is possible to verify the inexistence of a paradigm which deals with the optimisation of water and energy resources (and, as such, with the consideration of water-using and energy-using processes) in a simultaneous and integrated manner, and therefore the complete promotion of the water-energy nexus character associated to a determinate site. The methodology of Combined Water and Energy Integration (CWEI) does (in fact) deal with the optimisation of hot/ cold utilities in a water system. However, it is limited in this same scope (it only deals with a specific set of energy resources that are inputted to allow the operation of water systems). In a tentative attempt to establish a pinch analysis-based optimisation procedure based on the minimization of water and energy resource supply, Prabhakar and Bandyopadhyay [118] developed a method for the promotion of the water-energy nexus character associated with the geographical region (through the determination of the minimum water and energy generation needed for the demand satisfaction of the approached region). The developed method is associated with the same aspects previously pointed out for pinch analysis-based methods, such as limitations with respect to simulation performance and a high number of modelling assumptions. In this prospect, the effective integration between the principles of both water-energy nexus and process integration shall subsist on a paradigm which may be adequate to developing mathematical programming-based computational models (not only at the level of optimisation models, but also simulation models).

The innovative paradigm of Water and Energy Integration Systems (WEIS) (which has been introduced by the authors in several past works [8, 17, 18, 107, 119–128]) has been developed as a simultaneous expansion and combination of the Combined Water and Energy Integration (CWEI) and Total Site Integration (TSI) methodologies in conceptual terms, considering categories of processes that are not considered in CWEI (it only considers water-using processes and associated consumption of material and energy-related utilities) and TSI (its scope is limited to energy-using processes and related utility systems). The CWEI is essentially encompassed within the general concept of WEIS, and as such, the improvement potential of water systems is the same for CWEI and WEIS. While the TSI does not only consider energy efficiency improvement based on energy recovery methods but also the adaptation/ substitution of energy inputs, the general paradigm of WEIS considers system-level improvements only through discharge stream recirculation (which would otherwise be classified as several types of wastes).

In addition, the general paradigm of WEIS considers several types of energy recovery methods in addition to heat recovery, which is the one primarily considered in TSI. For instance, the general paradigm of WEIS encompasses the opportunity for the implementation of waste-to-energy technologies, given that determinate material streams within a certain site exist to be valorised in terms of energy content. Examples of these technologies include anaerobic digestion (to produced biogas), gasification (to produce a synthetic gas), electrolysis and thermochemical water splitting (both to produce hydrogen fuel). The produced fuels in these units may be then supplied to the combustion-based processes, thus signifying an additional reduction of energy input in relation to the one already achieved by heat recovery. These are particularly relevant in the case that the material stream which serves as an input on these waste-to-energy units is water or wastewater. This inherent characteristic of the WEIS of encompassing this type of energy recovery units in its most general paradigm does not only represents an advancement in respect to the TSI methodology but also to the integration between energy use optimisation-based methodologies with water system-based ones such as CWEI (given the existence of water and wastewater streams which may be appreciably valorised in terms of energy content).

In the case that the set-to-be-considered energy recovery technologies of this type are based on biofuel production (which is the case of anaerobic digestion), such paradigm is particularly important, since it promotes the production of appreciable quantities of biofuels with the compensation of water and energy requirements of the newly installed units (the general concept of WEIS itself already predicts the new requirements of both these resources brought by the installation of new technologies, these the necessity to compensate these by stream recirculation). The production of biofuels is a practice which entails significant uses of both water and energy, and thus the analysis of the water-energy nexus character associated to this practice has been subject of analysis of several authors [129–132]. The WEIS paradigm effectively conjoins the potential integration of units set to produce this type of fuels with systems which overall promote the water-energy nexus character of real-life installations.

## 17.4 The Paradigm of Water and Energy Integration Systems (WEIS)

The Water and Energy Integration Systems (WEIS) are generally characterized by the following aspects:

- The object of study is a set of water-using and combustion-based thermal processes within a plant;
- It is based on the assessment of all potential recirculation points of discharge water and heat streams;
- Heat streams may be recirculated to all thermal processes (independently from the process of origin), thermodynamic cycles to be installed, the heaters of the water system and the heat-driven wastewater treatment units;

- Discharge water streams (with variable concentration of contaminants, according to the point of origin) may be recirculated from one unit operation of one water-using line to the inlet of the others within the same line and to the inlet of all the processes within the other water-using lines.
- The practical method is based on the valorisation of output streams, rather than on the rationalization of input streams:
- It does not completely disregard the consideration of other energy inputs (such as renewable resources and alternative hot and cold utilities), it only focuses on waste stream valorisation. Such a conceptual option is set to maximise solely the valorisation of otherwise waste streams generated in a plant (so as to minimise energy inputs only through the maximization of the recirculation of to-be-valorised streams);
- In terms of technological integration, it overall encompasses the same technologies as CWEI and TSI, and in addition wastewater-to-energy units;

### *17.4.1 General Concept for Energy Recovery Systems and Water Systems*

The conceptualization of energy recovery systems (ERS) subsists on the application of energy management principles to plan the installation of several technologies requiring and promoting the recirculation of streams with an associated energy valorisation potential, to obtain savings in energy use [133, 134]A specific type of energy recovery system is heat recovery systems (HRS), which subsist on the practice of recirculation of heat streams (which would otherwise be regarded as waste heat streams) and the implementation of waste heat recovery (WHR) technologies.

The recirculation of water streams has a similar principle to that of heat recovery, in which the water stream at the outlet of a particular process is recirculated in order to produce savings on freshwater consumption [135]. Nonetheless, in the case of the planning of a water recirculation system, it is generally necessary to first treat the wastewater from a specific process to recirculate it [136]. In this sense, these retrofitted systems are referred to in this work as water treatment and recirculation systems. In the context of the more complex conceptual simultaneous water recirculation and energy recovery systems, these will be referred to as water systems for convenience. As mentioned afore, a designation used by authors for water systems which are conceptualized to achieve an almost non-existent discharge is near-zero liquid discharge (NZLD) systems [137, 138]. Since a completely non-existent discharge is difficult, the term near-zero discharge systems is used instead of zero liquid discharge [139]. Another designation of water systems encompassing a set of heat exchangers, hot utility input, and cold utility input is water allocation and heat exchanger networks (WAHEN).

A generic heat recovery system may be planned considering the following opportunities:

- Reduction of fuel use in the combustion-based process from which the waste heat stream is emitted [134];
- Reduction of fuel use in other combustion processes [140];
- Reduction of electric energy use by the implementation of a thermodynamic cycle to produce electric energy from a waste heat stream [133].
- Additional energy sources from the thermal energy that is generated by renewable energy integration (such as solar thermal [141]);
- Additional energy sources from the thermal energy supplied by a thermal energy storage (TES) unit [141, 142];
- Additional energy sources from waste-to-energy technologies integration [143].

In the context of the conceptualization of water systems (systems in which energy-using units are present and in whose associated energy consumption is set to be reduced, in addition to freshwater):

- Several points of stream recirculation may exist, such as the streams at the outlet of heaters, coolers and the water-using process itself (these streams are set to be recirculated whereas limiting contaminant concentration in determinate points of the water system are not surpassed) [7, 20–22, 31, 68, 96–101, 103–106, 108–114, 144–147];
- The discharge streams from the water system (either the discharge water stream from the main water-using line or the sludge stream resulting as a by-product from the wastewater treatment unit) may be furtherly valorised in a context of recovery, through the installation of waste-to-energy (WtE) technologies [143].

The general conceptualization of a heat recovery system and a water recirculation and treatment system is represented in Fig. 17.1.

### *17.4.2 The Innovative Methodology of the Water and Energy Integration Systems (WEIS)*

The conceptualization of Water and Energy Integration Systems (WEIS) subsists on the combination of the two categories of conceptual systems (energy recovery systems and water recirculation and treatment systems), being overall characterized by the following sequence of steps:

- The use of the waste heat potential from a discharge stream from a combustion-based process for the purpose of reducing fuel consumption in a set of combustion-based processes [8];
- The use of the waste heat potential from a discharge stream from a combustion-based process for electricity generation through the installation of thermodynamic cycles [8];
- The recirculation of a set of water streams to produce savings in freshwater consumption [107];

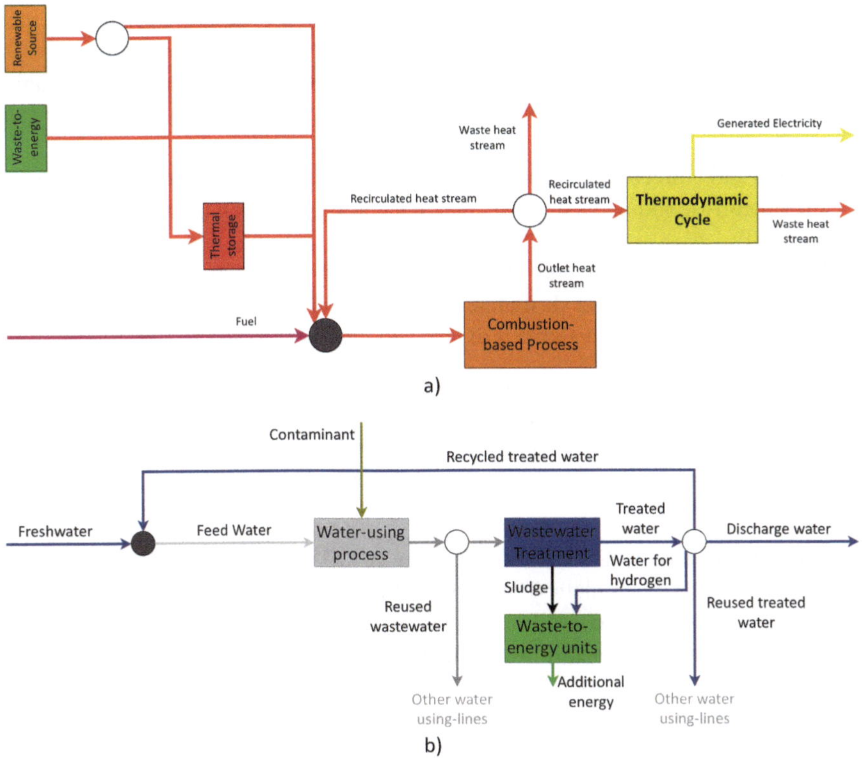


**Fig. 17.1** Basic approach of **a** a heat recovery system for a combustion-based process considering the recycling of waste streams and the input of additional energy streams and **b** a water treatment and recirculation system (representing a single water-using line constituted by one water-using process, one wastewater treatment unit and two wastewater-to-energy units for respectively the sludge and the discharge water streams)

- The recirculation of a set of water streams in the context of the use of its waste heat potential to produce savings in the hot utilities used in specific points of a water system (heaters) [20];
- The recirculation of a set of water streams produces savings in the cold utilities used in specific points of a water system (coolers) [20];
- The use of the waste heat potential from a discharge stream from a combustion-based process to set the operation of a water treatment and recirculation system (in the case of heat-driven water treatment [148]);
- The use of the discharge water streams and discharge sludge streams from a wastewater treatment unit for the production of additional fuels in wastewater-to-energy units [143].

In relation to water systems (standalone), the following concepts are regarded as analogous:

- **Wastewater treatment** and **desalination** (in this case, desalination represents a specific wastewater treatment unit in which the contaminants are salts);
- **Treated water** and **desalinated water** stream (in the case of desalination units, the set-to-be produced stream is a water stream with a significant quantity of salts removed);
- **Sludge** and **concentrate** stream (in the case of desalination units, the by-product stream corresponding to sludge is the concentrate, in which the removed salts are retained);
- **Wastewater-to-energy** and **energy recovery from water** (in the case of the former term, wastewater designates all types of discharge streams from a water system, which may be valorised in terms of energy recovery).

The paradigm of Water and Energy Integration Systems (WEIS) may be further expanded and adapted to constitute a universal methodology which deals with the optimisation of any category of water and energy resources in a determinate and defined site. For instance, it may be furtherly extended and adapted to include the following aspects:

- **Carbon capture, utilization and storage (CCUS) technologies**, serving for the purpose of the recirculation of carbon dioxide-rich streams with the ultimate objective of reducing overall water and energy inputs;
- **Energy-using processes in addition to combustion-based processes**, such as electric-based processes;
- **Direct electrification-based (Power-to-Heat) technologies**, mainly for purposes of complementing the existing paradigm (for instance, in the case of the input of electricity originating from an external renewable energy source, such as a photovoltaic (PV) system, which is then supplied to the thermal energy-based processes).

In Fig. 17.2 the abridged pictorial representation of a Water and Energy Integration System (WEIS) is presented. In Fig. 17.3, it is the abridged pictorial representation of a Water and Energy Integration System (WEIS) encompassing the proposed expansion to include CCUS technologies.

## 17.5 The ThermWatt Computational Tool

The analysis of the implementation of these types of Water and Energy Integration Systems (WEIS) may be performed in a virtual basis through the use of computational models. For this purpose, the ThermWatt computational tool was created, subsisting in a digital twin-based tool for the analysis of set-to-be-implemented WEIS and the further sustainability assessment of associated Engineering projects.

For the virtual implementation of WEIS in a determinate end-use sector sit, the first step is the development and use of simulation models (system-level models which encompass several interconnected component-level models corresponding to

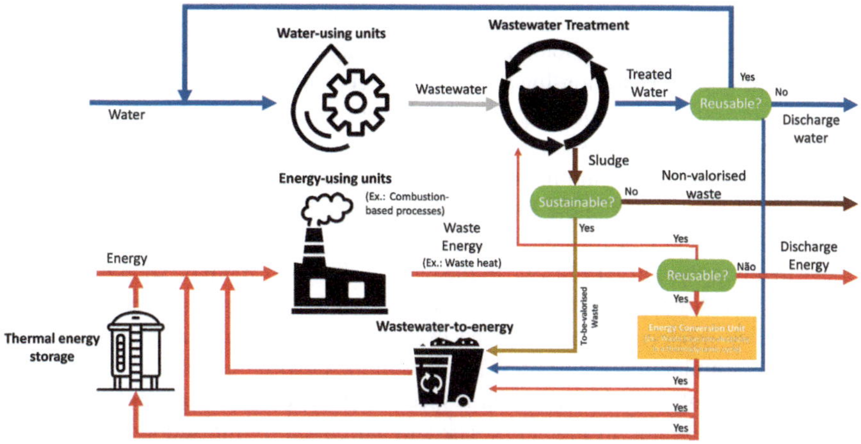


**Fig. 17.2** Abridged representation of Water and Energy Integration Systems (WEIS)

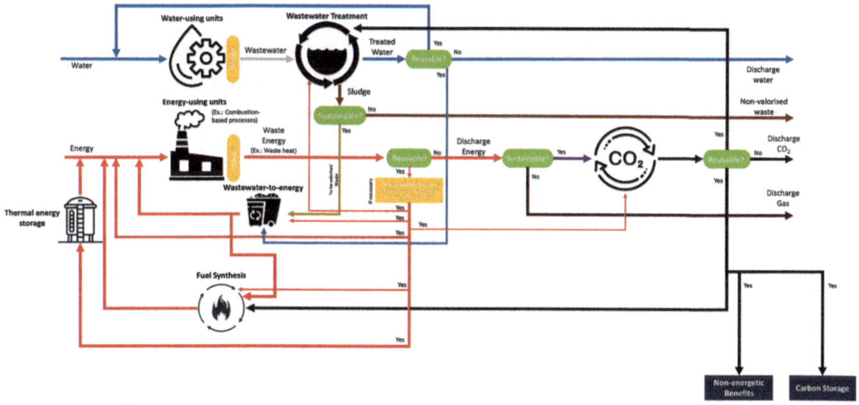


**Fig. 17.3** Abridged representation of a Water and Energy Integration System (WEIS) encompassing a carbon capture unit and a fuel synthesis unit

the industrial processes and improvement technologies), which is performed using the object-oriented modelling language Modelica. Several scenarios may be created based not only on several different points of stream recirculation but also on the installation of several analogous technologies. Such takes part in the overall scenario analysis procedure, which allows for the selection of the most favourable scenarios for each targeted case study.

The obtained results are nonetheless still associated with an uncertainty regarding their consideration as the optimal scenarios (defined by the minimisation of the water and energy-input related operational costs). As such, an optimisation procedure is undertaken, subsisting on the development of an optimisation model which serves as a counterpart to the previously developed simulation models. According to the specific modelling requirements, these models are developed using two different languages: the Python language for steady-state perspective ones (namely the ones using the linear, non-linear, mixed-integer linear, mixed-integer non-linear and multi-objective programming methodologies) and Modelica for transient-based ones (namely through the use of the OpenModelica *optimization* solver for the development of a model applying the dynamic programming methodology). In the case of the models developed in the Python language, the respective optimisation results are set to be integrated in the counterpart Modelica simulation models using the OpenModelica Python API [149]. In the case of the latter models, such direct integration is not directly possible, due to inexistence of a proper API that connects the Modelica models developed to be used with the *optimization* solver and the counterpart simulation models (developed with the DASSL solver).

The aforementioned concept with respect to the ThermWatt computational tool is pictorially presented in Fig. 17.4. The simulation and optimisation model-based achievements associated with the ThermWatt computational tool in light of the requirements to fully build a digital twin for a WEIS are characterized in Table 17.3 (which takes into account all the modelling work presented across the previous works by the authors [8, 17, 18, 107, 119–128]).

As may be verified by the analysis of Table 17.3, the ThermWatt computational tool (in its current development stage at the moment the present work is being elaborated) contains all the necessary components to serve for the purpose of building counterpart simulation and optimisation models for a same WEIS, with only the technologies associated to the expansion and adaptation purposes having not yet been modelled. The achievements only regarded as reasonable in the context of optimisation models may be considered irrelevant to the overall modelling framework of ThermWatt, as these do not appreciable affect the procedure associated to the obtention of optimised results (the main objective of ThermWatt is the obtention of a final

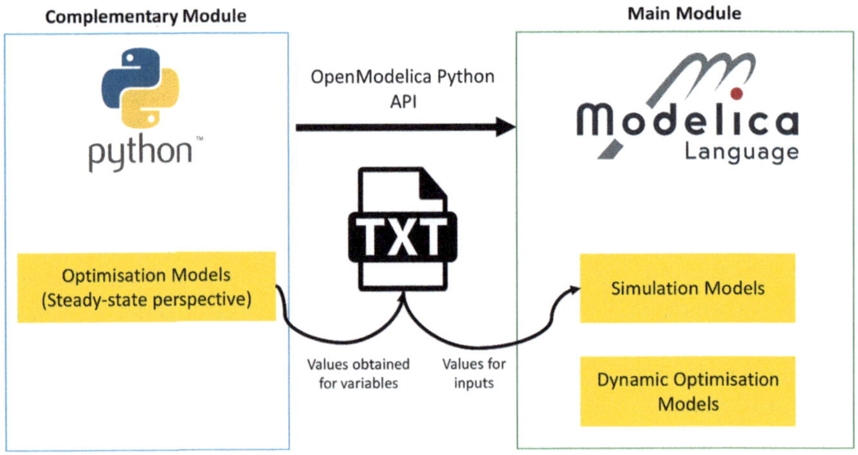


**Fig. 17.4** General concept for the ThermWatt computational tool

**Table 17.3** Achievements associated to the ThermWatt computational tool at the light of the paradigm of Water and Energy Integration Systems

| Aspect | Simulation models | Optimisation models |
|---|---|---|
| General | | |
| *Modelica language* | Achieved | Achieved (Dynamic Programming) |
| *Python language* | Not-applicable | Achieved (Steady state-based methodologies) |
| *Component-level Models* | | |
| Combustion-based processes | Achieved | Achieved |
| Water-using processes | Achieved | Achieved |
| Basic categories of heat exchangers (air preheaters and economisers) | Achieved | Achieved |
| Electricity Generation Thermodynamic cycles (e.g. Organic Rankine Cycle) | Achieved | Reasonably achieved (Through an assumption at the level of thermal-to-electric conversion factor) |
| Thermal energy storage technologies | Achieved | Reasonably achieved (Simplification to the finite difference method) |
| Wastewater treatment technologies (e.g. Desalination through Multi-effect distillation) | Achieved | Achieved |
| Energy recovery from water/ wastewater technologies (e.g. Electrolysis) | Achieved | Achieved |
| Electrification technologies | Non-achieved | Non-achieved |
| Carbon capture, utilization and storage technologies | Non-achieved | Non-achieved |

simulation models integrated with the results generated by the counterpart optimisation models, which are associated to limitations at the level of slight simplifications and assumptions).

In Fig. 17.5 an example of a WEIS model built using the capacities of the ThemWatt computational model is presented. The WEIS configuration given in the referred model has been obtained in the follow-up of an extensive simulation and optimisation process, taking into account the concept presented in Fig. 17.4. The conceptualized WEIS, which is presented in Fig. 17.5, considers a set of combustion-based and water-using processes (namely two combustion-based processes operating in a continuous mode, two combustion-based processes operating in a batch mode and three water-using processes), all newly set-to-be installed technologies and stream recirculation points. The system selected to be presented in the context of this work is one corresponding to the a dynamic WEIS configuration, in which a transient state and a dynamic perspective have to be considered for the overall analysis of the

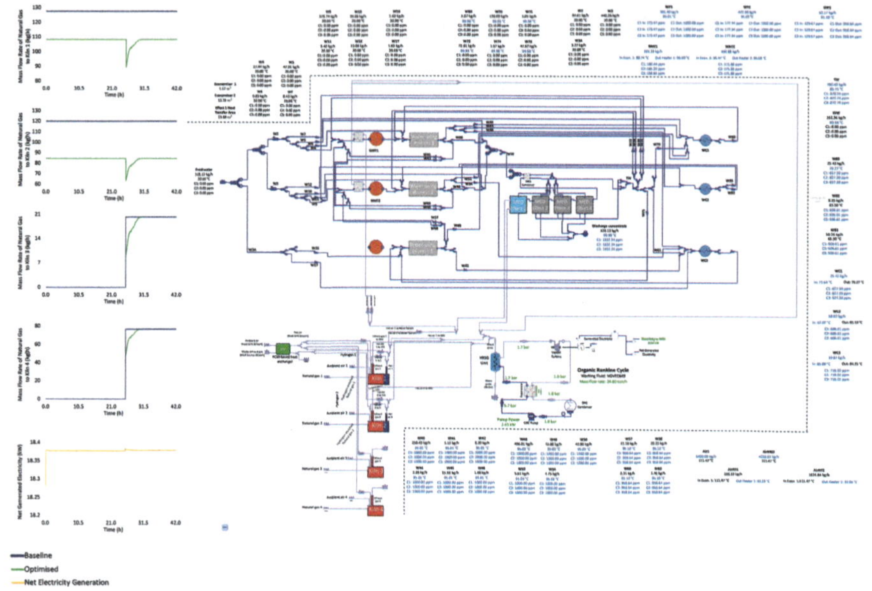


**Fig. 17.5** Simulation model and results (after an extensive simulation and optimisation process) obtained by the use of the capacities of the ThermWatt computational tool (adapted from [123])

operation of the system (owing to the installation of a thermal energy storage which justifies the analysis of the variation of determination parameters along with time).

## 17.6 Conclusions

This work exploits the fundamental aspects in respect to the implementation of the principles of process integration (PI) in the context of the promotion of the interdependencies of water and energy resources encompassed by the water-energy nexus in a combined social, economic and technical perspective. This exploitation is performed by the analysis of the existing PI methodologies (from the simpler to the most complex ones), followed by the analysis of the existing simulation and optimisation modelling methods implemented to develop digital twins for real-life physical systems implementing the concept of process integration for the purpose of improving the overall use of water and energy resources. Such culminates on the presentation of the Water and Energy Integration Systems (WEIS) as an ultimate paradigm to promote water and energy resources interdependencies in any site (a process industry plant or any other end-use sector site) with the aim to improve the use of these resources (thus allowing the promotion of the water-energy nexus and process integration principles in a determinate context in a fully simultaneous manner). Furthermore, it is presented and detailed a computational tool designated

ThermWatt, which subsists in a digital twin-based counterpart of the WEIS paradigm (the tool serves for the creation of the simulation and optimisation models of these types of systems). Both these assets (WEIS and ThermWatt) have been developed by the authors of this work and have served for several academic, industrial and entrepreneurial purposes.

In respect to the existing process integration methodologies and associated modelling methods, the following affirmations may be performed:

- The most complex and encompassing exiting methodologies of total site integration (TSI) and combined water and energy integration (CWEI) are adequate for the analysis of water systems and energy systems, respectively, in standalone perspectives, although each one suffers from being extensively oriented to the only one of the involved resources (CWEI only consists of the improvement of the use of hot and cold utilities encompassed by a water system, not being oriented to the improvement of any other energy resource);
- The existing graphic methods for process integration-based optimisation modelling are adequate for the implementation of a holistic perspective on water and energy resource use improvement in a specific site. However, these are associated with several limitations in terms of physical phenomena modelling, which ultimately leads to the inclination to mathematical programming-based methods as preferable (in the prospect of simultaneously combining the accuracy related to physical phenomena modelling and the consideration of a holistic perspective associated with the analysis of water and energy systems).

With respect to the two more innovative assets of Water and Energy Integration Systems (WEIS) and the ThermWatt computational tool:

- The paradigm of Water and Energy Integration Systems (WEIS) is effective in respect to the achievement of the objective to simultaneously improve the use of water and energy resources within a system encompassing a high number of water-using and energy-using processes (in the limit, all processes of these two categories in determinate site). In this sense, it completely makes use of all the potential interdependencies of water and energy (as encompassed by the concept of water-energy nexus) with the aim to improve the use of these two categories of resources (making extensively use of the principles of process integration);
- The ThermWatt computational tool is effective in terms of the achievement of the overall objective of creating digital twins of WEIS (with several models for each one of the involved components of these types of systems and for these systems as a whole having been already developed in the scope of previous works by the authors);
- The paradigm of WEIS (and the ThermWatt tool as its digital twin-based counterpart) may be furtherly extended to include other concepts (such as the one of carbon capture, utilization and storage) to be gradually updated to the emerging sustainability promotion technologies (namely those which completely handle the use of water and energy resources or that simply reduce the use of these resources).

## References


1. 'Environmental Performance and Information Division Environment Directorate O (2017) The economic significance of natural resources: key points for reformers in Eastern Europe, Caucasus and Central Asia. Environ Perform Inf Div Environ Dir OECD 5:1–8
2. Gennitsaris S, Castro Oliveira M, Vris G, Bofilios A, Ntinou T, Frutuoso AR, Queiroga C, Giannatsis J, Sofianopoulou S, Dedoussis V (2023)Energy efficiency management in small and medium-sized enterprises: current situation, case studies and best practices. Sustain 15
3. IEA (2020) Energy and water (2020) Exploring the interdependence of two critical resources. Int Energy Agency
4. Onar OC, Khaligh A (2015) Energy sources. Altern Energy Power Electron 81–154
5. Pure Water for the World: Fresh Water: A Precious Natural Resource (2023)
6. Immerman G (2021) Production and process optimization in manufacturing. Mach
7. N Ibrić E Ahmetović Z Kravanja 2016 Mathematical programming synthesis of non-isothermal water networks by using a compact/reduced superstructure and an MINLP model Clean Technol Environ Policy 18 1779 1813
8. M Castro Oliveira M Iten HA Matos 2022 Simulation and assessment of an integrated thermal processes and Organic Rankine Cycle (ORC) system with Modelica Energy Rep 8 764 770
9. L Miró J Gasia LF Cabeza 2016 Thermal energy storage (TES) for industrial waste heat (IWH) recovery: a review Appl Energy 179 284 301
10. P Royo L Acevedo VJ Ferreira T García-Armingol AM López-Sabirón G Ferreira 2019 High-temperature PCM-based thermal energy storage for industrial furnaces installed in energy-intensive industries Energy 173 1030 1040
11. Nyamsi SN, Tolj I, Lototskyy M (2019) Metal hydride beds-phase change materials: dual mode thermal energy storage for medium-high temperature industrial waste heat recovery. Energies 12
12. Chan MHE, Chu KK, Chow HF, Tsang CW, Ho CKD, Ho SK (2019) Improving the energy efficiency of petrochemical plant operations: a measurement and verification case study using a balanced wave optimizer. Energies 12
13. A Gherghel C Teodosiu S Gisi De 2019 A review on wastewater sludge valorisation and its challenges in the context of circular economy J Clean Prod 228 244 263
14. Yukesh Kannah R, Kavitha S, Preethi, Parthiba Karthikeyan O, Kumar G, Dai-Viet NV, Rajesh Banu J (2021) Techno-economic assessment of various hydrogen production methods—a review. Bioresour Technol 319
15. Bauer D (2014) The water-energy nexus: challenges and opportunities
16. Naveed U, Mohammad Rozali NE, Mahadzir S (2022) Energy–water–carbon nexus study for the optimal design of integrated energy–water systems considering process losses. Energies 15
17. Castro Oliveira M, Iten M, Matos HA (2022) Review on water and energy integration in process industry: water-heat nexus. Sustain 14
18. M Castro Oliveira M Iten HA Matos 2023 Simultaneous optimisation of energy recovery and water recirculation in a ceramic plant Comput Aided Chem Eng 52 2785 2790
19. European Commission. Powering a climate-neutral economy: an EU strategy for energy system integration. https://eur-lex.europa.eu/legal-content/EN/TXT/HTML/?uri=CELEX:52020DC0299&from=EN
20. Savulescu LE, Alva-Argaez A (2013) Process integration concepts for combined energy and water integration. Handb Process Integr Minimisation Energy Water Use Waste Emiss 461–483
21. HH Chin DCY Foo HL Lam 2019 Simultaneous water and energy integration with isothermal and non-isothermal mixing—a P-graph approach Resour Conserv Recycl 149 687 713
22. E Ahmetović N Ibrić Z Kravanja IE Grossmann 2015 Water and energy integration: a comprehensive literature review of non-isothermal water network synthesis Comput Chem Eng 82 144 171

23. European Commission: 2030 Climate & Energy Framework. http://data.consilium.europa.eu/doc/document/ST-169-2014-INIT/en/pdf
24. European Commission: A European Green Deal | European Commission. https://ec.europa.eu/info/strategy/priorities-2019-2024/european-green-deal_en
25. Commission E: 2050 Long-term strategy. https://ec.europa.eu/clima/policies/strategies/2050_en
26. Horowitz CA. Paris Agreement. https://ec.europa.eu/clima/policies/international/negotiations/paris_en
27. Oliveira MC, Iten M, Matos HA, Michels J (2019) Water-energy nexus in typical industrial water circuits. Water (Switzerland) 11
28. Klemeš JJ, Varbanov PS, Ocłoń P, Chin HH (2019) Towards efficient and clean process integration: utilisation of renewable resources and energy-saving technologies. Energies 12
29. Dunn RF, Ristau JS (2016) Using process integration technology to retrofit chemical plants for energy conservation and wastewater minimization
30. Widmann D, Mader H, Friedrich H, Heywang W, Müller R. Process integration. https://www.ceas.manchester.ac.uk/research/themes/process-integration/
31. Kermani M, Kantor ID, Maréchal F (2019) Optimal design of heat-integrated water allocation networks. Energies 12
32. HL Lam JJ Klemeš Z Kravanja PS Varbanov 2011 Software tools overview: process integration, modelling and optimisation for energy saving and pollution reduction Asia-Pacific J Chem Eng 6 696 712
33. Iten, M, Fernandes U, Oliveira MC (2021) Framework to assess eco-efficiency improvement: case study of a meat production industry. SSRN Electron J
34. F Thibault A Zoughaib S Pelloux-Prayer 2015 A MILP algorithm for utilities pre-design based on the pinch analysis and an exergy criterion Comput Chem Eng 75 65 73
35. Zhou C, Zhuang Y, Zhang L, Liu L, Du J, Shen S (2020) A novel pinch-based method for process integration and optimization of Kalina cycle. Energy Convers Manag 209
36. Y Abdelouadoud E Lucas P Krummenacher D Olsen B Wellig 2019 Batch process heat storage integration: a simple and effective graphical approach Energy 185 804 818
37. TK Zhelev KA Semkov 2004 Cleaner flue gas and energy recovery through pinch analysis J Clean Prod 12 165 170
38. VE Araújo FP Bernardo CM Reis FG Martins 2017 Cluster analysis of process operational data to identify representative scenarios for pinch analysis and energy optimisation studies Comput Aided Chem Eng 40 2539 2544
39. MN Hamsani TG Walmsley PY Liew SR Wan Alwi 2018 Combined pinch and exergy numerical analysis for low temperature heat exchanger network Energy 153 100 112
40. X Feng XX Zhu 1997 Combining pinch and exergy analysis for process modifications Appl Therm Eng 17 249 261
41. AIA Salama 2006 Determination of the optimal heat energy targets in heat pinch analysis using a geometry-based approach Comput Chem Eng 30 758 764
42. A Ghannadzadeh M Sadeqzadeh 2017 Exergy aided pinch analysis to enhance energy integration towards environmental sustainability in a chlorine-caustic soda production process Appl Therm Eng 125 1518 1529
43. PM Pereira MC Fernandes HA Matos 2016 FI2EPI—a freeware tool for performing heat integration based on pinch analysis Comput Aided Chem Eng 38 1815 1820
44. N Angsutorn K Siemanond R Chuvaree 2014 Heat exchanger network synthesis using MINLP stage-wise model with pinch analysis and relaxation Comput Aided Chem Eng 33 139 144
45. SG Yoon J Lee S Park 2007 Heat integration analysis for an industrial ethylbenzene plant using pinch analysis Appl Therm Eng 27 886 893
46. J Geldermann M Treitz O Rentz 2006 Integrated technique assessment based on the pinch analysis approach for the design of production networks Eur J Oper Res 171 1020 1032
47. Elias AM, Giordano R, Secchi AR, Furlan FF (2019) Integrating pinch analysis and process simulation within equation-oriented simulators. Comput Chem Eng 130

48. JC Bonhivers A Moussavi A Alva-Argaez PR Stuart 2016 Linking pinch analysis and bridge analysis to save energy by heat-exchanger network retrofit Appl Therm Eng 106 443 472
49. S Valiani N Tahouni MH Panjeshahi 2017 Optimization of pre-combustion capture for thermal power plants using pinch analysis Energy 119 950 960
50. T Han C Wang C Zhu D Che 2018 Optimization of waste heat recovery power generation system for cement plant by combining pinch and exergy analysis methods Appl Therm Eng 140 334 340
51. PS Roychaudhuri V Kazantzi DCY Foo RR Tan S Bandyopadhyay 2017 Selection of energy conservation projects through financial pinch analysis Energy 138 602 615
52. A Manizadeh A Entezari R Ahmadi 2018 The energy and economic target optimization of a naphtha production unit by implementing energy pinch technology Case Stud Therm Eng 12 396 404
53. D Olsen Y Abdelouadoud P Liem B Wellig 2017 The role of pinch analysis for industrial ORC integration Energy Proc 129 74 81
54. SM Khezri F Lotfi S Tabibian Z Erfani 2010 Application of water pinch technology for water and wastewater minimization in aluminum anodizing industries Int J Environ Sci Technol 7 281 290
55. ZY Liu YZ Yang Y Zhang 2007 Determining the pinch point and calculating the freshwater target for water-using systems with single contaminant Chem Eng Res Des 85 1485 1490
56. X Jia L Zhang Z Li RR Tan J Dou DCY Foo F Wang 2019 Pinch analysis for targeting desalinated water price subsidy J Clean Prod 227 950 959
57. YL Tan ZA Manan DCY Foo 2007 Retrofit of water network with regeneration using water pinch analysis Process Saf Environ Prot 85 305 317
58. M Sorin S Bédard 1999 The global pinch point in water reuse networks Process Saf Environ Prot 77 305 308
59. G Skouteris S Ouki D Foo D Saroj M Altini P Melidis B Cowley G Ells S Palmer S O'Dell 2018 Water footprint and water pinch analysis techniques for sustainable water management in the brick-manufacturing industry J Clean Prod 172 786 794
60. M Bavar MH Sarrafzadeh H Asgharnejad H Norouzi-Firouz 2018 Water management methods in food industry: corn refinery as a case study J Food Eng 238 78 84
61. Wan Alwi SR, Manan ZA (2013) Water pinch analysis for water management and minimisation: an introduction. Handb Process Integr Minimisation Energy Water Use Waste Emiss 353–382
62. S Mohammadnejad GRN Bidhendi N Mehrdadi 2011 Water pinch analysis in oil refinery using regeneration reuse and recycling consideration Desalination 265 255 265
63. ZA Manan YL Tan DCY Foo 2004 Targeting the minimum water flow rate using water cascade analysis technique AIChE J 50 3169 3183
64. Zoller F (2017) Integration of energy networks and the water cycle with surface water energy as connecting element. In: 12th IEA heat pump conference
65. Tan RR, Foo DCY (2013) Pinch analysis for sustainable energy planning using diverse quality measures. Handb Process Integr Minimisation Energy Water Use Waste Emiss 505–523
66. XY Lim DCY Foo RR Tan 2018 Pinch analysis for the planning of power generation sector in the United Arab Emirates: a climate-energy-water nexus study J Clean Prod 180 11 19
67. Silori GK, Khanam S (2017) Simultaneous water and energy integration techniques: a review, vol 15
68. M Souifi A Souissi 2019 Simultaneous water and energy saving in cooling water networks using pinch approach Mater Today Proc 13 1115 1124
69. JJ Klemeš PS Varbanov TG Walmsley X Jia 2018 New directions in the implementation of pinch methodology (PM) Renew Sustain Energy Rev 98 439 468
70. AH Tarighaleslami TG Walmsley MJ Atkins MRW Walmsley PY Liew JR Neale 2017 A Unified total site heat integration targeting method for isothermal and non-isothermal utilities Energy 119 10 25
71. Klemeš JJ, Varbanov PS (2012) From HEN to total site, to energy supply chains. Chalmers.Se

72. AH Tarighaleslami TG Walmsley MJ Atkins MRW Walmsley JR Neale 2017 Total site heat integration: utility selection and optimisation using cost and exergy derivative analysis Energy 141 949 963
73. Klemeš JJ, Varbanov PS, Škorpík J, Pospíšil J (2020) Total site integration. Sustain Util Syst 11:347–408
74. OE Prun AB Garyaev 2020 Method for optimization of heat-exchange units working in heat recovery systems Therm Eng 67 560 566
75. Oluleye G, Jobson M, Smith R (2015) Optimisation-based design of site waste heat recovery systems. In: ECOS 2015—28th international conference efficient cost, optimisation simulation environment impact energy system
76. A Ganjeh Kaviri MN Mohd Jafar ML Tholudin 2012 Modeling and optimization of heat recovery heat exchanger Appl Mech Mater 110–116 2448 2452
77. MH Yang 2016 Optimizations of the waste heat recovery system for a large marine diesel engine based on transcritical Rankine cycle Energy 113 1109 1124
78. D Li Q Sun K Sun G Zhang S Bai G Li 2021 Diesel engine waste heat recovery system comprehensive optimization based on system and heat exchanger simulation Open Phys 19 331 340
79. Tanasic N, Ivosevic M, Simonovic T (2021) Optimisation of waste heat recovery system in paper machine dryer section. In: Proceedings of 2021 6th international symposium environment energies applications EFEA 2021
80. X Wang M Jin W Feng G Shu H Tian Y Liang 2018 Cascade energy optimization for waste heat recovery in distributed energy systems Appl Energy 230 679 695
81. Castelli AF, Elsido C, Scaccabarozzi R, Nord LO, Martelli E (2019) Optimization of organic rankine cycles for waste heat recovery from aluminum production plants. Front Energy Res 7
82. Laouid YAA, Kezrane C, Lasbet Y, Pesyridis A (2021) Towards improvement of waste heat recovery systems: a multi-objective optimization of different organic Rankine cycle configurations. Int J Thermofluids 11
83. G Legorburu AD Smith 2019 Demonstrating the benefit of multi-objective optimization and clustering for the design of waste heat recovery systems ASHRAE Trans 125 436 443
84. MM Sani A Noorpoor MS Motlagh 2020 Multi objective optimization of waste heat recovery in cement industry (a case study) J Therm Eng 6 604 618
85. Valencia G, Núñez J, Duarte J (2019) Multiobjective optimization of a plate heat exchanger in a waste heat recovery organic rankine cycle system for natural gas engines. Entropy 21
86. Zhang L, Wang Y, Feng X (2021) A framework for design and operation optimization for utilizing low-grade industrial waste heat in district heating and cooling. Energies 14
87. Feru E, Goyal S, Willems F (2018) Optimal sizing of waste heat recovery systems for dynamic engine conditions. Org Rank Cycle Technol Heat Recover
88. Powell KM, Hedengren JD, Edgar TF (2013) Dynamic optimization of a solar thermal energy storage system over a 24 hour period using weather forecasts. In: Proceedings of American control conference, pp 2946–2951
89. J Li Y Zhang P Ding E Long 2021 Experimental and simulated optimization study on dynamic heat discharge performance of multi-units water tank with PCM Indoor Built Environ 30 1531 1545
90. C Hoffmann H Puta 2006 Dynamic optimization of energy supply systems with Modelica models IFAC Proc 1 51 56
91. Marton S, Langner C, Svensson E, Harvey S (2021) Costs vs. Flexibility of process heat recovery solutions considering short-term process variability and uncertain long-term development. Front Chem Eng 3
92. Benvenuto G, Trucco A, Campora U (2016) Optimization of waste heat recovery from the exhaust gas of marine diesel engines. In: Proc Inst Mech Eng Part M J Eng Maritime Environ 230:83–94
93. JM Joe AM Rabiu 2013 Retrofit of the heat recovery system of a petroleum refinery using pinch analysis J Power Energy Eng 01 47 52

94. M Cassanello KP Yeoh Y Liang MS Pamudji CW Hui 2018 Modelling and optimization of multistream heat exchanger with area targeting Chem Eng Trans 70 673 678
95. Biyanto TR, Tama NE, Permatasari I, Sabillah MG, Napitupulu DH, Anda AR (2019) Optimization heat transfer coefficient in retrofit heat exchanger network using pinch analysis and killer whale algorithm. In: AIP conference proceedings, vol 2088
96. HG Dong CY Lin CT Chang 2008 Simultaneous optimization approach for integrated water-allocation and heat-exchange networks Chem Eng Sci 63 3664 3678
97. F Yan H Wu W Li J Zhang 2016 Simultaneous optimization of heat-integrated water networks by a nonlinear program Chem Eng Sci 140 76 89
98. X Hong Z Liao B Jiang J Wang Y Yang 2017 Targeting of heat integrated water allocation networks by one-step MILP formulation Appl Energy 197 254 269
99. E Ahmetović Z Kravanja 2013 Simultaneous synthesis of process water and heat exchanger networks Energy 57 236 250
100. Ibrić N, Ahmetović E, Kravanja Z, Grossmann IE (2021) Simultaneous optimisation of large-scale problems of heat-integrated water networks. Energy 235
101. X Dong C Zhang X Peng C Chang Z Liao Y Yang J Sun J Wang Y Yang 2022 Simultaneous design of heat integrated water allocation networks considering all possible splitters and mixers Energy 238 121916
102. M Bagajewicz H Rodera M Savelski 2002 Energy efficient water utilization systems in process plants Comput Chem Eng 26 59 79
103. HP Zhao Y Yang ZY Liu 2019 Design of heat integrated water networks with multiple contaminants J Clean Prod 211 530 536
104. X Hong Z Liao J Sun B Jiang J Wang Y Yang 2018 Energy and water management for industrial large-scale water networks: a systematic simultaneous optimization approach ACS Sustain Chem Eng 6 2269 2282
105. SY Alnouri P Linke M El-Halwagi 2014 Water integration in industrial zones: a spatial representation with direct recycle applications Clean Technol Environ Policy 16 1637 1659
106. Caballero JA, Pavão LV, Costa CBB, Ravagnani MASS (2021) A novel sequential approach for the design of heat exchanger networks. Front Chem Eng 3
107. Oliveira MC, Vieira SM, Iten M, Matos HA (2022) Optimisation of water-energy networks in process industry: implementation of non-linear and multi-objective models. Front Chem Eng 3
108. Zhelev, T.K., Zheleva, S.R.: Combined pinch analysis for more efficient energy and water resources management in beverage industry. Waste Manage Environ 623–632
109. XS Zheng X Feng DL Cao 2003 Design water allocation network with minimum freshwater and energy consumption Comput Aided Chem Eng 15 388 393
110. Y Hou W Xie Z Duan J Wang 2017 A conceptual methodology for simultaneous optimization of water and heat with non-isothermal mixing Front Chem Sci Eng 11 154 165
111. Kamat S, Bandyopadhyay S (2021) A hybrid approach for heat integration in water conservation networks through non-isothermal mixing. Energy 233
112. BS How Á Orosz SY Teng JY Lim F Friedler 2021 Heat integrated water regeneration network synthesis via graph-theoretic sequential method Chem Eng Trans 88 49 54
113. X Hong Z Liao J Sun B Jiang J Wang Y Yang 2018 Heat transfer blocks diagram: a novel tool for targeting and design of heat exchanger networks inside heat integrated water allocation networks ACS Sustain Chem Eng 6 2704 2715
114. GC Sahu S Bandyopadhyay 2010 Energy conservation in water allocation networks with negligible contaminant effects Chem Eng Sci 65 4182 4193
115. Nemati-Amirkolaii K, Romdhana H, Lameloise ML (2019) Pinch methods for efficient use of water in food industry: a survey review. Sustain 11
116. Buabeng-Baidoo E, Majozi T (2019) Simultaneous optimization of water and energy in integrated water and membrane networks. Water Manag 513–535
117. Teles J (2012) New global optimization algorithm for water-using networks. In: 2012 AIChE annual meeting

118. S Prabhakar S Bandyopadhyay 2023 Graphical optimisation of supplying water and energy in a water-energy nexus system Chem Eng Trans 103 307 312
119. M Castro Oliveira M Iten PL Cruz H Monteiro 2020 Review on energy efficiency progresses, technologies and strategies in the ceramic sector focusing on waste heat recovery Energies 13 6096
120. M Castro Oliveira M Iten 2021 Modelling of a solar thermal energy system for energy efficiency improvement in a ceramic plant Renew Energy Environ Sustain 6 31
121. Castro Oliveira M, Iten M, Matos HA (2022) Review of thermochemical technologies for water and energy integration systems: energy storage and recovery. Sustain 14
122. Castro Oliveira M, Matos HA (2024) Sustainability and strategic assessment of water and energy integration systems: case studies of the process industry in Portugal. Energies 17
123. Castro Oliveira M (2023) Simulation and optimisation of water and energy integration systems (WEIS): an innovative approach for process industries. https://scholar.tecnico.ulisboa.pt/records/KwfdB28dqObhPEiL7RtB3VJ6YxviyXFEhWnt
124. Castro Oliveira M (2023) Sustentabilidade, nexus água-energia e escassez de recursos. Público
125. Oliveira MC, Iten M, Matos HA (2021) Assessment of energy efficiency improvement in ceramic industry through waste heat recovery modelling. In: Computer aided chemical engineering. Elsevier, pp 1653–1658
126. M Castro Oliveira P Coelho M Iten HA Matos 2022 Modelling of heat-driven water treatment systems: multi-effect distillation (MED) model in Modelica Comput Aided Chem Eng 51 397 402
127. M Castro Oliveira R Castro Oliveira HA Matos 2024 Dynamic simulation and optimisation of water and energy consumption in a ceramic plant: application of the customised ThermWatt computational tool Comput Aided Chem Eng 53 2377 2382
128. Oliveira MC, Oliveira RC, Matos HA (2024) Optimisation model for a closed-loop industrial water-energy system within a ceramic industry plant. In: 7th international Congress water, waste energy management
129. M Martín IE Grossmann 2015 Water-energy nexus in biofuels production and renewable based power Sustain Prod Consum 2 96 108
130. DC López-Díaz LF Lira-Barragán JM Ponce-Ortega MM El-Halwagi 2019 Optimization of biofuel supply chain design via a water-energy-food nexus framework Comput Aided Chem Eng 46 1567 1572
131. Strapasson A, Lee H, Schnettler J (2021) Biofuels and the water-energy nexus: perspectives for the United States. Environ Nat Resour Progr Pap
132. Seabra JEA (2021) Roadmap to 2050: the land-water-energy nexus of biofuels. Sustain Dev Solut Netw Fond Eni Enrico Mattei 164
133. Semmari H, Filali A, Aberkane S, Feidt R, Feidt M (2020) Flare gas waste heat recovery: assessment of organic rankine cycle for electricity production and possible coupling with absorption chiller. Energies 13
134. A Mezquita J Boix E Monfort G Mallol 2014 Energy saving in ceramic tile kilns: cooling gas heat recovery Appl Therm Eng 65 102 110
135. JJ Klemes 2012 Industrial water recycle/reuse Curr Opin Chem Eng 1 238 245
136. M Salgot M Folch 2018 Wastewater treatment and water reuse Curr Opin Environ Sci Heal 2 64 74
137. K Lu Y Lü Y Bai J Zhang N Bie Y Ren Y Ma 2020 Experimental investigation and theoretical modeling on scale behaviors of high salinity wastewater in zero liquid discharge process of coal chemical industry Chinese J Chem Eng 28 969 979
138. R Xiong C Wei 2017 Current status and technology trends of zero liquid discharge at coal chemical industry in China J Water Process Eng 19 346 351
139. SY Alnouri P Linke MM El-Halwagi 2018 Accounting for central and distributed zero liquid discharge options in interplant water network design J Clean Prod 171 644 661
140. Monteiro H, Cruz PL, Oliveira MC, Iten M (2020) Technical and economical assessment of waste heat recovery on a ceramic industry. In: Wastes: solutions, treatments and opportunities III Sel P from 5th International Conference Wastes Solut Treat Oppor, pp 524–530

141. L Kumar M Hasanuzzaman NA Rahim 2019 Global advancement of solar thermal energy technologies for industrial process heat and its future prospects: a review Energy Convers Manag 195 885 908
142. MH Mahfuz MR Anisur MA Kibria R Saidur IHSC Metselaar 2014 Performance investigation of thermal energy storage system with Phase Change Material (PCM) for solar water heating application Int Commun Heat Mass Transf 57 132 139
143. Nazari L, Sarathy S, Santoro D, Ho D, Ray MB, Xu CC (2018) 3—recent advances in energy recovery from wastewater sludge. Direct Thermochem Liq Energy Appl 67–100
144. M Boix L Pibouleau L Montastruc C Azzaro-Pantel S Domenech 2012 Minimizing water and energy consumptions in water and heat exchange networks Appl Therm Eng 36 442 455
145. Z Chijin R Congjing L Zuwei S Jingyuan W Jingdai Y Yongrong 2021 Recent progresses on optimal design of heat integrated water allocation network China Pet Process Petrochemical Technol 23 69 75
146. Gundersen T (2013) Heat integration: targets and heat exchanger network design. Handb Process Integr Minimisation Energy Water Use Waste Emiss 129–167
147. Kermani M, Kantor ID, Maréchal F (2018) Synthesis of heat-integrated water allocation networks: a meta-analysis of solution strategies and network features. Energies 11
148. Rahimi B, Chua HT (2017) Low grade sensible heat-driven distillation. Low Grade Heat Driven Multi-Effect Distill Desalin 19–26
149. Kalaiarasi Ganeson A, Fritzon P, Rogovchenko O, Asghar A, Sjölund M, Pfeiffer A (2012) An OpenModelica python interface and its use in PySimulator. In: Proceedings of 9th international modelling conference, 3–5 Sept 2012, Munich, Germant, vol 76, pp 537–548

# Chapter 18
# Transitioning to Sustainability: Challenges and Innovations in Biorefinery Process Design and Control

**Massimiliano Errico and Stefania Tronci**

**Abstract** The transition from fossil fuel-based production to sustainable biomass utilization is critical for addressing environmental concerns and achieving a circular economy. Biorefineries, utilizing renewable biomass feedstocks, provide an alternative to traditional crude oil refineries, reducing greenhouse gas emissions and wastes. Despite the extensive knowledge and research in conventional oil refineries, biorefineries face unique challenges, particularly in feedstock variability and process complexity. The diverse nature of biomass necessitates accurate thermodynamic data and component properties for effective biorefinery design. Additionally, advanced separation techniques and tailored process control methodologies are essential to manage the inherent variability of bio-derived streams and ensure efficient operations. In this paper, critical aspects of biorefinery development are discussed, starting with a discussion of how the physicochemical properties of biomass influence design and operation. The paper then examines the role of process intensification in significantly improving processes by reducing equipment size, lowering energy consumption, and enhancing sustainability. Additionally, the importance of process control is highlighted as a key factor in optimizing biorefinery operations. Effective process control ensures the consistent performance of unit operations, enables real-time adjustments, and improves overall efficiency and product quality. Through these integrated approaches, biorefinery development can move toward more sustainable, cost-effective, and efficient processes.

**Keywords** Green transition · Process synthesis · Process control

M. Errico (✉)
University of Southern Denmark, Odense, Denmark
e-mail: maer@igt.sdu.dk

S. Tronci
Università Di Cagliari, Cagliari, Italy
e-mail: stefania.tronci@unica.it



A. Angelini et al. (eds.), *Security Enhancement for Climate Change Impacting Urban Resources - SECCURe*, NATO Science for Peace and Security Series C: Environmental Security, https://doi.org/10.1007/978-94-024-2345-7_18

## 18.1 Introduction

The cornerstone of our society ended up reconciling economic growth with sustainability. This challenge has become a driving force for many countries, pushing them to reconsider economic models that rely heavily on non-renewable resources and to accelerate the transition toward sustainable alternatives. The European Union has embraced this vision through the Green Deal, a landmark initiative in which all 27 Member States have pledged to make the EU the world's first climate-neutral continent by 2050 [1]. This ambitious goal is set to be achieved by reducing emissions by at least 55% by 2030 compared to 1990 levels, marking a decisive step toward a more sustainable future.

One of the possible ways to shift to a bio-based economy is to embrace biorefining as a new production standard. The International Energy Agency defined biorefining as the sustainable processing of biomass into a spectrum of marketable products and energy [2]. This simple definition encompasses a wide range of processes and technologies designed to optimize biomass utilization and align its composition with the desired product portfolio. About 300 biorefineries at commercial or demonstration scale are located in the EU [3], highlighting unlocked opportunities to approach the 2050s goal.

Oil-based refineries have the unquestionable historical advantage of an amount of know-how and research performed from the time of the Industrial Revolution and the use of coal. On the other side, biorefineries were developed across the concept of green chemistry in the 90s [4]. It may be claimed that unit operations, process synthesis, design principles, optimization strategies, and analytical techniques are still the same and completely transferrable in a bio-based production. Still, new challenges are hindering a wider diffusion of biorefineries. The first point that arises is the feedstock. Crude oil contains hydrocarbon compounds such as paraffins, naphthenes, and aromatics, small amounts of organic compounds of sulfur, oxygen, and nitrogen, and small amounts of metallic compounds. Despite the different proportions between the classes of compounds, the elemental analysis varies within narrow limits with a typical distribution of 84.5% carbon, 13% hydrogen, 1–3% sulfur, and less than 1% each of nitrogen, oxygen, and metals [5].

In the case of biomass, the classification becomes more complex and different parameters can be considered for the definition of classes with a certain degree of similarity. A possibility is to classify biomasses according to their original compositions, source, and possible conversion technology. For example, the following groups represent biomass with a common origin:

- Wood and wood processing waste
- Forestry biomass and waste materials
- Biogenic materials in municipal solid waste
- Animal manure and human sewage.

Among these biomass classes, there is significant variation in composition. For example, waste cooking oils, categorized under waste materials, primarily consist of fatty acids and triglycerides, whereas forestry biomass is composed mainly of cellulose, hemicellulose, and lignin in varying proportions. Due to this diversity, numerous valorization pathways exist, each involving distinct treatment methods, reaction mechanisms, intermediates, and end products. Consequently, accurate thermodynamic data and component thermo-properties are crucial for designing unit operations, evaluating biorefinery outputs, and developing reliable process models for optimization and dynamic analysis. The first challenge addressed in this chapter is the availability of primary data necessary for the design and development of biorefineries.

Another key point of discussion is the fundamental differences between the processes in a conventional oil refinery and the unit operations involved in biomass treatment. Oil refineries primarily rely on processes such as distillation, hydrotreating, reforming, isomerization, and cracking to convert crude oil into fuels and chemicals. In contrast, biomass processing requires a broader range of technologies, including recovery of added-value compounds through solvent-free extraction methods such as pressurized fluid extraction, microbial fermentation of sugars for lignocellulosic biomass, thermochemical and catalytical pathways together with different separation technologies to purify the products to the market level.

Nevertheless, while conventional distillation remains the standard separation method in oil refineries, the separation of bio-derived compounds presents additional challenges. The presence of diluted streams containing alcohols, proteins, and polar compounds often leads to complex mixtures with multiple azeotropes. Consequently, alternative and more advanced separation techniques are required, such as membrane separations, hybrid processes incorporating liquid–liquid extraction, or enhanced distillation methods.

The last point discussed is extending process control methodologies to bio-based production. In such complex manufacturing systems, controllability becomes a critical factor in ensuring consistent and efficient operation [6]. The necessity of different and interconnected separation techniques makes maintaining stable operations more challenging than in conventional processes. The presence of disturbances in one stage, such as upstream reactions or feed variations, can cascade downstream, impacting the performance of separation and purification units [7]. Processes showing the best ability to achieve acceptable control performance despite unknown but bounded variations, can better respond to input variations and maintain stable operating conditions, ensuring consistent product quality and process efficiency. Unlike conventional systems, where standard techniques like distillation are robustly optimized, biomanufacturing requires control solutions tailored to the variability of bio-derived streams. Designing a proper controller helps obtain a balance between energy efficiency, scalability, and operational stability. However, it is important to note that quality variables are often not measured due to limitations in measuring devices, presenting further challenges in the implementation of process control.

## 18.2 The Need for Fundamental Properties in Bio-processing Design

The modern process design relies on powerful simulation software where the unit operation models are already implemented and interfaced with the component/system data necessary to run the thermodynamic model and to complete the design. In the component definition and thermodynamic method selection stage, the user's input is usually limited to identifying the components from the databank and selecting the appropriate thermodynamic model based on their interaction and processing conditions. Nevertheless, the source of primary data is often not so obvious to the user as well as the possibility that a property is estimated and not measured. The amount of experimental data and estimation methodologies developed for oil-derived compounds is extensive, building up a certain confidence in the user that the necessary data is "available and reliable". Nevertheless, there is not the same research effort and historical advantage to extend this confidence when dealing with biorefining. However, it should be noted that essential references such as the DETHERM database provide thermophysical property data for about 85,310 pure compounds and 202,540 mixtures. The database is updated yearly and grows continually by around 5.7% [8].

Literature review becomes the first source of updated measurements with contributions in the characterization of different biomasses, intermediates and products.

Yahya et al. [9] reported the characterization of corncob, groundnut shell, cashew leaves, lemongrass, sawdust, and *Ixora coccinea* (jungle geranium) using different techniques. In particular, the authors provided proximate and ultimate analyses, functional group, microstructure, and elemental determination, thermogravimetric analysis, different fuel indexes, compositional analysis, and extractive contents, giving valuable information to the designer for the definition of possible process and reaction pathways.

Essential physicochemical, elemental, and biological biomass characterization techniques were discussed by Shandangi et al. [10] to provide a basic path to follow when approaching research in bioderived compounds.

Limiting the analysis to lignocellulosic biomass, Cai et al. [11] discussed the influence of physicochemical properties on design and operation related to its conversion. The authors provided a summary of the properties, the standard method or the analytical procedure, the advantages and disadvantages of the methodology, and their influence in engineering applications. For example, density is related to the design of handling, storage, and transportation facilities, while chemical composition is correlated to conversion efficiency. However, examples are missed to give the designer a clear sense of the extent of variation of the design due to properties uncertainty. A different approach was followed by Lin et al. [12] that related the biomass characterization to the composition of bio-oils obtained through thermal conversion. This represents an important step to analyze also the influence on the unit operations required for the bio-oil upgrade.

Many other relevant contributions highlight the importance of biomass characterization and collecting compounds' properties and analytical methodologies, but specific examples of design cases are often overlooked. To bridge this gap and demonstrate the relationship between properties and design reliability, the separation of levulinic acid has been analyzed.

Levulinic acid (4-oxopentanoic acid) is a bifunctional chemical containing a ketone and a carboxylic acid group that gives the molecule a wide range of possible applications as component in polymer resins, animal feed, food as well as flavor and fragrance industry products, solvents, textile dyes, additives, extenders for fuels, antimicrobial agents, herbicides, and plasticizers [13].

Levulinic acid (LA) can be obtained by lignocellulosic biomass acid hydrolysis. A typical hydrolyzate contains 3–8 wt% LA, 1–5 wt% formic acid (FA), and 1–5 wt% furfural (F) [14]. Levulinic acid does not form any azeotrope with the mixture components, the pair water-furfural forms a minimum-boiling, heterogeneous azeotrope at 97.8 °C and a furfural content of 64.5 wt.%, while the pair formic acid–water forms a maximum-boiling heterogeneous azeotrope at 106.8 °C making the separation based on phase equilibria challenging. Errico et al. [15] proposed different hybrid schemes based on liquid–liquid extraction combined with distillation. Those alternatives were reconsidered and extended by Tronci et al. [16] discussing the impact of the components′ properties on the separation schemes.

Differences between critical properties and normal boiling point reported in the literature and those included in the Aspen Plus data base are summarized in Table 18.1.

The authors further extended their analysis to the binary interaction parameters of the LA-water system, derived by fitting the NRTL model to T-x-y experimental data published in 1968. These data are also included in the NIST database within Aspen Properties. However, when subjected to standard thermodynamic consistency tests, the overall data quality coefficient was found to be close to zero, indicating poor data reliability.

Comparing the design of hybrid systems of liquid–liquid extraction followed by ordinary distillation performed with the standard setting of Aspen Plus and by updating the property database with experimental data, an offset of 2% in the FA purity with respect to the design specification was observed. This result proved the necessity to keep an active research interest in providing fundamental data for new systems that could represent a future industrial standard.

**Table 18.1** Critical properties available in the literature and difference (Δ) with respect to the Aspen database value

| Component | Tc (K) | Pc (bar) | Tb (K) | ΔTc (%) | ΔPc (%) | ΔTb (%) |
|---|---|---|---|---|---|---|
| LA | 766.00 | 53.10 | 518.70 | 3.8 | 32.1 | 2.3 |
| FA | 581.20 | 58.10 | 373.71 | 1.2 | 0.0 | 0.0 |
| F | 670.15 | 58.90 | 434.85 | 0.0 | 4.1 | 0.0 |

### *18.2.1 Future Trends*

Biomass processing encompasses a wide range of alternatives and possibilities regarding the components obtainable, reaction schemes, and intermediates. Among these, biomass pyrolysis is emerging as a promising technique for producing bio-oil, which can be refined and utilized as a substitute for traditional fuels.

A growing body of literature highlights the interest in testing various biomass feedstock to assess key performance metrics, such as optimized process conditions, overall yield, and bio-oil characterization. For instance, Bouaik et al. [17] investigated the effects of temperature, heating rate, and residence time on the yield and quality of pyrolysis products from different macroalgae species. While feasibility studies provide valuable insights, a significant challenge arises when attempting to model the process and design unit operations due to the lack of thermophysical property data for pyrolysis components.

To address this, Gorensek et al. [18] developed a property-based model to integrate the pyrolysis of lignocellulosic biomass into process simulators. The authors acknowledged that many components involved in pyrolysis were absent from existing data banks, necessitating the use of estimation methods. They calculated estimates for the standard solid enthalpy of formation, heat capacity, and density based on the limited experimental data available. Their study covered six biomass constituents and nine pyrolysis intermediates and end-products, including glyoxal, acrolein, furfural, xylo-san, levoglucosan, and 5-hydroxymethylfurfural.

Among these compounds, levoglucosan—an anhydrous sugar and a major product of biomass pyrolysis—stands out as particularly important. Experimental data on its properties are crucial for improving process modeling accuracy. Several researchers have contributed to filling this gap. Kabo et al. [19] measured its heat capacity over a temperature range of 5–370 K using adiabatic calorimetry and also reported data on its solid-phase transitions and enthalpy of fusion. Additionally, Shoji et al. [20] determined its boiling point, further expanding the available property data.

To enhance the reliability of pyrolysis modeling, these newly available experimental data should be integrated into process simulator databanks. Furthermore, existing databanks should be critically reviewed to determine the origin of the available property values and whether they are experimentally determined or estimated. Improving the accuracy of these properties will be instrumental in refining biomass pyrolysis models and advancing the field of bio-based fuel production.

## 18.3 Biorefining Processes

In 2014, Kiss defined distillation as a technology that is "still young and full of breakthrough opportunities" [21]. The author fully embraced as a mature technology as distillation can still offer opportunities when coupled with process intensification principles. In general terms, process intensification includes technologies that allow

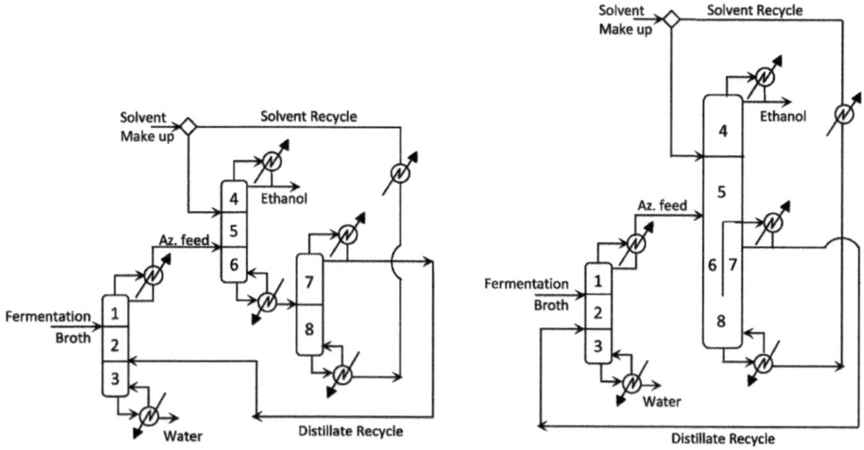


**Fig. 18.1** Simple column extractive distillation configuration (left), intensified alternative with divided wall column (right)

a substantial enhancement in the process. This can be achieved through a reduction of the equipment size, a reduction of the energy consumption and all possible ways to increase the process sustainability.

Intensified alternatives have found interesting applications in the separation of biofuels.

Errico et al. [22] proposed an intensified scheme obtained from the traditional configuration reported in Fig. 18.1 (left) to separate bioethanol from fermentation broths. This scheme, based on extractive distillation includes three columns, the first called prefractionator is used to approach the ethanol azeotropic composition. In the second column, ethanol is separated at the desired purity using a high boiling point solvent. The third column is used for solvent recovery and to recycle the water with ethanol residues.

In the intensified alternative (Fig. 18.1, right), the first column remains unchanged while the second and third columns are combined in a divided wall column that separates ethanol as distillate, the solvent is obtained as residue while a side stream is recycled back to increase the overall ethanol recovery. The two configurations achieved the same energy consumption however, the intensified alternative reached over 20% savings in the capital costs.

For the same separation, Errico et al. [23] proposed a combination of forward osmosis membranes and different distillation alternatives. In particular, reactive distillation was considered as an alternative to overcome the azeotropic limitations through simultaneous reaction and separation in the same unit [24].

Analogous to the previous case, in the configuration reported in Fig. 18.2 (left), the first column is the fractionator that separates a stream with a composition close to the azeotropic composition. In the second column, the reaction between ethylene

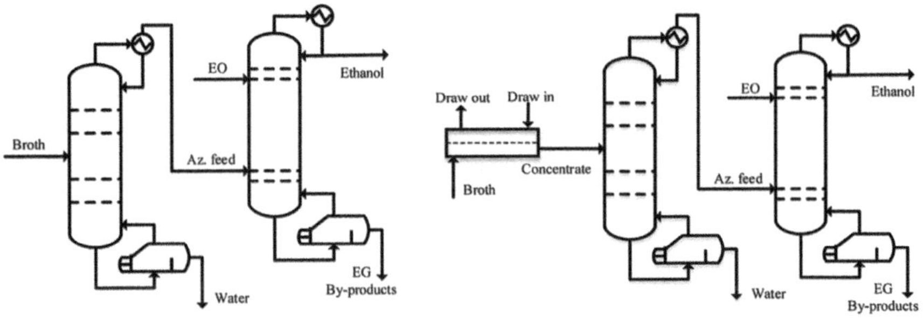


**Fig. 18.2** Preconcentration plus reactive distillation (left), membrane-assisted configuration (right)

oxide and water is used to produce pure ethanol as distillate and the reaction product, ethylene glycol, as residue.

In the alternative reported in Fig. 18.2 (right), part of the water removal is performed through a membrane separation. This configuration can achieve substantial savings in the prefractionator column's reboiler duty. However, a direct comparison is not completely fair due to the difference in the ethanol recovery due to a partial loss in the draw solution. The reactive distillation option still represents an interesting alternative for the bioethanol separation, but the ethyl oxide cost, together with the marketability of the reaction products, needs to be carefully evaluated in the overall evaluation.

Another intensified cases applied in a biorefinery context are the production and separation of biodiesel from waste cooking oil through reactive distillation [25], the separation of acetone, butanol, ethanol (ABE mixture) from fermentation broth through hybrid schemes including liquid–liquid extraction and different distillation alternatives [26], or heat pump-assisted vacuum distillation for the separation of butanediols from fermentation [27], just to mention a few.

## 18.4 Process Control

The previous sections highlighted that one of the main aspects related to obtaining bio-chemicals from sustainable but difficult-to-treat feedstock is the necessity of non-conventional separation technologies, and the importance of process intensification for improving efficiency, reducing energy consumption, and enhancing the overall scalability and feasibility of the production processes [28]. Automation can give a valid help in this direction, because it can improve the efficiency of the process and guarantee the maintenance of the optimal conditions. However, this is not an easy task, as process intensification can lead to operations that are difficult to manage due to strong interactions among the control loops [29]. Additionally, quality-related variables are often not measured, posing a challenge for implementing effective control strategies. It is important to highlight that for addressing the mentioned process control issues, it is fundamental to have a dynamic model of the process, that is used for assessing controllability, designing and implementing control strategies, and evaluating control performance. The following subsections will discuss the challenges associated with dynamic modeling, the analysis of controllability, and the development of advanced control strategies for biorefineries.

## *18.4.1 Dynamic Modeling of Biorefinery Processes*

Design strategies centered on process intensification and integration typically rely on steady-state models, which are invaluable for optimizing systems improving energy efficiency and minimizing waste. However, effective process control needs a deep understanding of the dynamic relationships between input and output variables. This makes dynamic models indispensable for accurately predicting system behavior, designing robust control strategies, and ensuring satisfactory performance in all the operation units in the plant [30]. Despite their importance, the biorefinery community still faces a significant gap in dynamic modeling efforts, which raises concerns [31]. One reason can be related to the biorefinery-specific conversion processes, being the understanding of the dynamic nature of many of these reactions still incomplete and limited. This is due to a combination of insufficient mechanistic understanding, a scarcity of experimental data and difficult chemical characterization of the feedstocks and products [32].

A review on biowaste refinery models was proposed by De Buck et al. [33], primarily focusing on pretreatment and conversion processes and neglecting the downstream separation units. Similarly, Pomeroy et al. [34] explored the application of neural network modeling across different stages of biorefinery processes, highlighting its potential for optimizing individual operations. However, they did not present a unified model encompassing all stages of a biorefinery, from pretreatment to separation. Recently, Gargalo et al. [35] discussed how hybrid modeling for biochemical processes integrates mechanistic and data-driven approaches to enhance process understanding and optimization. This approach can offer a valid tool for advancing digitalization in biochemical processes, enabling real-time decision-making, sustainability, and efficient resource use.

Several detailed models for biorefineries focus on the production of bioethanol, which is the most common biofuel derived from biowaste. Prunescu et al. [30] developed a plant simulator implemented in modules for each conversion step, i.e. pretreatment, enzymatic hydrolysis and fermentation. A notable advancement in comprehensive biorefinery modeling was made by Isimite [36], where a training simulator for a biorefinery producing ethanol was proposed, incorporating detailed mathematical representations of pre-treatment, liquefaction, and distillation units. The simulator integrates operator input to address realistic training needs and features an emulated distributed control system for improved usability. A different process was investigated in Sbarciog et al. [37] where the model for a small size biorefinery, converting biowaste into oligosaccharides, lignin, fibres, biogas, fertiliser and compost, was developed. The integrated model combines processes like steam refining, anaerobic digestion, ammonia stripping, and composting to simulate and improve the performance of a small-scale biorefinery.

While these efforts mark significant progress, there is still the need to expand modeling activities to encompass larger, more complex biorefinery systems. This will require a concerted effort to deepen mechanistic understanding, enhance experimental data collection, and develop innovative modeling frameworks capable of addressing the unique challenges posed by biomass conversion and valorization.

### *18.4.2 Controllability Analysis for Robust Operational Performance in Biorefineries*

According to Skogestad and Postlethwaite [38], input–output controllability is defined as the system's ability to achieve acceptable control performance. This involves maintaining the system's outputs within specified bounds or close to desired reference values, despite unknown but bounded variations such as disturbances and changes in the plant, using the available inputs and measurements. In biorefineries, which consist of a complex network of interconnected units for pretreatment, conversion, and separation processes, achieving input–output controllability is crucial to meet operational targets. In particular, controllability metrics—such as the minimum singular value, the condition number of the transfer matrix, and relative gain arrays (RGA) to quantify interactions within the control loops—should be incorporated into the design process. These metrics enable the evaluation and differentiation of various solutions, ensuring optimal dynamic performance and robust control under feedback conditions [29, 39]. It is worth noting that input–output controllability is also influenced by the placement of sensors and actuators, making it imperative to carefully evaluate these parameters during the design phase, using a robust dynamic model that accurately mimics the real behavior of the processes occurring in the biorefinery. Another important aspect is related to the required operational flexibility to handle variations in feedstock composition and process conditions. Addressing both flexibility and controllability in process synthesis and design leads to a system that can adapt to varying conditions while maintaining good dynamic performance. Essentially, flexibility is not solely determined by the chosen design and operating point but is also influenced by the type of controller and control structure implemented [40].

A potential limitation of the controllability analysis usually performed is its reliance on linear measures, which may appear restrictive when applied to biorefinery processes. In many cases, this is not a significant problem, because nonlinearities such as input constraints can be effectively managed through linear analysis, and controllability can be assessed at multiple operating points. Numerous case studies in chemical engineering have shown that linear methods are often highly effective [39]. However, the complexity of biorefinery processes—particularly the conversion of biomass and separation of non-ideal mixtures—can introduce nonlinearities that linear methods may struggle to address. In these instances, more advanced techniques such as the ones based on Lie-algebra, can be used to assess global reachability and controllability, determining the minimal number of control inputs required to control the system effectively [41, 42].

Controllability analysis methods are extensively documented in the literature for both linear and nonlinear systems. However, it is essential to extend their application to biorefineries due to the distinctive challenges these facilities face. One aspect is model uncertainty, which can arise from various factors such as the inherent complexity of biochemical processes, variability in feedstock composition, separation of nonideal mixture, and operational conditions. Additionally, the lack of comprehensive sensors in biorefineries means that not all phenomena can be accurately captured and related in the model. This gap in measured data can lead to incomplete or imprecise models, further complicating this type of analysis.

### *18.4.3 Process Control as a Tool for Improving Biorefineries Efficiency and Sustainability*

Given the challenges highlighted in the previous sections, it naturally follows that control strategies are crucial for ensuring the efficient operation of biorefineries. While conventional control techniques provide a starting point, the complexity of biorefinery operations may necessitate advanced approaches tailored to these systems.

An investigation where the effect of a proper tuning of traditional PI controller is important to improve the behavior of the process was given by Arenas-Grimaldo et al. [39]. The authors implemented PI controllers in the separation sequence for ethanol production from sotol bagasse and found that higher controller gain values were associated with a reduction in the integral of absolute error, suggesting an improvement in controllability. This correlation highlighted that dynamic system responses can give useful insights to optimize control strategies for better overall system performance.

Most contributions in the literature on process control for bioprocesses primarily focus on bioreactors, where fermentation occurs. A seminal study by Nagy [43] investigated the application of feedforward neural networks for dynamic modeling and temperature control in a continuous yeast fermentation bioreactor. This research used a quite detailed model to describe the dynamic behavior of the reacting system and compared neural network-based model predictive control (MPC) with linear MPC and PID control, highlighting the superior robustness of the neural network approach in handling setpoint changes and noisy temperature measurements. Nagy's work has since inspired a range of studies on bioreactor control using advanced control techniques [44–48]. One of the most recent is the paper by Rajasekhar et al. [49] where a reinforcement learning algorithm was exploited to achieve precise temperature control in a fermentation bioreactor, with the goal of optimizing ethanol production. Their study tackles challenges such as model-process mismatches in model-based control systems, focusing on maintaining reactor temperature within a narrow range to promote optimal microbial growth and maximize ethanol yield.

It is worth noting that the fermentation reactor used in ethanol production serves as a representative model for bioreactors, providing a solid foundation for developing control strategies within the field of biomanufacturing. However, it is essential to advance further by designing controllers for more complex biowaste feedstocks and integrating all process units [36]. This approach will help identify optimal control solutions that enhance the overall performance and efficiency of the entire biorefinery system. Additionally, the challenge of unmeasurable variables complicates the implementation of effective control strategies [47]. Therefore, there is a pressing need to develop innovative online sensors, state estimation methods, or soft sensors to address these gaps, ensuring that all critical parameters can be accurately monitored and controlled in real-time for optimal biorefinery performance.

## 18.5 Conclusions

The design of plants for converting waste biomass into valuable products requires dedicated research to develop efficient, competitive systems capable of handling diverse feedstocks and adapting products to market demands and feedstock characteristics. This paper highlights three key aspects to achieve these goals. First, the importance of enhancing property databases for biomass conversion components is emphasized, with levulinic acid production serving as an illustrative example. Modern process design depends heavily on simulation software that integrates unit operation models and thermodynamic data, but gaps in experimental data for bio-derived compounds hinder accurate modeling and design. Continued research on characterizing biomass and its derivatives is crucial for improving property databases, refining simulation tools, and progressing in the development of biofuel technologies. Second, the application of process intensification principles is explored showing that it is a necessary tool for enhancing efficiency, sustainability, and cost-effectiveness. Innovations in biofuel separation processes, including bioethanol recovery and biodiesel production, benefit from advanced techniques such as divided wall columns, reactive distillation, and hybrid membrane-assisted systems, offering cost and energy savings while addressing typical challenges, such as the presence of azeotropes. Finally, the paper highlights the critical role of process control, demonstrating that controllability measures are essential for designing optimized process configurations. Advanced process control strategies are discussed, highlighting challenges and contributions in this area to further enhance the reliability and performance of biomass conversion processes.

## References


1. European Union. https://reform-support.ec.europa.eu/what-we-do/green-transition_en. Last accessed 31 Jan 2025
2. The International Energy Agency. https://www.ieabioenergy.com/wp-content/uploads/2013/10/Task-42-Booklet.pdf. Last accessed 31 Jan 2025
3. EU Biorefinery Outlook to 2030. https://op.europa.eu/en/home. Last accessed 31 Jan 2025
4. Anastas P, Warner J (2000) Green chemistry: theory and practice. Oxford University Press
5. Aitani AM (2004) Oil refining and products. In: Encyclopedia of energy. Elsevier, pp 715–729
6. Velazquez-Samano TE et al (2025) Assessment of control properties in intensified processes for levulinic acid purification. In: Separation and purification technology, vol 361
7. Gao J et al (2024) Reinforcement learning-based control for waste biorefining processes under uncertainty. Commun Eng 3
8. DECHEMA. https://dechema.de/en/detherm.html. Last accessed 2025/02/04
9. Yahya et al (2023) Comprehensive characterization of some selected biomass for bioenergy production. ACS Omega 8(46):43771–43791
10. Shandangi KP (2023) Characterization techniques of biomass: physico-chemical, elemental, and biological. Elsevier, pp 51–66
11. Cai J et al (2017) Review of physicochemical properties and analytical characterization of lignocellulosic biomass. Renew Sustain Energy Rev 76
12. Lin F et al (2015) Relationship between biomass composition and liquid products formed via pyrolysis. Frontiers 3
13. Pyo S-H et al (2020) Clean production of levulinic acid from fructose and glucose in salt water by heterogeneous catalytic dehydration. ACS Omega 5(24)
14. Di Bucchianico DDM et al (2022) Production of levulinic acid and alkyl levulinates: a process insight. Green Chem 2
15. Errico M et al (2021) Novel intensified alternatives for purification of levulinic acid recovered from lignocellulosic biomass. Processes 9(3)
16. Tronci S et al (2023) Analysis of hybrid separation schemes for levulinic acid separation by process intensification and assessment of thermophysical properties impact. In: Separation and purification technology, vol 310
17. Bouaik H et al (2025) Pyrolysis of macroalgal biomass: a comprehensive review on bio-oil, biochar, and biosyngas production. Sustain Chem One World 5
18. Gorensek M et al (2019) Development of a thermophysical properties model for flowsheet simulation of biomass pyrolysis processes. ACS Sustain Chem Eng 7(9)
19. Kabo GJ et al (2015) Experimental and theoretical study of thermodynamic properties of levoglucosan. J Chem Thermodynamics 85
20. Shoji T et al (2014) Boiling point of levoglucosan and devolatilization temperatures in cellulose pyrolysis measured at different heating area. J Anal Appl Pyrolysis 109
21. Kiss AA (2014) Distillation technology—still young and full of breakthrough opportunities. J Chem Technol Biotechnol 89
22. Errico M et al (2013) Optimal synthesis of distillation systems for bioethanol separation. Part 2. Extractive distillation with complex columns. Ind Eng Chem Res 52(4)
23. Errico M et al (2020) Membrane assisted reactive distillation for bioethanol purification. Chem Eng Process Process Intensification 157
24. Kiss AA et al (2018) Reactive distillation: stepping up to the next level of process intensification. Ind Eng Chem Res 58(15)
25. Grosmann MT et al (2022) Hydrated metal salt pretreatment and alkali catalyzed distillation: a two-step production of waste cooking oil biodiesel. Chem Eng Process Process Intensification 176
26. Errico M et al (2027) Multiobjective optimal acetone-butanol-ethanol separation systems using liquid-liquid extraction-assisted divided wall columns. Ind Eng Chem Res 56(40)
27. Jankovic T et al (2025) Adaptable downstream processing design for recovery of butanediols after fermentation. Chem Eng Res Des 213

28. Pandey K et al (2024) Bio-processing 4.0 in biomanufacturing: paving the way for sustainable bioeconomy. Syst Microbiol Biomanufacturing 4(2):407–424
29. Velazquez-Samano TE et al (2025) Assessment of control properties in intensified processes for levulinic acid purification. In: Separation and purification technology, vol 131457
30. Prunescu RM et al (2017) Model-based plantwide optimization of large scale lignocellulosic bioethanol plants. Biochem Eng J 124:13–25
31. Pujan R, Preisig HA (2022) Biorefinery modelling is in tatters, and here is why. In: Computer aided chemical engineering, vol 51. Elsevier, pp 295–300
32. Pérez PAL, Lopez RA, Femat R (2020) Control in bioprocessing: modeling, estimation and the use of soft sensors. John Wiley & Sons
33. De Buck V et al (2020) Modeling biowaste biorefineries: a review. Frontiers Sustain Food Syst 4
34. Pomeroy B et al (2022) Artificial neural networks for bio-based chemical production or biorefining: a review. Renew Sustain Energy Rev 153:111748
35. Gargalo CL et al (2024) Navigating industry 4.0 and 5.0: the role of hybrid modelling in (bio) chemical engineering's digital transition. Frontiers Chem Eng 6:1494244
36. Isimite JO (2024) Development of a biorefinery training simulator with automatic operating procedure adaptation. Doctoral dissertation, University College London
37. Sbarciog M et al (2022) Design, implementation and simulation of a small-scale biorefinery model. Processes 10(5):829
38. Skogestad S, Postlethwaite I (1996) Multivariable feedback control. John Wiley & Sons
39. Arenas-Grimaldo C et al (2024) Design and control of a distillation sequence for the purification of bioethanol obtained from sotol bagasse (Dasylirium sp.). In: Chemical engineering research and design, vol 203, pp 11–17
40. Bahri PA et al (1997) Integrated flexibility and controllability analysis in design of chemical processes. AIChE J 43(4):997–1015
41. Drexler DA et al (2018) Controllability and reachability of reactions with temperature and inflow control. Fuel 211:906–911
42. Lisci S et al (2021) Different control strategies for a yeast fermentation bioreactor. IFAC-PapersOnLine 54(3):306–311
43. Nagy ZK (2007) Model based control of a fermentation bioreactor using optimally designed artificial neural networks. Chem Eng J 127:95–109
44. Pachauri N et al (2017) Bioreactor temperature control using modified fractional order IMC-PID for ethanol production. Chem Eng Res Des 122:97–112
45. Kumar M et al (2019) Temperature control of fermentation bioreactor for ethanol production using IMC-PID controller. Biotechnol Rep 22:e00319
46. Petre E et al (2021) Advanced nonlinear control strategies for a fermentation bioreactor used for ethanol production. Biores Technol 328:124836
47. Lisci S et al (2020) A geometric observer-assisted approach to tailor state estimation in a bioreactor for ethanol production. Processes 8(4):480
48. Chai et al (2022) Fermentation process control and optimization. Chem Eng Technol 45(10):1731–1747
49. Rajasekhar N et al (2024) Reinforcement learning based temperature control of a fermentation bioreactor for ethanol production. Biotechnol Bioeng 121(10):3114–3127

# Part V
# Energy/Food/Water Nexus: Technical and Social Perspective for a Safer Management of Resources

# Chapter 19
# Using Systems Thinking Approaches for 'Nexus' Understanding and Analysis

**Virginia Rosa Coletta, Stefania Santoro, Alessandro Pagano, Wendy Francesconi, Lucila Marcela Beltrán, Martín Londoño-Behaine, and Raffaele Giordano**

**Abstract** Human-natural systems are characterized by complex interactions across different domains, challenging sustainable resource management. Holistic approaches, such as the Nexus approach, are required to address mutual interdependencies across sectors and to explore synergies and trade-offs. This study applies a System Thinking approach, implemented through qualitative System Dynamics modelling based on Causal Loop Diagrams (CLDs), to identify and prioritize Nexus challenges, i.e., cross-sectoral issues that affect system sustainability. Structural and descriptive analyses of the CLDs are proposed to support this prioritization, while stakeholder engagement throughout the modelling process enhances system understanding. The modelling approach was applied to two Nexus hotspots: the Nima River Basin in Colombia, a Water-Ecosystem-Food (WEF) system facing water scarcity

V. R. Coletta · S. Santoro (✉) · R. Giordano
Water Research Institute, National Research Council, Bari, Italy
e-mail: stefaniasantoro@cnr.it

V. R. Coletta
e-mail: virginiarosacoletta@cnr.it

R. Giordano
e-mail: raffaele.giordano@cnr.it

A. Pagano
Department of Civil, Environmental, Land, Construction and Chemistry, Polytechnic University of Bari, Bari, Italy
e-mail: alessandro.pagano@poliba.it

W. Francesconi
Alliance of Bioversity International and International CIAT, Rome, Italy
e-mail: w.francesconi@cgiar.org

L. M. Beltrán · M. Londoño-Behaine
The Alliance of Bioversity International and CIAT, Palmira, Colombia
e-mail: m.beltran@cgiar.org

M. Londoño-Behaine
e-mail: m.londono@cgiar.org



A. Angelini et al. (eds.), *Security Enhancement for Climate Change Impacting Urban Resources - SECCURe*, NATO Science for Peace and Security Series C: Environmental Security, https://doi.org/10.1007/978-94-024-2345-7_19

and pollution driven by intensive agriculture; and in the Ucayali Region in Peru, a Biodiversity-Climate-Society (BCS) system impacted by deforestation linked to cocoa production and weak infrastructure. Key findings highlight the central role of variables such as 'water demand' in Nima and 'deforestation' in Ucayali, which influence multiple sectors and therefore need attention. The approach revealed systemic trade-offs and sets the stage for identifying priority areas of the systems (i.e., leverage points) for integrated governance. The modelling process proves versatile, offering a robust, participatory framework applicable to diverse Nexus systems and contexts.

**Keywords** System dynamics modelling · Nexus challenges · Stakeholder engagement

## 19.1 Introduction

Human-natural systems are composed of a complex set of interacting physical, technological, ecological and socio-economic subsystems, deeply conditioned by human intervention [1]. Their sustainability is increasingly in question, especially because of the unsustainable use of the resources on which they rely (e.g., water, soil, energy, food) and the effects of climate change (e.g., droughts and floods) [2]. Managing these systems requires integrated approaches that support the understanding of all key aspects, considering the complex and non-linear interactions of various social and ecological processes across multiple spatial and temporal scales [3]. The literature presents different integrated approaches for conceptualising and analysing complex human-natural systems. These approaches aim to achieve integration and coordination for efficient, equitable and sustainable natural resource management [4]. One of the best-known examples is Integrated Water Resources Management (IWRM). However, existing integrated approaches tend to assume a water-centered perspective, considering other sectors (e.g., food, energy and ecosystem) merely as users of water resources [5]. Recent reports, such as the 2023 World Water Council publication, call for broadening the IWRM paradigm to explicitly include interlinkages with other sectors (e.g., energy, food, health and education) [6]. This hinders an in-depth understanding of the complexity and dynamics of interconnections and trade-offs across different resources [7], thus limiting governmental institutions in practically implementing an integrated resource management approach into policy and decision-making [8]. All these considerations highlight the need for a shift from integrated approaches to holistic approaches, i.e., approaches that emphasize the intertwined nature of human-natural systems and reveal the complex linkages and feedbacks that occur on different temporal and spatial scales and between different levels, sectors and groups [9]. The idea behind holistic approaches is that the analysis of individual parts of complex systems (such as human-natural systems) is not sufficient for fully understanding them because their behaviours come from the interactions among their different components [10]. One holistic approach that is receiving increasing attention is the "Nexus" approach [11], which seeks to comprehensively address the mutual interdependencies between different sectors of a complex system (e.g., water-energy/ecosystem-food (WEF), biodiversity-climate-society (BCS), etc.) also emphasising synergies and trade-offs [12]. The concept of Nexus was originally framed in the influential World Economic Forum (2011) report [13], which introduced the Water-Food-Energy-Climate Nexus as a foundation for integrated sustainability governance. Since then, the WEF Nexus has become the most widely studied and applied configuration. However, in response to evolving environmental and societal challenges, alternative Nexus configurations, such as the Biodiversity-Climate-Society (BCS) Nexus, are now emerging to expand the scope and relevance of Nexus thinking. However, regardless of the specific sectors involved, the central idea of the Nexus approach remains unchanged: it is a systems perspective that reveals cross-sectoral linkages, feedbacks and trade-offs to support more integrated and sustainable resource management. Despite the theoretical benefits of the Nexus approach, its operationalization within policies is making slow progress mainly due to the difficulty of effectively identifying the key Nexus challenges of a complex system using the majority of traditional analytical frameworks [14]. The term "Nexus challenges" refers to the key cross-sectoral issues that affect the sustainability of the Nexus and need to be addressed in an integrated way across sectors [15]. System Thinking, operationalized through System Dynamics (SD) modelling, has recently been identified in scientific literature as a holistic approach to understanding and addressing Nexus challenges (see e.g., [16]). In fact, while the traditional thinking paradigm tends to isolate problems without considering their broader context, System Thinking provides methods for accounting for feedback loops, interconnections, and time delays, leading to a more comprehensive understanding of complex systems' behaviour [17]. Specifically, SD modelling allows for a transdisciplinary perspective, combining inputs from different domains and taking into account people's different perceptions of the real-world system, particularly the causal relationships and feedback between components of a system [18]. It helps describe the dynamic patterns, feedback loops, trade-offs, and synergies that shape interactions between systems, leading to more effective decision making [19]. Among the available SD modelling tools, Causal Loop Diagrams (CLDs) have the ability, in a qualitative way (i.e., without having to deal with computational mathematical calculations) to (i) support the mapping of the feedback structure of the system; (ii) make preliminary assumptions about the system's behaviour, taking into account how changes in one part may propagate and affect other elements; (iii) incorporate both scientific and stakeholder knowledge, thus expanding understanding of the system [20]. Based on that, these qualitative modelling tools prove to be reliable in identifying key variables/interactions in the system [21], being functional in prioritizing Nexus challenges and elements that can strongly influence their dynamic evolution [22]. This prioritization obtained from qualitative modelling can then be functional in the construction and simulation of SD quantitative models (e.g., Stock and Flow models), which is useful in supporting policy and decision-making.

Starting from these considerations, the present work applies the principles of System Thinking, implemented through a participatory SD modelling approach based on CLDs, to support the analysis of complex interactions in Nexus systems and identify Nexus challenges (crucial and cross-sectoral issues that impact the sustainability of the Nexus and require an integrated approach). Although the proposed modelling approach is versatile and applicable to different Nexus contexts, regardless of the specific components and sectors involved, this work focuses on WEF and BCS Nexus systems.

The research question this paper aims to answer is: to what extent does the proposed modelling approach, and in particular the use of CLDs, enable the prioritization of Nexus challenges thus increasing our understanding of complex systems? To address this question, the specific objectives of the study are: (i) to design a participatory qualitative SD modelling process based on CLDs; (ii) to apply this process to two case studies representing different Nexus configurations; and (iii) to identify and prioritize cross-sectoral Nexus challenges through structural and descriptive analysis.

The proposed modelling process is applied to two case studies from EU-funded projects. The first case study is the Nima River Watershed (Colombia), an example of the WEF Nexus, examined within the REXUS (Managing Resilient Nexus Systems Through Participatory Systems Dynamics Modelling, [23]) and LENSES (LEarning and Action Alliances for NexuS EnvironmentS in an Uncertain Future, [24]) projects. The second case study is the Ucayali Region (Peru), an example of the BCS Nexus, analysed within the BIOTRAILS (Nexus Framework for biodiversity-relevant transformative change, [25]) project.

The paper is structured as follows: Sect. 19.2 outlines the proposed modelling process; Sect. 19.3 presents the results of its application to the WEF and BCS Nexus systems; Sect. 19.4 discusses the main implications of the modelling approach.

## 19.2 Modelling Process

To support the identification of Nexus challenges within complex human-natural systems, the participatory SD modelling process is structured into three phases: (i) system conceptualization, (ii) system mapping, (iii) system analysis (see Fig. 19.1).

The *System Conceptualization* phase aims to define the key elements and interconnections of Nexus systems. Activities such as literature review and insights from case study leaders contribute to the description of the Nexus interlinkages baseline. These activities are enriched by information collected through Semi-Structured Interviews (SSIs) with case study stakeholders. SSIs are chosen for their ability to capture individual experiences, perceptions, and attitudes, thereby bridging gaps in the literature [26]. For the SSIs, a content analysis is carried out to transform the "interview quotes" into a Causal Loop Diagram (CLD) [27] following three phases: (i) text coding, i.e., SSIs are analysed to identify key terms in stakeholders' arguments, which are then translated into variables within the CLD; (ii) identification of the links, i.e., causal variables, effect variables, and their interrelationships are determined; (iii) CLD model building, i.e., the system is represented by depicting

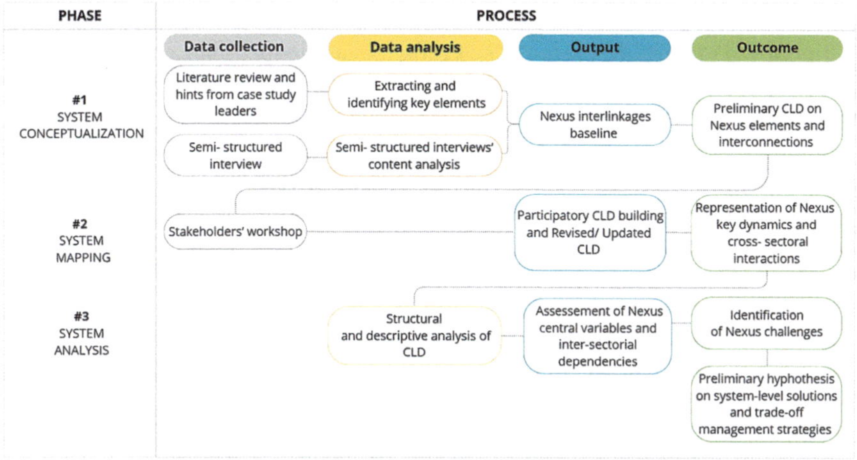


**Fig. 19.1** Outline of the proposed modelling process

the cause-and-effect relationships among key variables, enabling the integration of scientific and stakeholders knowledge [28]. Causal links between variables are illustrated with directed arrows, marked with polarity signs to denote positive (+) or negative (–) relationships [29]. These interconnections frequently form feedback loops that can impact the system's stability, potentially causing either imbalances or equilibrium states. Feedback loops (see Fig. 19.2) can be either reinforcing (R), driving exponential growth or decline, or balancing (B), helping to maintain stability or induce cyclical fluctuations within the system [30]. The outcome of this phase is a preliminary CLD focused on Nexus elements and their connections.

The *System Mapping* phase aims to validate and expand the preliminary version of the CLD. It takes place through a participatory exercise conducted during a workshop with local stakeholders. The participatory exercise is a tool for collecting local

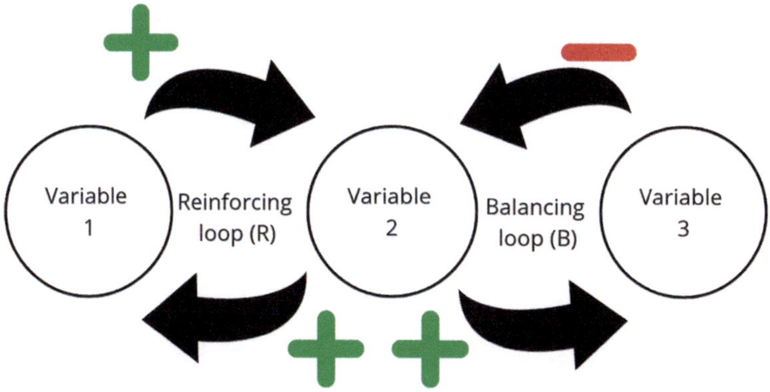


**Fig. 19.2** Graphical representation of feedback loops within a causal loop diagram

knowledge, bridging gaps in the literature, and integrating missing information [31]. During the participatory exercise, stakeholders are divided into groups, and each group is asked to connect key aspects of the Nexus system in the form of a CLD. The discussion is facilitated by an analyst. After that, using Synthesis Theory, the preliminary CLD is validated and enriched by incorporating the information from the CLDs produced by each stakeholder group. Synthesis theory is generally applied to models characterized by high complexity [32]. The CLDs from individual groups are analysed to identify key concepts, and subsequently, to draw main relationships. The outcome of this phase is a more precise representation of the Nexus's key dynamics and cross-sectoral interactions in the form of a causal map.

The *System Analysis* phase aims to identify the key Nexus challenges, i.e., the variables characterized by a high centrality within the Nexus system and having cross-sectoral interactions.

To this end, both structural and descriptive analyses of the CLD are conducted. The structural analysis employs calculating Centrality Measures based on Graph Theory to examine the system's structure. Two measures are used: (i) Degree Centrality, which measures the direct influence of a node by counting the number of connections it has within the network. A node with a high number of connections occupies a prominent position in communication and information access and is therefore considered the most central [33]; (ii) Betweenness Centrality, which counts the number of shortest paths through the network that pass through a given node, indicating its role as a communication hub. The Kumu [34] Social Network Analysis module can be used to automatically calculate these measures. The combination of different Centrality Measures enables to make some preliminary hypothesis on the Nexus challenges.

The descriptive analysis of the CLD focuses on identifying the inter-sectoral impacts of the variables selected during the structural analysis to formulate hypotheses on their potential dynamic evolution due to their position within the system [35].

The outcome of this phase is twofold: on the one hand, it supports the identification of Nexus challenges based on the structural relevance and inter-sectoral role of system variables; on the other hand, it enables the formulation of preliminary hypotheses on potential system-level solutions and strategies to manage trade-offs across sectors. These insights contribute to laying the groundwork for future decision-making and policy development.

## 19.3 Case Studies

### *19.3.1 Nima River Watershed as an Example of the Water-Ecosystem-Food Nexus*

The Nima River Basin is located in the southeastern part of the Cauca Valley Department in the Colombian Andes (Fig. 19.3). It covers an area of 16,739 ha and is impacted by two rainy seasons throughout the year, separated by drier periods [36]. The catchment is divided into two areas with elevations ranging from 1050 to 4100 m above sea level: a mountainous upstream area, fundamental for the provision of water-related ecosystem services provisioning to downstream users; a flat downstream area, dominated by agricultural activities thanks to the water provided by the upstream region. About 8700 ha are protected from an environmental point of view [37]. One of the most important functions of the Nima watershed is supplying water to the municipality of Palmira (312,519 inhabitants), to a hydropower plant, and to an irrigation district dedicated mainly to sugarcane cultivation. During the dry season, water availability is significantly affected, mainly with reference to the agricultural part. The basin has different types of asymmetries (i.e., hydrological, biophysical and socioeconomic) that require conservation, restoration and sustainable land use interventions. Key issues include the need to improve water use efficiency, exploration of alternative agricultural systems, and implementation of incentive mechanisms to align land management with shared environmental and socioeconomic goals.

### *19.3.2 Ucayali Region as an Example of the Biodiversity-Climate-Society Nexus*

The Ucayali Region (Peru) (Fig. 19.4) is currently experiencing one of the highest deforestation rates in the country with the aim of expanding cocoa production areas. In fact, Peru is the world's second-largest cocoa producer, with the Ucayali region serving as a key production hub. In 2017, the region's cocoa exports were valued at $2.7 million [39]. While the cocoa industry makes a significant contribution to the country's GDP, it also poses substantial environmental and social challenges [40]. Among these challenges, deforestation is one of the most pressing concerns. Cocoa cultivation often leads to forest clearing, which threatens biodiversity and intensifies climate change. Additionally, deforestation accelerates soil erosion and degradation, reducing long-term fertility and agricultural productivity. Another critical issue is the widespread use of chemical fertilizers and pesticides, which contaminate soil and water resources, harming both the ecosystem and local communities' health [41].

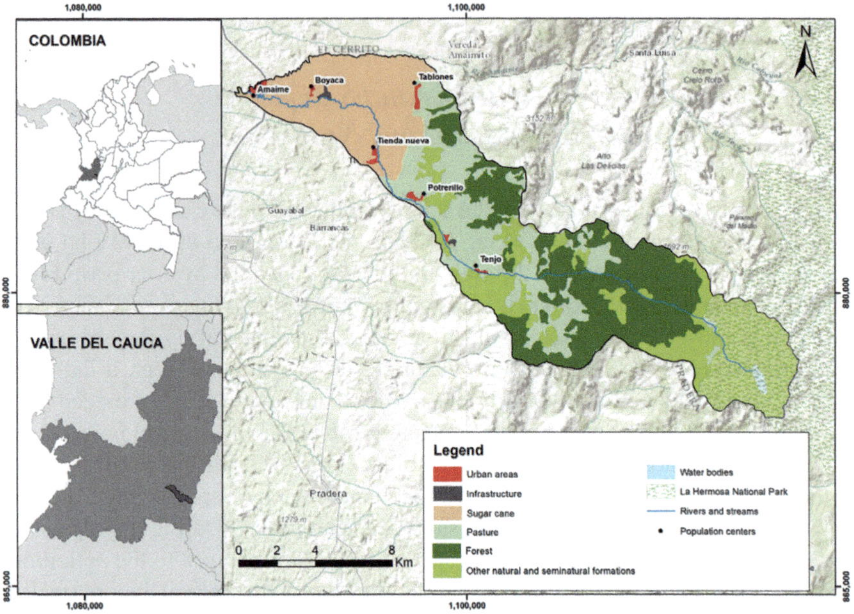


**Fig. 19.3** Nima River Basin location and land covers. Reproduced from [38], licensed under CC BY 4.0

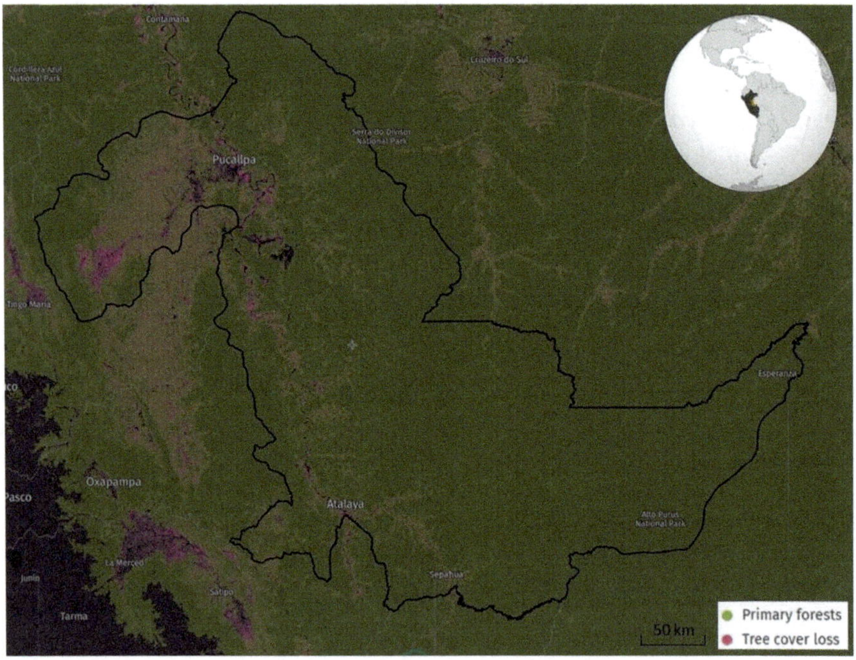


**Fig. 19.4** Ucayali region location and cover [42]

## 19.4 Results

### *19.4.1 Application of the Modelling Process to the Nima River Watershed*

**System Conceptualization**

The information provided by both the baseline description and the Pilot leader (CIAT - International Center for Tropical Agriculture) was useful to build a very simple version of the Nima CLD. This map, developed using the online platform Kumu [34], represented the main aspects of the area, with reference to WEF resources security, as well as key cross-sectoral issues and interdependencies. A round of SSIs conducted between February and May 2022 helped improve the CLD, thus obtaining a more complex preliminary version. In particular, the interviews highlighted, on one hand, the important role agriculture plays for local community well-being and the limited sustainability of sugarcane production over the long term. On the other hand, the stakeholders emphasised the impact of deforestation on natural areas, biodiversity, water availability and regulation functions. Information on key socio-institutional dynamics also emerged. These included: (i) the influence of the low community awareness on the main environmental issues (e.g., poor waste disposal, deforestation, ecosystem degradation); (ii) weak territorial planning and control, exacerbated by fragmented regulations, negatively affecting soil and water pollution and enabling harmful activities such as illegal mining, which increasingly threatens riverbanks and wetlands. Stakeholders perceive a conflict between local communities (which take care of forest conservation and riverine restoration) and big companies (which pursue economic profit but are also involved in environmental restoration and ecotourism programs). In this context, the lack of public authority oversight is seen as a major barrier, despite the existing opportunity presented by PES (Payment for Ecosystem Services) schemes.

**System Mapping**

The workshop in the Nima pilot area was organized on August 3, 2022, in Palmira, with the participation of 18 stakeholders. To simplify the discussion, stakeholders were divided into two groups: one focusing on the 'upstream area' of the basin, characterized by natural landscapes and conflicts related to water and land use; the other focusing on the 'downstream area', the most productive area, mainly dealing with water pollution issues caused by agro-industrial crops, urban development and agricultural waste dumping. A participatory mapping exercise was performed in each group for properly locating different classes of variables in the basin (i.e., resources, socio-economic activities, technologies and pressures). After that, stakeholders were asked to indicate the main cause-effect connections among variables (e.g. what pressures are exerted by economic activities and technologies on resources, what pressures economic activities receive and what are additional pressures that may impact resources). These interconnections were reported on a blank paper on the wall, where four quadrants (dedicated to the specific classes of variables) were drawn. An example of the outcome of the process is represented in Fig. 19.5. The preliminary CLD built in the System Conceptualization phase was then revised to include the new knowledge that emerged from the workshop, also considering that the stakeholders highlighted two key issues for the area: the increasing level of water pollution caused by several productive activities and the limited availability of water. Therefore, the data used to inform this System Mapping phase included both qualitative information collected during the stakeholder workshop and prior knowledge derived from the baseline analysis, semi-structured interviews, and expert insights. These data supported the identification of key variables and their interconnections within the Nexus system. Figure 19.6 presents the final version of the Nima CLD representing key Nexus dynamics and cross-sectoral interactions.

**System Analysis**

Based on the final version of the CLD, both the structural and descriptive analyses were performed by the analysts. The structural analysis was oriented to identify the most central variables, useful to make some preliminary hypothesis on the Nexus challenges within the system. Using the Kumu [34] Social Network Analysis module a list of high-ranked variables in terms of centrality in the CLD was obtained (see Table 19.1).

The results in Table 19.1 indicate that 'Contamination' and 'State of ecosystem conservation' may have a strong influence on the overall system because of their highest Degree Centrality values (10). Other highly connected variables include 'Demand for human water' (9), 'Irrigation areas' (8), 'Deforestation' (8), and 'Water availability' (8). Furthermore, 'Water availability' (0.295) and 'Livestock' (0.225) show the highest Betweenness Centrality values, indicating their role as key intermediaries in system interactions. Additional variables such as 'Urban development', 'Energy demand', and 'Community awareness' (Degree Centrality = 7), along with 'Agricultural crop yields', 'River state', 'Agricultural water supply', and 'Wastewater production' (high Betweenness Centrality values between 0.064 and 0.081), also emerged as relevant.

The variables identified as central in Table 19.1 were then examined within the CLD to assess the cross-sectoral interactions in which they are involved. Figure 19.7 shows an example of this analysis, focusing on 'Water demand (anthropogenic)' and its interconnections with multiple sectors.

As shown in Fig. 19.7, 'Water demand (anthropogenic)' is influenced by key economic activities, such as 'Industrial activities', 'Extensive cattle ranching', 'Livestock farming', and 'Agricultural water demand'. These interactions amplify pressures on 'Water availability' and 'Water security'. At the same time, increasing 'Water demand (anthropogenic)' negatively impacts 'Ecosystem conservation', highlighting the trade-offs between economic activities and environmental sustainability.

Following this in-depth analysis of the CLD, considering all variables listed in Table 19.1, a set of Nexus challenges was identified based on their cross-sectoral relevance and systemic impact (see Table 19.2).

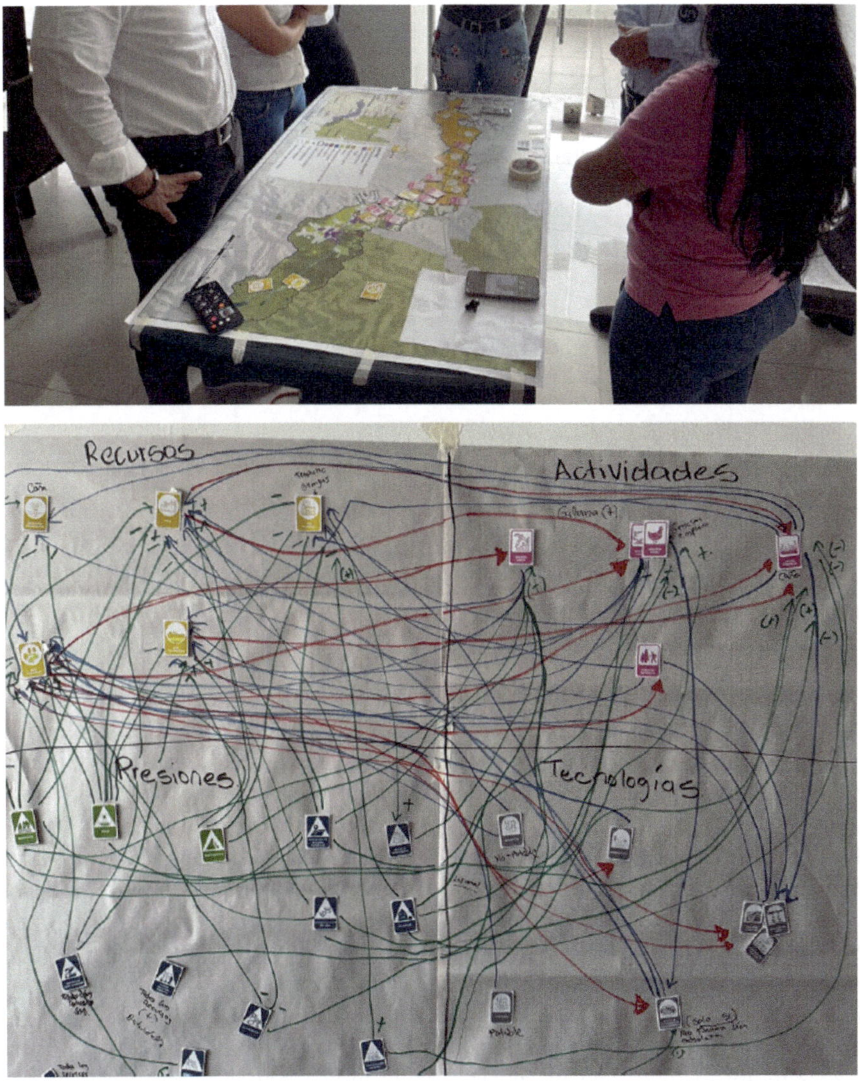


**Fig. 19.5** An example of participatory mapping exercise to support causal loop diagram development in the Nima case study

## *19.4.2 Application of the Modelling Process to the Ucayali Region*

**System Conceptualization**

During March and April 2024, nine SSIs were conducted. The stakeholders involved included distributors, retailers, and artisans to obtain a diverse perspective on the BCS

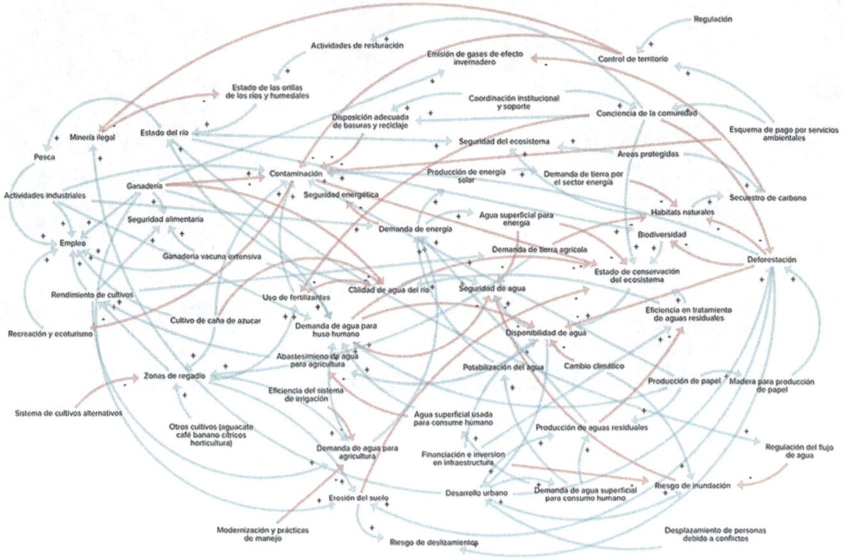


**Fig. 19.6** Causal loop diagram of the Nima case study

Nexus dynamics. The SSIs addressed two main themes: on one hand, the dynamics of the cocoa value chain and their impacts on biodiversity, climate change and human activities; on the other hand, the obstacles and needs related to stakeholder interactions, existing policies, and their implementation. The results of the SSIs and scientific information from the baseline, were used to build a preliminary CLD using Kumu [34] based on: (i) key variables of the system, classified in biodiversity loss, climate change and social dynamics, as well as key dynamics and impacts along the cocoa value chain. The key variables were also classified considering the Ecosystem Services categories established by the Millennium Ecosystem Assessment [43]; (ii) interrelationships between the key variables of the BCS Nexus. The variables identified and integrated into the preliminary CLD primarily relate to access to resources, both economic and infrastructural. Economic access was represented by the variable 'accessibility to funds and incentives', while infrastructural access was captured by 'quality of infrastructure'. The bureaucratic complexity, which hinders business formalization, was included in the model as 'formalization of business'. Challenges in the agricultural sector, such as low innovation and limited technical training, were summarized in the variable 'technical assistance and technology'. Among the identified social issues, land invasion associated with forest fires was integrated into the model as 'forest burning', while various illegal activities (including land trafficking, illegal crop cultivation, illegal logging, and illegal hunting) were also included in the CLD.

**Table 19.1** Variables with highest values of centrality measures within Nima causal loop diagram. The last column reports the centrality values calculated using the Kumu Social Network Analysis module. Higher degree centrality indicates stronger direct influence within the system, while higher betweenness centrality reflects the variable's role in mediating interactions between other variables

| Variable | Metric | Metric value |
|---|---|---|
| Contamination | High Degree Centrality | 10 |
| Ecosystem conservation | High Degree Centrality<br>High Betweenness Centrality | 10<br>0.168 |
| Water demand (anthropogenic) | High Degree Centrality<br>High Betweenness Centrality | 9<br>0.078 |
| Irrigated agricultural areas | High Degree Centrality | 8 |
| Deforestation | High Degree Centrality<br>High Betweenness Centrality | 8<br>0.090 |
| Water availability | High Degree Centrality<br>High Betweenness Centrality | 8<br>0.295 |
| Livestock farming | High Degree Centrality | 7 |
| Urban development | High Degree Centrality<br>High Betweenness Centrality | 7<br>0.225 |
| Community awareness | High Degree Centrality | 7 |
| Energy demand | High Degree Centrality<br>High Betweenness Centrality | 7<br>0.065 |
| Agricultural crop yields | High Betweenness Centrality | 0.081 |
| River state | High Betweenness Centrality | 0.074 |
| Agricultural water supply | High Betweenness Centrality | 0.073 |
| Wastewater production | High Betweenness Centrality | 0.064 |

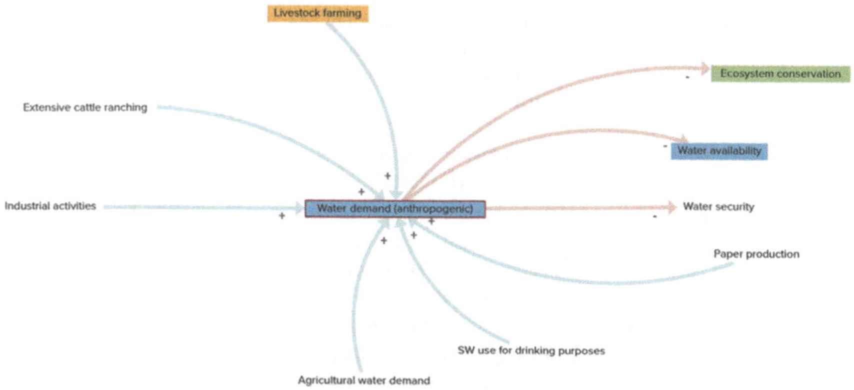


**Fig. 19.7** Focus on 'Water demand (anthropogenic)' and its cross-sectoral interactions within the causal loop diagram. Positive (+) and negative (−) feedback relationships are represented by blue and red arrows, respectively. Possible Nexus challenges are color-coded according to the sector they belong to: blue for water-related issues, orange for food-related aspects, and green for ecosystem-related factors

**Table 19.2** Nexus challenges and their corresponding cross-sectoral interactions for the Nima causal loop diagram. Each Nexus challenge is linked to variables involved in interactions across different sectors: water (W), food (F), and ecosystem (E). The last column details the sectoral classification of each variable contributing to the identified challenge, allowing the cross-sectoral nature of the issue to be explicitly recognized

| Nexus challenges | Cross-sectoral interactions |
|---|---|
| Water demand (anthropogenic) | Water availability (W sector)<br>Water security (W sector)<br>Agricultural water demand (W sector)<br>Industrial activities (F sector)<br>Extensive cattle ranching (F sector)<br>Livestock farming (F sector)<br>Ecosystem conservation (E sector) |
| Irrigated agricultural areas | Sugarcane cultivation (F sector)<br>Other crops (avocado, coffee, bananas, citrus, horticulture) (F sector)<br>Agricultural land demand (F sector)<br>Soil erosion (E sector)<br>Use of fertilizers (F sector)<br>Alternative crop systems (F sector)<br>Agricultural water demand (W sector) |
| Water availability | Water resources planning (W sector)<br>Ecosystem conservation (E sector)<br>Deforestation (E sector)<br>Climate change<br>Water demand (anthropogenic) (W sector)<br>Community aqueducts (W sector)<br>Funding and investment in infrastructure<br>Agricultural water supply (F sector)<br>Urban development |
| Livestock farming | Water demand (anthropogenic) (W sector)<br>Food security (F sector)<br>Employment and well-being<br>GHG emissions<br>River water quality (E sector)<br>Soil erosion (E sector)<br>Pollution (E sector) |
| Agricultural crop yields | Irrigated agricultural areas (F sector)<br>Agricultural land demand (F sector)<br>Food security (F sector)<br>Employment and well-being<br>Agricultural water supply (W sector) |
| River state | Illegal mining (E sector)<br>River slope encroachment (E sector)<br>Fishing (E sector)<br>Ecosystem security (E sector)<br>Biodiversity conservation (E sector)<br>Territorial planning (E sector)<br>Environmental awareness (E sector)<br>Water flow regulation—flood (W sector)<br>River water quality (E sector) |

**Fig. 19.8** An example of participatory mapping exercise to support causal loop diagram development in the Ucayali case study

## System Mapping

The stakeholder workshop took place in October 2024 and involved approximately 30 stakeholders. Stakeholders were organized into three different groups. Participants were provided with cards representing all the variables identified in the preliminary CLD. They were asked to draw the relationships between these variables or add new ones. To consider the spatial dimension of the system, participants were asked to place the cards on a geographical map, as illustrated in Fig. 19.8. Based on their perceptions and knowledge, they carried out the exercise by selecting and geographically locating the most significant biodiversity losses, the key processes/activities in the cocoa value chain that influenced these losses, the impacts of climate change and the social dynamics.

The cause-effect relationships identified by each group were subsequently aggregated using Synthesis Theory to create an aggregated CLD. The result of the aggregation process is presented in Fig. 19.9.

The CLD represents color-coded variables that highlight different stages of the cocoa value chain (i.e., consumption, production, transformation, export) and influencing factors related to the BCS Nexus. The consumption stage is marked in red, focusing on market demand and consumer trends, while the production stage is represented in yellow. The transformation stage, shown in green, encompasses processing activities, including the conversion of raw beans into cocoa butter, powder and chocolate. The export stage, shown in blue, covers logistics and international trade. Additionally, variables related to ES, marked in black, are included in the map. These include environmental factors such as soil fertility, water management, and biodiversity support. Variables related to the BCS Nexus are: (i) biodiversity and ecosystem degradation, highlighted in yellow, (ii) climate change, represented in blue, (iii) social dimension, in purple. Variables without colour characterization represent aspects suggested by stakeholders during the participatory exercise.

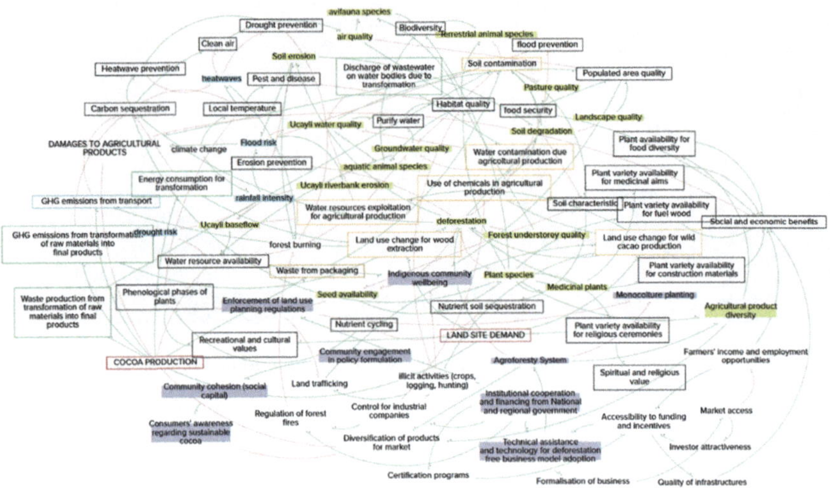


**Fig. 19.9** Causal loop diagram of the Ucayali region

## System Analysis

The CLD was explored through a structural and descriptive analysis to identify the BCS Nexus challenges. The structural analysis was performed by selecting variables with a high Degree Centrality and high Betweenness Centrality, as shown in Table 19.3.

**Table 19.3** Variables with highest values of centrality measures within Ucayali causal loop diagram

| Variable | Degree | Betweenness |
|---|---|---|
| Deforestation | 12 | 0.106 |
| Plant species | 10 | 0.103 |
| Soil contamination | 9 | 0.062 |
| Cocoa production | 9 | 0.187 |
| Carbon sequestration | 8 | 0.105 |
| Social and economic benefits | 8 | 0.189 |
| Land use change | 6 | 0.110 |
| Flood risk | 6 | 0.102 |
| Accessibility to funding | 5 | 0.013 |
| Quality of infrastructures | 5 | 0.110 |
| Illegal activities | 5 | 0.040 |
| Soil erosion | 5 | 0.104 |
| Forest burning | 5 | 0.071 |
| Heatwaves | 5 | 0.069 |

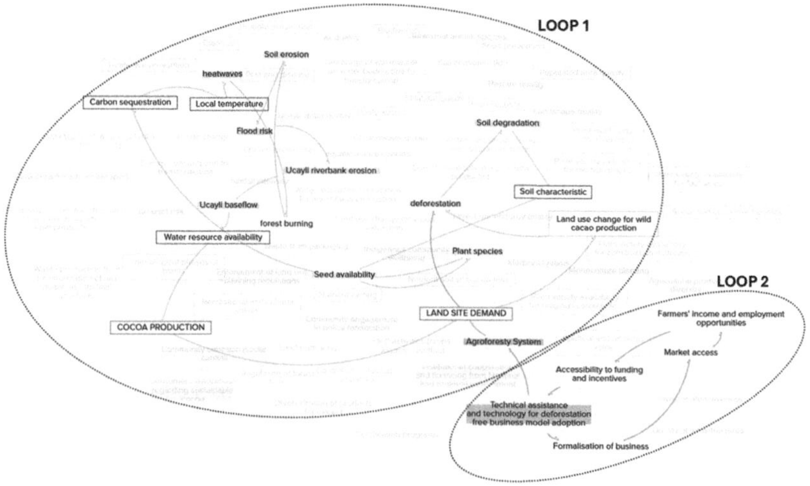


**Fig. 19.10** Loop of Cocoa produced in Peru

The same variables were subjected to a descriptive analysis to define the main cross-sectoral interactions and define the Nexus challenges. An example related to two feedback loops of the system involving some of the central variables is shown in Fig. 19.10.

Loop 1 highlights the effects of market demand on land use change for cocoa production. Expansion of cultivation leads to 'deforestation', which in turn affects the availability of 'Plant species', 'Soil degradation' and the characteristics on which ecosystem services depend such as 'Water resources availability'. A further effect of 'deforestation' is the reduction of 'Carbon sequestration', which contributes to the increase in 'Local temperatures' and the intensification of 'heat waves'. These climatic changes amplify the risk of soil erosion and land degradation ('Ucayali riverbank erosion').

Loop 2, closely linked to Loop 1, highlights the economic and social aspects of cocoa production, paying particular attention to the importance of infrastructure, investment and access to markets. Improving marketing opportunities and encouraging business formalization allows producers to gain a more competitive position, increasing income and incentivizing investments in more sustainable practices, such as agroforestry techniques. This process not only facilitates the transition to environmentally responsible production models but also generates new job opportunities and economic growth. Another key element is training and technical assistance: without the right technological and innovative support, many producers may continue to resort to deforestation to expand their crops. In addition, adequate infrastructure facilitates the marketing and distribution of products, strengthening the entire agricultural sector and promoting more balanced and sustainable economic development.

The same analysis was performed for other feedback loops within the Ucayali CLD. A set of Nexus challenges was identified based on their cross-sectoral relevance

**Table 19.4** Nexus challenges for the Ucayali causal loop diagram

| Group | Variable | Nexus challenge | Description—motivation |
|---|---|---|---|
| 1 | Deforestation<br>Plant species<br>Carbon sequestration<br>Soil contamination<br>Soil erosion | Preserve forest ecosystem | Intensive agricultural expansion causes deforestation, reducing biodiversity and carbon sequestration, accelerating climate change. In addition, unsustainable practices contaminate soils, impair their fertility and increase the risk of erosion, threatening the productivity and resilience of ecosystems |
| 2 | Flood risk<br>Heatwaves | Managing the effects of climate change | Climate change can affect both the availability and quality of resources, as well as the living environments of indigenous communities and natural habitats |
| 3 | Cocoa production<br>Land use change | Manage value chain phases | Develop production models that allow the expansion of cocoa production without compromising ecosystems, incentives for soil regeneration and more responsible land-use strategies |
| 4 | Accessibility to funding<br>Social and economic benefits<br>Quality of infrastructures<br>Forest burning | Develop an economic and infrastructural support system that promotes sustainable practices | Access to finance is essential to promote sustainable agricultural practices and reduce the need for forest fires. However, a lack of infrastructure hinders economic growth and the transition to sustainable models. Investing in infrastructure and financing can strengthen local communities, generating socio-economic benefits and reducing pressure on forests |

and systemic impact (see Table 19.4). Given the strong correlation between the different variables, they were grouped, and the relative "macro" challenges were identified.

## 19.5 Discussion and Conclusions

This study proposed a System Thinking modelling approach based on qualitative SD principles for the identification of Nexus challenges (and their cross-sectoral interactions), using two complex human-natural systems (i.e., the WEF Nexus in the Nima River Basin (Colombia) and the BCS Nexus in the Ucayali Rainforest (Peru)) for its practical implementation.

The findings confirmed the role of System Thinking, operationalized through CLDs, in providing a structured and participatory modelling approach to select and prioritize Nexus challenges. In fact, while CLDs effectively captured the multi-dimensional interactions across sectors, which are often overlooked in traditional modelling approaches [18, 21], the combination of structural analysis (using Centrality Measures) and descriptive analysis (considering interdependencies and feedback loops) allowed for a systematic selection of key Nexus challenges, based on their influence and connectivity within the system. In both case studies, this approach revealed the underlying dynamics driving resource management conflicts, enabling a more holistic understanding of interdependencies between sectoral processes. In order to enhance the legitimacy and applicability of the results, local stakeholders were engaged throughout the modelling process [44]. Therefore, scientific insights were improved (or expanded) by integrating local knowledge into the modelling framework.

Within the Nima CLD, 'Water demand (anthropogenic)' emerged as a highly connected and influential variable, linking agricultural expansion, urban water consumption, and industrial activities. The key trade-off highlighted in the system is between sugarcane production and water security: while irrigation supports economic productivity, it exacerbates water scarcity and pollution, affecting both ecosystems and human needs. Deforestation further undermines hydrological regulation, amplifying risks of seasonal water shortages. These results confirm the urgency of integrating land-use and water governance, moving beyond sectoral policies that prioritize agricultural productivity over long-term resource sustainability.

Within the Ucayali CLD, 'Deforestation' emerged as a highly connected and influential variable, linking cocoa production expansion, biodiversity loss, and climate risks. The key trade-off highlighted in the system is between agricultural expansion and ecosystem stability: while land conversion for cocoa cultivation drives economic activity, it accelerates carbon sequestration loss, soil degradation, and heatwave intensification, ultimately threatening long-term agricultural productivity. Additionally, limited financial access and poor infrastructure reinforce unsustainable practices, preventing a shift toward more resilient farming systems. These results confirm the urgency of integrating economic incentives, agroforestry strategies, and stronger land-use governance to decouple economic development from environmental degradation.

As emphasized by recent studies, identifying trade-offs within Nexus systems is essential for decision-making. Trade-offs represent the tensions that arise when actions taken to enhance one sectoral goal (e.g., food production or economic growth) result in unintended negative outcomes in other sectors (e.g., water quality, biodiversity). As such, the early identification of trade-offs can inform more coherent and anticipatory governance responses. This dimension is often underexplored in traditional modelling practices, yet it is critical to resource management in contexts marked by high complexity and limited institutional coordination. In this regard, the findings of this study resonate with recent work by [45], who reinforced the notion that trade-offs are not only biophysical but also embedded in economic and institutional systems, requiring integrated solutions that are sensitive to cross-sectoral dynamics. Incorporating trade-off thinking into participatory modelling tools such as CLDs enables for the visualization and discussion of these tensions early in the planning process.

In addition to the identification of Nexus challenges and trade-offs, the proposed modelling process sets the stage for acting on them. To be effective, interventions should focus on 'sensitive' areas and processes within the system, i.e., the so-called 'leverage points', rather than dispersing efforts across multiple disconnected initiatives. By explicitly identifying trade-offs alongside leverage points, the proposed Systems Thinking approach supports the co-design of strategies that balance competing priorities, avoid unintended consequences, and promote systemic coherence, especially in contexts where governance structures are fragmented and sector-specific policies risk being misaligned.

Despite the effectiveness of the proposed qualitative modelling approach, its practical impact on Nexus system management strongly depends on its integration with a subsequent quantitative modelling phase. In particular, the CLDs developed through stakeholder engagement serve as a conceptual foundation to inform the structure of quantitative Stock and Flow Diagrams (SFDs). This transition enables the quantification of system dynamics and the simulation of long-term trajectories under different policies and external scenarios. As demonstrated in recent applications (e.g., [19, 46]), the knowledge elicited during the qualitative phase, particularly regarding variable interdependencies, feedback structures, and leverage points, guides the mathematical formalization of system components and relationships within the SFD framework. The integration of stakeholder-informed CLDs into SFDs allows analysts to define key state and rate variables, identify appropriate equations based on available data or expert knowledge, and distinguish between sectors where quantitative modelling is feasible and those that require qualitative or semi-quantitative representations. This blended approach improves model robustness while maintaining the participatory essence of the process. It also supports the implementation of tools such as sensitivity analysis and scenario testing, enabling policymakers to assess the system's responsiveness to different strategies and explore future trajectories under uncertainty. Therefore, the qualitative phase not only supports shared understanding but also provides the essential structure and narrative logic that make quantitative modelling operational and policy-relevant.

**Acknowledgements** The research activities described in this work were conducted as part of the REXUS Project (H2020—Grant Agreement No. 101003632, https://www.rexusproject.eu), the LENSES Project (PRIMA programme—Grant Agreement No. 2041, https://www.lenses-prima.eu) and the BIOTRAILS Project (Horizon Europe—Grant Agreement No. 101082008, https://biotrails project.eu/). The authors would like to thank both projects' teams for their invaluable contributions and inspiring discussions.

## References


1. Razavi S, Duffy A, Eamen L, Jakeman AJ, Jardine TD, Wheater H et al (2025) Convergent and transdisciplinary integration: on the future of integrated modeling of human-water systems. Water Resources Res 61:e2024WR038088
2. Biggs R, de Vos A, Preiser R, Clements H, Maciejewski K, Schlüter M (eds) (2021) The Routledge handbook of research methods for social-ecological systems. Routledge, Taylor & Francis, London
3. Smajgl A, Ward J, Pluschke L (2016) The water-food-energy Nexus: realising a new paradigm. J Hydrol 533:533–540
4. Bazilian M, Rogner H, Howells M, Hermann S, Arent D, Gielen D, Steduto P, Mueller A, Komor P, Tol RS, Yumkella KK (2011) Considering the energy, water and food nexus: towards an integrated modelling approach. Energy Policy 39(12):7896–7906
5. Di Baldassarre G, Sivapalan M, Rusca M, Cudennec C, Garcia M, Kreibich H et al (2019) Sociohydrology: scientific challenges in addressing the sustainable development goals. Water Resour Res 55:6327–6355
6. World Water Council (2023) Opening up integrated water resource management to include energy, food, health and education. World Water Council Report
7. Blöschl G et al (2019) Twenty three unsolved problems in hydrology (UPH)—a community perspective. Hydrol Sci J 64(10):1141–1158
8. Stringer LC, Quinn CH, Le VHT, Msuya F, Pezzuti J, Dallimer M et al (2018) A new framework to enable equitable outcomes: resilience and nexus approaches combined. Earth's Future 6(6):902–918
9. de Amorim WS, Valduga IB, Ribeiro JMP, Williamson VG, Krauser GE, Magtoto MK, de Andrade Guerra JBSO (2018) The nexus between water, energy, and food in the context of the global risks: an analysis of the interactions between food, water, and energy security. Environ Impact Assess Rev 72:1–11
10. Cilliers P (2000) What can we learn from a theory of complexity? Emergence 2(1):23–33
11. Teutschbein C, Jonsson E, Todorović A, Tootoonchi F, Stenfors E, Grabs T (2023) Future drought propagation through the water-energy-food-ecosystem nexus—a nordic perspective. J Hydrol 617:128963
12. Pahl-Wostl C (2019) The role of governance modes and meta-governance in the transformation towards sustainable water governance. Environ Sci Policy 91:6–16
13. World Economic Forum (2011) An integrated sustainability index for effective water policy. In: Waughray D (ed) Water security: the water-food-energy-climate nexus. Island Press, Washington, Covelo, London, p 271
14. Liu J, Hull V, Godfray HCJ, Tilmans D, Gleick P, Hoff H et al (2018) Nexus approaches to global sustainable development. Nature Sustain 1:466–476
15. Wu L, Elshorbagy A, Pande S, Zhuo L (2021) Trade-offs and synergies in the water-energy-food nexus: the case of Saskatchewan, Canada. Resour Conserv Recycl 164:105192
16. Giordano R, Osann A, Enao E, Lopez LM, Piquera JS, Nikolaidis NP, Lilli M, Coletta VR, Pagano A (2025) Causal loop diagrams for bridging the gap between Nexus thinking and Nexus doing: evidence from two case studies. J Hydrol 650:132571
17. Preiser R, Biggs R, De Vos A, Folke C (2018) Social-ecological systems as complex adaptive systems: organizing principles for advancing research methods and approaches. Ecol Soc 23(4):46
18. Sterman JD (2000) Systems thinking and modeling for a complex world. McGraw-Hill, New York
19. Coletta VR, Pagano A, Zimmermann N, Davies M, Butler A, Fratino U, Giordano R, Pluchinotta I (2024) Socio-hydrological modelling using participatory system dynamics modelling for enhancing urban flood resilience through Blue-Green infrastructure. J Hydrol 636:131248
20. Coletta VR, Pagano A, Pluchinotta I, Zimmermann N, Davies M, Butler A, Fratino U, Giordano R (2024a) Participatory causal loop diagrams building for supporting decision-makers integrating flood risk management in an urban regeneration process. Earth's Future 12:e2023EF003659

21. Meadows DH (2008) Thinking in systems—a primer. Wright D (ed). Chelsea Green Publishing
22. Murphy RJA, Jones P (2021) Towards systemic theories of change: high-leverage strategies for managing wicked problems. Des Manage J 16(1):49–65
23. REXUS Project Homepage. https://www.rexusproject.eu/
24. LENSES Project Homepage. https://www.lenses-prima.eu/
25. BIOTRAILS Project Homepage. https://biotrailsproject.eu/
26. Ahlin EM (2019) Semi-structured interviews with expert practitioners: Their validity and significant contribution to translational research. SAGE Publications Ltd, London
27. Newberry P, Carhart N (2024) Constructing causal loop diagrams from large interview data sets. Syst Dyn Rev 40(1):e1745
28. Dhirasasna N, Sahin O (2019) A multi-methodology approach to creating a causal loop diagram. Systems 7(3):42
29. Coletta VR, Pagano A, Pluchinotta I, Fratino U, Scrieciu A, Nanu F, Giordano R (2021) Causal loop diagrams for supporting nature based solutions participatory design and performance assessment. J Environ Manage 280:111668
30. Haraldsson HV (2004) Introduction to system thinking and causal loop diagrams. Lund University, pp 3–4
31. Hemmerling SA, Barra M, Bienn HC, Baustian MM, Jung H, Meselhe E et al (2020) Elevating local knowledge through participatory modeling: active community engagement in restoration planning in coastal Louisiana. J Geogr Syst 22:241–266
32. Ryan E, Pepper M, Munoz A (2021) Causal loop diagram aggregation towards model completeness. Syst Pract Action Res 34(1):37–51
33. Freeman LC (1979) Centrality in social networks: conceptual clarification. Soc Netw 1(3):215–239
34. Kumu Homepage. https://www.kumu.io/
35. Egerer S, Cotera RV, Celliers L, Costa MM (2021) A leverage points analysis of a qualitative system dynamics model for climate change adaptation in agriculture. Agric Syst 189:103052
36. Corporación Autónoma Regional del Valle del Cauca (CVC): Balance Oferta—Demanda De Agua, Dirección Técnica Ambiental Grupo De Recursos Hídricos. https://www.cvc.gov.co/sites/default/files/2018-09/Balance_Amaime_0.pdf. Accessed 13 Aug 2024
37. Sepúlveda P. CVC recomienda reubicación de asentamientos en márgenes del río Nima. https://www.cvc.gov.co/carousel/2740-cvc-recomienda-reubicacion-de-asentamientos-en-margenes-del-rio-nima. Accessed 13 Aug 2024
38. Restrepo JD, Bottaro G, Barci L, Beltrán LM, Londoño-Behaine M, Masiero M (2024) Assessing the impacts of nature-based solutions on ecosystem services: a water-energy-food-ecosystems Nexus approach in the Nima River Sub-Basin (Colombia). Forests 15(11):1852
39. Ivanova Y, Tristán Febres MC, Romero Sánchez MA, Charry A, Lema S, Choy JS et al (2020) Moving towards a deforestation-free cacao and chocolate value chain with low greenhouse gas emissions. CIAT Publication No 502. International Center for Tropical Agriculture (CIAT), Cali, Colombia
40. Dreoni I, Schaafsma M, Matthews Z (2021) The social impacts of cocoa production: a systematic review. UKRI GCRF TRADE Hub
41. Afrane G, Ntiamoah A (2011) Use of pesticides in the cocoa industry and their impact on the environment and the food chain. In: Pesticides in the modern world—risks and benefits. InTech, pp 51–68
42. Global Forest Watch: Forest monitoring and analysis platform. https://www.globalforestwatch.org/. Accessed 20 Feb 2025
43. Millennium Ecosystem Assessment—MEA (2001) Vol 2. Island Press
44. Santoro S, Pluchinotta I, Pagano A, Pengal P, Cokan B, Giordano R (2019) Assessing stakeholders' risk perception to promote nature based solutions as flood protection strategies: the case of the Glinščica river (Slovenia). Sci Total Environ 655:188–201
45. Lee SH, Assi AT, Mohtar RH, Hamane M, Yoon PR, Yoo SH (2023) Development of WEF-P Nexus based on product-supply chain: a case study of phosphorous fertilizer industry in Morocco. Sci Total Environ 857:159520

46. Malamataris D, Pisinaras V, Pagano A, Baratella V, Vanino S, Bea M, Babakos K et al (2025) Managing water-ecosystem-food Nexus using participatory approaches: insights from an innovative methodological approach developed in two Mediterranean areas. Frontiers Water 7

# Chapter 20
# Securing Our Future: A Water-Energy-Food Nexus Approach to Climate Risks, Resource Scarcity, and Human Mobility in the Middle East

**Konstantinos Pappas, Rabi Mohtar, Bassel Daher, and Zainab Ashkanani**

**Abstract** Global challenges, including disasters escalating from extreme weather, population growth, urbanization, and resource scarcity, strain critical systems. Projected supply–demand gaps by 2030, coupled with migratory crises driven by conflict, famine, and resource deprivation, threaten human security and global stability. The Arab region faces acute water, energy, and food system pressures intensified by resource limitations, population growth, migration, and climate change. A holistic, systems-based Water-Energy-Food (WEF) Nexus approach is essential for regional sustainability. This chapter examines the WEF Nexus as a strategy for resource security in the Arab context, using Jordan and Kuwait as case studies. Jordan,

K. Pappas (✉) · R. Mohtar · B. Daher · Z. Ashkanani
Texas A&M Energy Institute, Texas A&M University, College Station, TX, USA
e-mail: kostis.pappas@tamu.edu

Department of Biological and Agricultural Engineering, Texas A&M University, College Station, TX, USA

R. Mohtar
e-mail: rabi.mohtar@ag.tamu.edu

B. Daher
e-mail: bdaher@tamu.edu

Z. Ashkanani
e-mail: ashkanani@tamu.edu

K. Pappas
Borders and Migration Program, Bush School of Government and Public Service, Mosbacher Institute for Trade, Economics, and Public Policy, Texas A&M University, College Station, TX, USA

R. Mohtar
Zachry Department of Civil and Environmental Engineering, Texas A&M University, College Station, TX, USA

 
A. Angelini et al. (eds.), *Security Enhancement for Climate Change Impacting Urban Resources - SECCURe*, NATO Science for Peace and Security Series C: Environmental Security, https://doi.org/10.1007/978-94-024-2345-7_20

severely water-scarce, illustrates the impact of resource constraints, regional instability, and refugee influxes, demanding integrated management strategies. Kuwait, resource-rich but environmentally challenged, highlights energy-intensive desalination and trade-offs between energy and water sustainability. These cases emphasize the need for context-specific WEF Nexus strategies and regional collaboration. This research advances WEF Nexus studies by integrating migration as a key factor and emphasizing socio-economic and geopolitical dimensions. It advocates evidence-based policies, participatory governance, and adaptive capacity-building, offering actionable insights for policymakers and stakeholders.

**Keywords** Water-Energy-Food (WEF) Nexus · Resource security · Migration · Sustainability

## 20.1 Introduction

The world is currently undergoing an energy transition and food system transformation amidst growing pressures on water resources, creating a critical need to better understand the interconnections between these systems and the challenges they face. Simultaneously, the convergence of climate change, resource scarcity, and human mobility presents a complex and multifaceted challenge to global security and sustainable development. The Water-Energy-Food (WEF) Nexus offers a framework to address these complexities, emphasizing the interdependencies and trade-offs inherent in managing these essential resources.

The Arab region finds itself at the forefront of these challenges, confronting acute issues at the intersection of water, energy, and food systems. The region's unique circumstances—characterized by resource scarcity, rapid population growth, significant migration pressures, and the escalating impacts of climate change—necessitate a holistic, systems-based approach to resource management. As the 2030 Agenda for Sustainable Development approaches its deadline, there is growing urgency to advance the Sustainable Development Goals (SDGs). These goals inherently recognize the interdependence of global priorities—poverty alleviation, improved health, sustainable systems, better water access, hunger elimination, and climate action—demanding a systems approach to ensure progress in one area does not come at the expense of another.

Understanding the interconnections between resource systems can help establish security and adaptability in the face of growing resource demand and compound shocks. While modeling these complexities, addressing their nuances, and translating insights into effective policies is a challenging endeavor, it holds the promise of significant economic benefits, enhanced resource efficiency, improved livelihoods, and better public health outcomes [1].

This chapter examines how the WEF Nexus framework can address resource management challenges in the Arab region, using Jordan and Kuwait as contrasting case studies. While existing research has often remained conceptual, our study applies systems thinking to provide evidence-based, context-specific recommendations for policymakers. By analyzing Jordan's water scarcity (exacerbated by refugee influxes) alongside Kuwait's energy-intensive desalination approaches, we demonstrate how integrated resource management strategies can enhance regional sustainability and resilience. The comparative analysis explores technological innovation and regional cooperation opportunities, identifying interdependencies, trade-offs, and synergies that yield actionable insights for stakeholders. This approach bridges the gap between theoretical frameworks and practical implementation, offering a template for similar analyses throughout water-stressed environments in the region and advancing the discourse on integrated resource security in the Arab context.

## 20.2 The "Nexus" Concept

The term 'Nexus,' by definition, refers to connections linking different elements [2]. The concept of tight interconnectedness across resources such as water, land, food, energy, and environment has gained increasing attention in research and decision-making communities [3]. While the Nexus concept initially focused on the security of water, energy, and food supply from the perspective of water [4], this approach can be traced to the World Economic Forum in 2008, where the global challenges related to economic development were recognized from the WEF perspective [5].

According to the World Economic Forum's Global Risks report, the WEF Nexus is highly influential in preserving human, social, and political security [6]. The necessity of identifying these interlinkages has grown since the announcement of the 2030 Agenda for Sustainable Development [7]. Despite the recognition of the importance of adopting a WEF Nexus approach, its application remains limited due to several barriers [8]. In addition, the Nexus approach often emphasizes interconnections among climate, ecosystems, society, and biodiversity, extending beyond the classical WEF approach [9–11]. This broader perspective highlights that, despite its holistic nature, certain boundaries must be considered in every study. Some researchers have reported that the introduction of the WEF Nexus concept is context-dependent and that generalized use of such terms can be overlapping and ambiguous [12, 13]. Further, the term "WEF" brings different meanings to different stakeholders [14]. It can move between Energy-Water-Food (EWF), Food-Energy-Water (FEW), or Water-Energy-Food (WEF), depending upon the focus of research or interest of stakeholders [15]. The scope can be broad, covering themes such as economics, resource security, and climate, or narrow, as in a defined system for which only a few interconnections among the resources are selected [16].

In this paper, we aim to build on the WEF Nexus literature and explore critical questions at the interfaces of the interconnected resource systems in the context of several case studies and to highlight both small- and large-scale security themes.

The Nexus approach emphasizes the interconnections among sectors, including climate, ecosystems, society, and biodiversity, extending beyond the classical WEF approach. This broader perspective highlights that, despite its holistic nature, certain boundaries must be considered in every study. The scope can be broad, covering themes such as economics, resource security, and climate, or narrow, as in a defined system for which only a few interconnections among the resources are selected [16].

## 20.3 Overview of the Water-Energy-Food (WEF) Nexus Approach

### *20.3.1 Historical Context and Evolution*

The WEF Nexus started becoming more mainstream since 2011, when greater attention was being paid to global economic issues, especially surrounding water resources and food scarcity [17, 18]. The interconnections between water, energy, and food sectors are many, but some key links highlight the extent of the overlap: globally, 72% of water resources are consumed by agriculture, and about 14% of water withdrawals go toward energy production [19]. Food systems use about 30% of globally available energy [20]. Adopting a systems approach to understanding and addressing the challenges facing these interdependent resource systems—each undergoing transitions and vulnerable to compound shocks—has become increasingly critical. Such an approach can guide evidence-based policymaking by highlighting trade-offs, forecasting long-term outcomes, and providing a valuable framework for policymakers and researchers to explore viable solutions.

Prior to the WEF Nexus's popularization, there was already substantial interest in the urgency and finiteness of water, energy, and food resources [2]. The Integrated Water Resources Management (IWRM) concept has long been instrumental in promoting sustainable water management practices by integrating social, economic, and environmental dimensions. It focused on creating policies and institutional frameworks that ensure the equitable distribution and efficient use of water resources, and address water-related issues and their direct impacts on various sectors. IWRM emphasizes the coordinated development and management of water, land, and related resources to maximize economic and social welfare without compromising the sustainability of vital ecosystems.

However, as global challenges have become more complex and interconnected, there has been a growing recognition that water management cannot be effectively addressed in isolation. This realization has given rise to the Water-Energy-Food (WEF) Nexus. Prior research primarily examined the interplay between energy production methods and their reliance on water resources, as well as how water management processes depend on energy [21]. Analysis of water and energy resources soon included how the agriculture sector utilizes energy and water [18, 22]. The World Economic Forum Water Initiative introduced the water-energy-food-climate nexus as a method of understanding and addressing water security in 2011, highlighting the complex position of water in the global economic system [23]. Interest in the water-energy-food nexus in its current form stemmed from the Bonn2011 Nexus Conference, titled "The Water, Energy and Food Security Nexus—Solutions for a Green Economy" [2, 24].

### *20.3.2 Nexus Frameworks and Modeling Approaches*

The core element of the WEF Nexus framework is its focus on the interdependencies and interconnections between resource systems. Other similar frameworks have sought to expand on the systems approach by including climate, security, ecosystems, soil, and waste [6, 22, 25, 26]. Recently, there has been a push to use the WEF Nexus to prioritize resource and system resilience, efficiency, and adaptability, in addition to productivity [27–30]. The WEF Nexus considers complex interconnections between resources and multidimensional resource systems, making it a growing area of research in the face of global emergent themes of climate change, mass migrations, and non-uniform resource distribution (see Table 20.1) [31].

The definition and structure of nexus modeling vary by context and research environment, but there are essentially two broad categories of consideration. The first category interprets the nexus as the interconnections among different resource subsystems. This approach aims to identify trade-offs and synergies of resource systems, internalize social and environmental impacts, and guide the development of cross-sectoral policies.

The interconnection process among these resources includes physical and chemical relations, input and output relations, and interactions with external factors such as markets, governing institutions, and security issues. The second approach presents the nexus as an analytical means to quantify the functionality of the link or identify the impact or influence of interlinkages. The FAO demonstrated how the coupled behavior of humans and nature can be integrated into the management of natural resources through the nexus approach. Despite methodological differences, both approaches can potentially provide a framework for better decision-making in resource governance by addressing trade-offs and identifying synergies in the WEF Nexus. The Nexus Approach extends beyond mere technological solutions: it necessitates collaborative efforts across governmental levels, ministries, and international partnerships. Effective governance under this framework is crucial and should incorporate participatory methods to ensure stakeholder involvement and community feedback in policy formulation and execution [27].

A comprehensive set of policy directives informed by the WEF nexus could substantially aid in understanding the interlinkages, trade-offs, and risks associated with these sectors, thereby facilitating the development of integrated solutions that minimize risks and maximize benefits. Embracing the nexus approach could significantly contribute to national objectives and align with several SDGs, including zero hunger (SDG2), clean water and sanitation (SDG6), affordable and clean energy (SDG7), and responsible consumption and production (SDG12) [32]. Insights from adopting the nexus approach would also support climate action (SDG13) and foster partnerships for achieving these goals (SDG17).

**Table 20.1** Summary of modeling approaches for WEF Nexus analysis

| Modeling approach | Key features | Advantages | Limitations | Applications in WEF Nexus |
|---|---|---|---|---|
| Scenario-based models [49, 50] | Multiple future pathways exploration<br>"What-if" analysis framework<br>Incorporates uncertainty<br>Often combines quantitative and qualitative elements<br>Temporal flexibility (short to long-term)<br>Can be integrated with other modeling approaches | Addresses future uncertainties<br>Supports robust decision-making<br>Tests policy effectiveness under different conditions<br>Facilitates stakeholder engagement and dialogue<br>Can identify critical thresholds and tipping points | Heavily dependent on scenario assumptions<br>May not capture complex feedbacks unless combined with other models<br>Can produce overwhelming number of scenarios<br>Challenge in scenario probability assignment<br>Often lacks spatial granularity | Climate change adaptation strategies in water-scarce regions<br>Long-term infrastructure planning for energy-water systems<br>Testing policy resilience under different socioeconomic futures<br>Evaluating system responses to extreme events<br>Food production planning under variable climate conditions |
| Tradeoff analysis models [50, 51] | Multi-criteria evaluation<br>Explicit consideration of competing objectives<br>Pareto optimization<br>Visual representation of tradeoff spaces<br>Balance between quantitative and qualitative indicators | Quantifies relationships between competing objectives<br>Supports balanced decision-making<br>Reveals opportunity costs of policy choices<br>Makes hidden tradeoffs explicit<br>Can incorporate stakeholder preferences | May oversimplify complex relationships<br>Challenges in weighting different objectives<br>Can be difficult to communicate to stakeholders<br>Data-intensive for comprehensive analysis<br>Static representation of dynamic systems | Balancing water allocation between agriculture and energy sectors<br>Evaluating energy-water tradeoffs in desalination<br>Analyzing food production vs. environmental impacts<br>Resource allocation under competing urban and agricultural demands<br>Optimizing reservoir operation for multiple benefits |

(continued)

**Table 20.1** (continued)

| Modeling approach | Key features | Advantages | Limitations | Applications in WEF Nexus |
|---|---|---|---|---|
| Optimization models [33, 34] | Linear or non-linear programming—Resource allocation focus<br>Objective function maximization/ minimization | Find optimal solutions for resource allocation<br>Quantify trade-offs systematically—allow incorporating multiple constraints | Require precise mathematical formulation<br>Often assume perfect information<br>May not capture dynamic feedback processes<br>Can oversimplify complex social factors | Optimal crop patterns to maximize yield while minimizing resource use<br>Water allocation between competing sectors<br>Energy portfolio optimization considering water constraints |
| System dynamics models [34, 36, 37] | Focus on feedback loops and time delays<br>Stock-and-flow structure<br>Causal loop diagramming | Capture complex system behavior over time<br>Visualize interdependencies and feedback loops<br>Incorporate non-linear relationships<br>Suitable for long-term scenario planning | Aggregated approach may miss spatial heterogeneity<br>Model structure highly dependent on modeler's mental models<br>Difficult to validate comprehensively –<br>Data-intensive for parameter estimation | Long-term impact assessment of population growth on resource demand<br>Policy intervention testing across linked resource systems<br>Analysis of cascading effects across nexus components |
| Input–output models [34, 40] | Matrix-based representation of economy<br>Track flows between economic sectors—Can be extended with physical units | Quantify direct and indirect sectoral interdependencies<br>Reveal hidden resource footprints in economic activities<br>Allow economic and environmental analysis integration<br>Good for system-wide impact assessment | Static representation (no dynamic changes)<br>Linear relationship assumptions<br>Homogeneous sector output assumption<br>Limited spatial resolution | Calculating virtual water/energy flows in trade<br>Assessing economy-wide impacts of WEF policies<br>Revealing cross-sectoral dependencies not visible in direct assessments |

(continued)

**Table 20.1** (continued)

| Modeling approach | Key features | Advantages | Limitations | Applications in WEF Nexus |
|---|---|---|---|---|
| Agent-based models [34] | Bottom-up approach—individual decision-making agents—emergent system behaviors—heterogeneous actors | Represent diverse stakeholder behaviors and preferences<br>Capture emergent phenomena from individual actions<br>Include spatial and social network interactions<br>Simulate adaptation to changing conditions | Computationally intensive<br>Difficult to calibrate and validate—results can be hard to interpret<br>Require detailed data on individual behaviors | Farmers' crop choices and irrigation practices<br>Household water and energy consumption patterns<br>Stakeholder responses to policy interventions<br>Land use change dynamics |
| Integrated assessment models (IAMs) [34] | Combine multiple model types<br>Link biophysical, economic, and social systems<br>Cross-scale integration—usually include climate change factors | Comprehensive analysis across domains<br>Connect human and natural systems<br>Support long-term policy planning—allow cross-scale assessment (local to global) | High complexity can reduce transparency<br>Uncertainty propagation across model components<br>Resource-intensive development and operation<br>Risk of compounding errors across integrated models | Climate change impacts on WEF systems<br>Policy intervention evaluations across sectors<br>Trade-offs between biofuel production and land/water use<br>Regional to global analysis of coupled human-natural systems |

Quantitative models play a vital role in WEF Nexus assessments by helping cross-sectoral decision makers evaluate trade-offs and synergies between water, energy, and food systems. Optimization models use linear or non-linear programming to allocate resources efficiently, such as determining optimal cropping patterns to maximize food production while minimizing water and energy use [33, 34]. System dynamics models capture feedback loops and inter-sectoral interactions, allowing for long-term impact assessments, such as how population growth drives resource demand and affects water availability [35–37]. Input–output models quantify interdependencies between sectors, assessing the direct and indirect effects of policies and revealing hidden water and energy footprints in economic activities [34, 38–40]. Agent-based models simulate individual decision-making, such as farmers' crop choices and irrigation practices, to understand how individual behaviors influence system-wide outcomes [34, 41, 42]. Integrated assessment models (IAMs) link biophysical, economic, and social models to assess climate change impacts on WEF systems and evaluate policy interventions [34, 43, 44]. These models provide valuable insights into policy impacts, such as the trade-offs between biofuel production, water withdrawals, and land use competition [45].

However, models have limitations. They require extensive data, and their accuracy depends on input quality and underlying assumptions. As simplified representations of reality, they may not capture all complexities [46]. Thus, they should be used as exploratory tools rather than definitive predictors, with sensitivity analyses ensuring robustness [47]. To enhance decision-making, models should be complemented with qualitative assessments and participatory approaches that incorporate stakeholder knowledge [48].

### *20.3.3 Research Gaps*

Existing WEF Nexus models and tools focus on better management of water, energy, and food resources within defined contextual factors such as local geography, climate, economics, resource demands, or national goals. However, few models assess the impact of sudden and significant disruptions to these resource systems, such as droughts, floods, cyclones, or other natural disasters. In Asia and Africa, flooding and storms resulting from disasters are major causes of agricultural loss [52]. Compared to uncertainty assessments, water, energy, and food cross-sectoral interactions have been understudied [53]. The loss of agricultural production due to sudden outbreaks of crop epidemics or disruptions in food demand–supply chains during a pandemic is understudied as part of integrated water, energy, and food security. Environmental degradation, identified as a major driver for migration, further complicates this issue [54–56].

Despite international institutions highlighting the interlinkage between water and migration, food security and migration, and energy and migration, no WEF model or framework includes migration as a thematic focus [56, 57]. Furthermore, there is a gap in understanding how technological innovations, such as renewable energy and water-efficient agriculture, can be integrated into the WEF Nexus framework to enhance sustainability. The importance of adopting a systems-thinking approach and greater 'nexus awareness' in policy and decision-making to address the complex interdependencies within the WEF Nexus is also crucial, as Susnik and Staddon (2022) have emphasized, and needs to be further studied for better understanding and mitigating the impacts of sudden disruptions and environmental changes on water, energy, and food security [58].

## 20.4 The WEF Nexus Approach in the Arab Countries

### *20.4.1 Context*

The Arab region is characterized by intricate interdependencies among water, energy, and food systems, a dynamic increasingly strained by climate change, population growth, urbanization, and migration. These pressures demand an integrated approach to resource security, where the WEF Nexus framework emerges as a critical tool. This framework not only highlights the interconnections between these sectors but also addresses trade-offs, synergies, and vulnerabilities across geographical and socio-economic contexts. Through systems thinking, it provides a pathway for developing sustainable, evidence-based policies to mitigate resource scarcity and foster resilience in a region grappling with multifaceted challenges.

The WEF Nexus framework is particularly relevant for Arab countries, where water scarcity, reliance on imported food, and energy-intensive resource production are defining characteristics. For instance, the region's water withdrawals for agriculture exceed global averages, with over 80% of water use in some countries allocated to this sector. Limited renewable water resources, making the region dependent on desalination, groundwater, and virtual water imports embedded in the food trade, further complicate this heavy reliance on water for food production. Energy systems are similarly strained, as water production processes such as desalination and wastewater treatment are highly energy-intensive, especially in countries like Kuwait, where fossil fuels dominate energy generation [59]. These interconnected systems illustrate the need for a holistic perspective to avoid siloed policies that exacerbate existing vulnerabilities.

### *20.4.2 The Role of Migration and Climate Change—Toward Integrated Resource Security*

In this context, migration presents an additional layer of complexity. Arab countries, particularly Jordan, have experienced significant influxes of refugees due to regional conflicts, straining already scarce resources. For example, Jordan's water scarcity has been intensified by increased demand from refugee populations, compelling the nation to overexploit finite groundwater reserves such as the Disi aquifer. This unsustainable pattern highlights the intersection of migration and resource security, where population shifts directly impact the water-energy-food balance. Migration-driven urbanization further amplifies pressures, as urban areas often demand more energy-intensive water and food systems to meet the needs of growing populations.

The systemic vulnerabilities of Arab countries are further influenced by climate change, which exacerbates water scarcity and alters energy and food production patterns. For example, rising temperatures and erratic precipitation threaten agricultural productivity and increase energy demand for cooling. The Gulf countries, including Kuwait, are particularly reliant on energy-intensive desalination to meet water needs, a process that underscores the region's dependency on fossil fuels and its exposure to global energy market fluctuations [31]. These vulnerabilities necessitate a shift toward integrated resource management that aligns with national and regional strategies for sustainability and resilience.

Addressing these interconnected challenges requires adopting a systems approach that integrates scientific tools, participatory governance, and capacity-building initiatives. The WEF Nexus framework, with its emphasis on intersectoral linkages and long-term planning, offers a foundation for Arab countries to navigate resource insecurity while accommodating the socio-economic and environmental realities of the region. Migration, as both a driver and consequence of resource challenges, must be central to this discourse, ensuring that policies are adaptive and inclusive.

## 20.5 Case Studies: Jordan and Kuwait

Jordan and Kuwait exemplify contrasting but complementary cases for understanding the application of the WEF Nexus framework. Jordan, one of the world's most water-scarce countries, faces severe constraints across all three nexus dimensions. Its reliance on shared and non-renewable water resources, combined with its dependency on imported energy and food, underscores the critical need for systems-based solutions. Kuwait, in contrast, leverages its wealth and energy resources to mitigate water scarcity through desalination and has integrated governance structures that align water and energy policies. Both nations highlight the importance of tailoring WEF strategies to local contexts while leveraging regional opportunities for cooperation.

Through the case studies of Jordan and Kuwait, the WEF Nexus approach demonstrates its potential to reconcile resource constraints with socio-economic needs, providing a pathway toward sustainable development and regional resilience. This perspective emphasizes the importance of multi-scalar and multi-sectoral strategies to address the pressing realities of water, energy, and food security in the Arab world, moving the Arab countries beyond reactive measures toward a proactive, integrated vision for resource sustainability and human security.

### 20.5.1 The Jordan Case Study: Resource Scarcity, Refugee Challenges, and the WEF Nexus Approach

The Hashemite Kingdom of Jordan represents a critical case study in resource management amid protracted refugee influxes. With an arid climate and limited natural resources, the country faces acute water scarcity, exacerbated by demographic expansion and climatic variability. The presence of large refugee populations, particularly from Syria, has further strained national infrastructure, necessitating shifts from emergency response strategies to sustainable, long-term solutions [60]. The complexities of transboundary water management, international donor funding, and national policy responses provide valuable insights into Jordan's ongoing efforts to address these challenges.

**Water Scarcity and Management**

Jordan's water scarcity is among the most severe globally, with renewable water resources estimated at only 100 cubic meters per capita annually, far below the 500-cubic-meter threshold for extreme water scarcity [61]. The over-extraction of groundwater from depleted aquifers has resulted in extraction rates exceeding recharge levels [62]. Additionally, declining precipitation and reduced surface runoff are expected to worsen under climate change scenarios. Transboundary water agreements with neighboring states, such as Syria and Israel, have been inconsistently upheld, limiting Jordan's access to shared water resources [62, 63].

Infrastructure inefficiencies further exacerbate water insecurity, with substantial losses occurring in the distribution network. Although most Jordanians have access to improved water sources, consumption remains below potential levels due to inefficiencies and regional disparities [64]. Sanitation infrastructure also presents challenges, particularly in rural areas, where many educational institutions lack basic facilities [65]. To address these issues, Jordan has invested in treated wastewater reuse, desalination projects, and capital improvements in northern governorates [66, 67].

**Refugee Crisis and Resource Strain**

Jordan has historically hosted refugee populations, but the Syrian Civil War led to a significant influx, placing immense pressure on national resources [68]. The government has developed response strategies, supported by international financial assistance, to ensure equitable resource distribution and infrastructure expansion [69, 70]. However, the refugee crisis complicates the achievement of Sustainable Development Goal 6 (SDG 6) on water and sanitation, as international organizations navigate the integration of unregistered refugees into needs assessments [71].

Jordan's sustainable development strategy has transitioned from emergency-based assistance to long-term resilience planning, ensuring both refugee and host communities benefit from infrastructure improvements [72]. The country's economic policies now emphasize sustainable resource use, with plans to institutionalize WEF Nexus management [73].

### The Water-Energy-Food (WEF) Nexus in Jordan

*Water Constraints and Their Impact.* Jordan receives approximately 8200 million cubic meters ($m^3$) of annual rainfall, but 80% is lost to evaporation, leaving only 1640 million $m^3$ available for use [74]. An additional 308 million $m^3$ comes from underground water basins, totaling 1948 million $m^3$ per year. This is well below the 500–1000 $m^3$ per capita recommended for sustainable water availability [75]. The scarcity of water has direct consequences on agriculture, industry, and domestic consumption, leading to higher reliance on costly desalination and wastewater reuse projects [76].

*Energy Challenges Linked to Water Scarcity.* Water scarcity directly affects Jordan's energy sector, as a significant portion of electricity consumption is used to pump water from the Jordan Valley [77]. Additional energy is required for water desalination, a necessity due to increased demand from urban centers and refugee camps, including Zaatari and Azraq [69, 70]. Jordan's dependence on imported petroleum and gas from Egypt has left it vulnerable to regional instability, such as disruptions to the Arab Gas Pipeline following the Arab Spring [78]. In response, the government has invested in shale oil and nuclear energy, though the latter remains controversial due to concerns over seismic activity and water requirements for cooling [79]. Jordan has also made substantial investments in solar and wind power to reduce reliance on fossil fuels and improve energy security [77].

*Food Security and Agricultural Constraints.* Jordan possesses one of the lowest levels of arable land per capita worldwide, limiting its agricultural capacity [80]. Given the country's constrained water and land resources, it imports a significant portion of its food supply [81]. Food security is influenced by availability, accessibility, and utilization [82]. While Jordan generally ensures food availability, poverty and unemployment limit access to nutritious food, exacerbating food insecurity challenges [83]. The influx of refugees has increased food demand, straining national food assistance programs [84].

Efforts to enhance food security through agricultural self-sufficiency initiatives have had mixed results. Increased reliance on irrigation has exacerbated water scarcity, leading to a cycle of rising food imports and economic burdens from subsidies [81]. Jordan's strategy now emphasizes regional cooperation and trade agreements to bolster food security while mitigating resource depletion.

*Regional Cooperation and Strategic Resource Management.* Given Jordan's resource limitations, regional collaboration is essential for long-term sustainability. Knowledge-sharing initiatives with neighboring Arab countries provide opportunities to develop more efficient WEF Nexus strategies. Bilateral agreements, such as those governing transboundary water resources, can improve Jordan's water security while fostering regional stability.

Jordan's case underscores the importance of integrated policy planning, technological innovation, and regional partnerships in managing resource scarcity. While challenges remain, the adoption of a structured WEF Nexus framework offers a pathway toward sustainable development, ensuring resilience in the face of environmental and geopolitical pressures.

### *20.5.2 The Kuwait Case Study: Enhancing Water Sustainability Through the WEF Nexus Approach*

**Environmental Contamination and Historical Context**

Kuwait faces significant water sustainability challenges due to historical environmental contamination, scarce natural water resources, and the exacerbating effects of climate change. The nation's heavy reliance on desalinated water, over-extraction of limited groundwater, and high per capita water consumption underscores the urgent need for effective water management strategies. Kuwait's environmental crisis stems from the severe impacts of the 1990–91 Gulf War, which led to the largest oil spill in history, releasing an estimated 6–8 million barrels of crude oil into the nation's terrestrial and marine ecosystems. This disaster resulted in extensive ecological damage, including the formation of oil lakes and dry oil lakes, known as "tarcrete," a solid mix of tar and sand [85, 86]. Despite significant efforts and support from a $460 million compensation fund to remediate and restore the environment, altered soil textures, decimated wildlife populations, and oil penetration affecting freshwater aquifers persist [86, 87]. These oil-induced damages are expected to affect the region and compromise soil and water quality for decades [87, 88]. In response, the United Nations Compensation Commission (UNCC) allocated $3 billion to Kuwait in what is considered the largest remediation project in history to cleanse the soil of oil contamination [85]. This situation underscores Kuwait's complexities in managing environmental contamination, water scarcity, limited agricultural capabilities, and the overarching threat of climate change.

**Water Scarcity and Hydrological Challenges**

Kuwait is identified by the World Resources Institute as one of the top seven countries at acute risk of extreme water stress by 2040 [89]. The nation's hydrological challenges are exacerbated by its geographical and climatic conditions: lack of permanent rivers, dams, or substantial natural reservoirs, and high evaporation rates. Coupled with an average annual rainfall of only 121 mm, these factors substantially impede the natural replenishment of groundwater reserves. Current groundwater extraction rates stand at an alarming 255 million cubic meters annually, starkly surpassing the minimal recharge from internal and transboundary inflows [90, 91]. To address these deficiencies, Kuwait has increasingly relied on desalination and treated wastewater to satisfy its water demands across multiple sectors, notably agriculture, which accounts for over half of the freshwater consumption. The nation's heavy reliance on desalinated water is a significant issue due to the high energy costs associated with desalination and environmental concerns such as thermal brine's impact on marine ecosystems and air pollution from burning fossil fuels. The situation is further strained by high water consumption rates that are among the highest globally and inefficiencies in water use, including substantial loss in the water distribution network [92].

**Food Security and Agricultural Challenges**

Kuwait's challenges in food security are compounded by limited agricultural potential due to harsh environmental conditions, leading to heavy reliance on food imports. This reliance makes the country vulnerable to global market fluctuations and geopolitical tensions that affect food prices and availability [93]. The ambition of Kuwait to attain food self-sufficiency by 2040 is critically hampered by its arid conditions, poor soil fertility, and pervasive scarcity of natural water resources. Sustainable agricultural practices are highly challenging due to substantial requirements for water and cooling energy [91]. The limited availability of water and adverse land conditions raise serious concerns about the feasibility and sustainability of pursuing agricultural self-sufficiency, a strategy that could potentially deplete all freshwater resources and necessitate heightened dependency on environmentally and economically costly desalination processes.

**Energy Sector and Sustainable Development**

The energy sector, while a substantial economic driver due to oil exports, faces its own set of challenges in terms of sustainable development. Kuwait has committed to significant decarbonization efforts aligned with global climate change mitigation goals. However, the transition from fossil fuels to renewable energy sources is complex and requires substantial investment and international cooperation [94].

**National Strategies and the New Kuwait Vision 2035**

Kuwait's efforts to address these issues include national strategies for environmental sustainability, enhanced water management practices, and renewable energy technologies. These efforts are part of a broader initiative to achieve environmental sustainability and reduce dependence on oil, aligning with the New Kuwait Vision 2035 [95]. A multifaceted approach integrating water, energy, and food considerations is urgently required to effectively address these interconnected challenges. This holistic perspective is crucial to understanding and mitigating the socio-economic and environmental impacts of such profound contamination and resource scarcity. It is imperative to optimize resource allocation across sectors to ensure environmental and production sustainability rather than solely focusing on achieving absolute food self-sufficiency.

**Climate Variability and the WEF Nexus**

Climate variability, marked by increasing occurrences of droughts and floods and shifting rainfall patterns, introduces additional complexity in managing the WEF nexus in Kuwait [96]. This nexus requires a comprehensive understanding of the interconnections of the water, energy, and food sectors to formulate sustainable policies effectively. This integrated approach is pivotal to maximizing resource efficiency and sustainability in urban development strategies within Kuwait and incorporates innovations such as energy generation from waste, reduced water use in energy production, the use of renewable energy for desalination, waste minimization, and improved supply chain management.

**The Role of KFAS in Addressing WEF Nexus Challenges**

Kuwait faces critical challenges in managing its interlinked WEF resources. These challenges are exacerbated by economic fluctuations, climate change, and reliance on external supply chains. Recognizing these issues, the Kuwait Foundation for the Advancement of Sciences (KFAS) has actively championed efforts to address the complexities of the WEF Nexus. KFAS focuses on fostering sustainable resource management through the integration of scientific research, policy innovation, and international collaboration. This approach aims to mitigate risks and enhance resilience [97]. KFAS's initiatives align with the key principles of the WEF Nexus. They emphasize integrated resource management by advocating strategies that address the interconnectedness of water, energy, and food systems to optimize resource use and minimize trade-offs. KFAS also adopts scientific data and empirical evidence by utilizing global databases to quantify and analyze resource interdependencies. Additionally, they highlight Kuwait's reliance on international trade for food imports, which exposes the country to vulnerabilities posed by external resource scarcities and climate change. KFAS promotes policy innovation and climate mitigation by stressing the importance of forward-looking policies to address the social cost of carbon and the global impacts of climate change on Kuwait's economy [97]. Domestically, Kuwait's nexus is energy-dominated, with high energy use for desalination and cooling, underscoring the need for water reuse and renewable energy adoption. Internationally, Kuwait's dependency on oil revenues and imported food highlights its exposure to risks in global supply chains, further emphasizing the importance of resilience and diversification.

**Migration's Impact on the WEF Nexus in Kuwait**

Unlike Jordan, Kuwait has not experienced large-scale refugee influxes. However, it hosts a substantial population of migrant workers, who constitute approximately 70% of the total population [98]. This significant demographic factor exerts considerable influence on Kuwait's WEF Nexus. The presence of a large migrant worker population substantially increases the demand for resources. In terms of water, the increased population places further strain on Kuwait's limited freshwater supplies, exacerbating the country's reliance on energy-intensive desalination processes and intensifying the interconnected challenges within the WEF Nexus [92]. Similarly, the energy needs of the migrant population, both for residential use and in various industries, contribute to Kuwait's already high per capita energy consumption [99]. This heightened demand adds pressure on the energy sector as it transitions towards sustainability and seeks to reduce its reliance on fossil fuels. Regarding food, meeting the requirements of a large migrant population further increases Kuwait's dependence on food imports, making the nation more vulnerable to global food price fluctuations and supply chain disruptions [100]. This underscores the importance of diversifying food sources and, where feasible, enhancing domestic food production.

Beyond resource demands, the large migrant population has significant labor force and economic implications. Migrant workers play a vital role in Kuwait's economy, particularly in sectors such as construction, manufacturing, and domestic services, contributing significantly to economic growth and development [101]. However, the remittances sent by these workers to their home countries represent a substantial outflow of funds, impacting Kuwait's balance of payments, even as they support the economies of the workers' home countries [102]. Furthermore, the availability of a large pool of migrant labor can influence wage levels and potentially create challenges for the employment of Kuwaiti nationals in certain sectors [103], necessitating careful management to ensure a balanced and equitable labor market.

Social and environmental considerations are also intertwined with the presence of a large migrant population. The concentration of migrant workers in urban areas contributes to urbanization pressures, increasing the demand for housing, infrastructure, and public services. This can exacerbate challenges related to waste management, traffic congestion, and urban planning [104, 105]. Moreover, the consumption patterns associated with a large migrant population contribute to Kuwait's overall environmental footprint, including carbon emissions, waste generation, and resource depletion [106], highlighting the need for sustainable consumption and production practices that encompass the entire population. The integration of a large migrant population into Kuwaiti society presents both challenges and opportunities, with issues related to cultural diversity, social cohesion, and access to services requiring attention to ensure a harmonious and inclusive society [102].

In light of these factors, Kuwait's national development plans, including the New Kuwait Vision 2035, need to explicitly address the implications of migration for the WEF Nexus. This includes investing in water and energy-efficient technologies and practices to reduce per capita resource consumption across the entire population, including migrant workers [107]. It also necessitates promoting sustainable urban planning and development that can accommodate the needs of a growing and diverse population while minimizing environmental impacts [108]. Diversifying the economy to reduce reliance on oil revenues and creating more opportunities for both Kuwaiti nationals and migrant workers in sustainable sectors is crucial [108]. Strengthening labor laws and regulations to protect the rights of migrant workers and ensure their fair treatment is also essential [102]. Finally, promoting social integration and intercultural dialogue will foster a cohesive and inclusive society [102]. By acknowledging the significant role of migration in shaping the WEF Nexus, Kuwait can develop more comprehensive and effective strategies for achieving long-term resource security and sustainability. This requires a holistic approach that considers the economic, social, and environmental dimensions of migration and its impact on water, food, and energy.

## 20.6 Policy Recommendations: Enhancing Water Sustainability Through the WEF Nexus Approach in the Middle East

The Middle East faces unprecedented water scarcity, exacerbated by climate change, population growth, and unsustainable consumption patterns. The WEF Nexus approach offers a holistic framework to address these interconnected challenges. These policy recommendations aim to operationalize the WEF Nexus, promoting water sustainability across the region while acknowledging the unique context of each nation.

### *20.6.1 Develop and Reform WEF Governance Structures*

A fundamental step toward effective management of WEF resources is through reforming governance structures at both national and regional levels. Many Middle Eastern countries face challenges due to fragmented governance, where water, energy, and food policies are managed independently, leading to inefficiencies and conflicts. To address this, each country should establish a National Water Council or equivalent high-level body chaired by a senior government official. These councils should go beyond sectoral silos and integrate energy and food security considerations into national water governance. Their primary mandate should include developing and implementing a National Integrated Water Resources Management (IWRM) plan with explicit WEF Nexus considerations, ensuring coordinated policymaking across ministries, research institutions, civil society, and the private sector [27].

Several countries have already taken steps toward strengthening water governance. Kuwait's proposed Supreme Water Council should explicitly incorporate regional coordination into its mandate [95]. Saudi Arabia could establish a National Water and WEF Nexus Council under the leadership of the Crown Prince or a designated minister to ensure Vision 2030's water-related goals are met within an integrated framework [109]. Qatar's Supreme Council for Water, Energy, and Food Security could align its National Vision 2030 with a WEF Nexus strategy, focusing on sustainable desalination and food production [110]. Similarly, the UAE could form a Federal WEF Nexus Authority to facilitate coordination across its emirates. Oman's National Council for Sustainable Resource Management could integrate water scarcity, energy diversification, and food security within a single governance mechanism [111].

At the regional level, a **Middle East Water and WEF Nexus Coordination Body (MEWNCB)** should be established under the Arab League or a newly dedicated entity. This body would promote regional collaboration through joint planning, harmonized standards, conflict resolution mechanisms, and knowledge exchange. It would also facilitate a regional database for water, energy, and food resources, enabling evidence-based policymaking across borders. Given the transboundary nature of many water resources in the region, regional coordination is essential for sustainable water governance and long-term resource security [27].

### *20.6.2 Address Barriers to Nexus Implementation*

Despite the potential of the WEF Nexus approach, several barriers hinder its implementation. Stakeholder participation remains a challenge due to competing interests between governmental agencies, private sectors, and local communities. Effective policies should promote inclusive decision-making processes, ensuring that all stakeholders are actively engaged in Nexus planning and implementation.

Uncertainty in data and information is another major constraint. Many Middle Eastern countries lack comprehensive, high-quality datasets on water, energy, and food interdependencies. Standardizing data collection, integrating remote sensing technologies, and establishing open-access databases would help mitigate these gaps and support informed policymaking.

Integration of scientific and stakeholder knowledge is crucial for operationalizing the Nexus approach. Traditional policy frameworks often rely heavily on technical expertise while neglecting local knowledge and socio-economic factors. By incorporating participatory modeling and transdisciplinary research, policymakers can create more effective and context-specific Nexus solutions.

### *20.6.3 Enhance the Role of Models and Modeling Approaches in Policy Support*

Models and modeling approaches play a crucial role in facilitating WEF Nexus implementation. They provide **scenario analysis**, enabling policymakers to evaluate the trade-offs between different water, energy, and food strategies. For instance, integrated assessment models can predict the long-term impacts of desalination on energy consumption and agricultural water availability.

Furthermore, models support **risk assessment**, helping governments anticipate and mitigate the effects of climate change, resource depletion, and policy interventions. Agent-based models and system dynamics models can simulate the behavior of multiple stakeholders and assess the consequences of different governance structures.

Finally, modeling tools enhance **policy coherence**, ensuring that strategies in one sector do not undermine objectives in another. For example, optimizing water allocation between agriculture and urban use can balance economic development and resource conservation. Governments should invest in Nexus modeling platforms and train policymakers in their application to support evidence-based decision-making.

### 20.6.4 Enhance Data Collection, Monitoring, and Evaluation

Reliable data is essential for evidence-based policymaking. Saudi Arabia should develop a national water information system integrating data from remote sensing, smart meters, and monitoring wells. Egypt should strengthen its Nile River monitoring to improve transboundary water management [112]. Lebanon should implement comprehensive water quality monitoring programs for its rivers and groundwater sources [113].

At the regional level, the MEWNCB should establish a standardized monitoring framework with shared indicators, enabling transparency and comparability across nations. High-quality data ensures accountability and improves policy adaptability in response to emerging challenges [46].

### 20.6.5 Strengthen Research, Innovation, and Capacity Building

Research and innovation are essential to addressing the region's WEF Nexus challenges. Kuwait should establish a dedicated WEF Nexus Research Center at KISR, while institutions like KAUST in Saudi Arabia and Masdar Institute in the UAE should take leading roles in advancing regional collaboration on water, energy, and food security research [89, 114].

Capacity building is equally important. Jordan's National Center for Agricultural Research and Extension (NCARE) is an example of an institution that could be expanded to train professionals in sustainable resource management [115].

The MEWNCB should facilitate joint training programs and regional workshops, ensuring knowledge-sharing across countries [27].

### 20.6.6 Enhance Energy Efficiency in the Water Sector

Energy consumption in the water sector is a major challenge across the Middle East, where desalination and water pumping require substantial electricity. To address this, countries should conduct regular energy audits for water infrastructure, identifying opportunities for efficiency improvements. Upgrading desalination plants with energy-efficient technologies and adopting smart water grids can optimize distribution, reducing energy waste. For example, Saudi Arabia and Kuwait should prioritize replacing outdated water pumps with modern, energy-efficient models, while the UAE could expand its smart metering systems to better regulate water use [89, 116]. Implementing strict building codes mandating the use of energy-efficient appliances, water heaters, and irrigation systems will also reduce unnecessary energy use in water management [117].

### *20.6.7 Strengthen the Water-Agriculture Nexus for Sustainable Food Production*

Agriculture is the largest consumer of water in the Middle East, making the adoption of efficient irrigation techniques and sustainable water management essential for long-term food security. Many countries still rely on inefficient flood irrigation, leading to significant water losses. Transitioning to modern irrigation systems such as drip and sprinkler irrigation must be prioritized. Saudi Arabia should enforce stricter groundwater extraction regulations while expanding water-saving irrigation technologies [118]. Egypt's irrigation modernization program, including the Tabteen initiative, aims to improve irrigation efficiency and minimize water losses in canals and fields [119]. Lebanon and Jordan should incentivize farmers to adopt precision irrigation, and Oman should expand treated wastewater use for agriculture to reduce reliance on freshwater sources [120].

Rainwater harvesting is another critical tool for improving agricultural resilience. Lebanon should promote farm-level rainwater collection and storage to supplement irrigation, particularly in areas with seasonal rainfall [113]. Jordan, facing severe water shortages, should implement community-based rainwater harvesting programs in rural farming communities. Expanding wastewater treatment and reuse in agriculture is also key—Egypt has increased its wastewater treatment capacity, but further investment is needed to ensure safe and effective reuse [121].

Investing in climate-resilient crops and innovative agricultural technologies will further strengthen the Water-Agriculture Nexus. Gulf countries such as Saudi Arabia, Qatar, and the UAE should scale up research into drought-resistant and salt-tolerant crops. The UAE has pioneered hydroponic and vertical farming, reducing water use in food production, and similar initiatives should be expanded regionally [122]. Precision agriculture technologies, including soil moisture sensors and automated irrigation systems, should be widely adopted to optimize water use on farms.

At a regional level, agricultural trade policies should be aligned to ensure food security without excessive strain on water resources. Middle Eastern countries import significant portions of their food, which mitigates local water demand but increases vulnerability to global supply chain disruptions. **The Middle East Water and WEF Nexus Coordination Body (MEWNCB)** should facilitate discussions on regional food reserves, trade agreements, and coordinated investments in sustainable agriculture.

Integrating water conservation into agricultural planning, promoting modern irrigation, and investing in resilient farming practices will ensure that food production does not come at the cost of long-term water sustainability. The Water-Agriculture Nexus must be at the heart of policies aiming to balance resource use and food security in the region.

### *20.6.8 Promote Public Awareness and Behavioral Change*

Sustainable resource management requires active public participation. Governments should implement nationwide awareness campaigns to educate citizens on water and energy conservation. In Kuwait, for example, a national campaign could highlight the link between water and energy conservation through social media, television, and public events [123]. Saudi Arabia and Jordan should expand conservation education in schools, integrating WEF Nexus topics into curricula. Egypt's campaigns to raise awareness about protecting the Nile from pollution should be expanded to include broader discussions on responsible water use [124]. In countries with high levels of water waste, such as Qatar, media campaigns should emphasize the high costs of desalination and encourage conservation measures like installing water-saving fixtures and reducing excessive outdoor water use [125].

## 20.7 Overall Discussion and Conclusion

The Middle East stands at a critical juncture, facing a confluence of challenges that threaten the sustainability of its water, energy, and food systems. The region's arid climate, coupled with rapid population growth, urbanization, and the impacts of climate change, has created a complex web of interconnected issues that demand a holistic and integrated approach. The WEF Nexus framework provides a powerful lens through which to understand these interdependencies and to develop effective strategies for long-term resource security and sustainable development.

The case studies of Jordan and Kuwait, while distinct in their specific circumstances, highlight the common challenges faced by many countries in the region. Jordan's experience underscores the profound impact of water scarcity, exacerbated by recurring refugee crises, on the WEF Nexus. The country's heavy reliance on groundwater, coupled with inadequate transboundary water cooperation and aging infrastructure, necessitates urgent reforms in water governance and management. The influx of refugees has further strained Jordan's limited resources, highlighting the need for integrated planning that considers the needs of both host communities and displaced populations. Kuwait, on the other hand, demonstrates the challenges and opportunities associated with leveraging wealth and energy resources to address water scarcity. While desalination has provided a technological solution, its energy intensity and environmental impacts underscore the need for a transition towards more sustainable practices, including renewable energy integration and enhanced water use efficiency.

These case studies, along with the broader regional context, reveal several key insights. The interconnected nature of the WEF Nexus necessitates a shift from siloed, sectoral approaches to integrated resource management. Policies and strategies must explicitly consider the trade-offs and synergies between water, energy, and food systems. For instance, promoting water-efficient agriculture not only conserves water but also reduces energy demand for pumping and can enhance food security. Similarly, investing in renewable energy for desalination can simultaneously address water scarcity and reduce greenhouse gas emissions.

Effective governance and institutional frameworks are crucial for successful WEF Nexus implementation. Establishing high-level national councils or coordination bodies, as recommended in the policy recommendations, can facilitate cross-sectoral collaboration, ensure policy coherence, and drive the development of integrated strategies. Regional cooperation, facilitated by a body like the proposed MEWNCB, is equally important for addressing transboundary resource issues, sharing best practices, and promoting joint investments in sustainable technologies.

Technological innovation plays a vital role in enhancing resource efficiency and sustainability within the WEF Nexus. Investing in research and development for advanced desalination technologies, renewable energy integration, climate-smart agriculture, and water-efficient irrigation methods can significantly improve the region's resource outlook. The examples of Kuwait's Shagaya Renewable Energy Park and the UAE's investments in solar energy demonstrate the potential for technological solutions to contribute to a more sustainable future.

Social equity and inclusivity must be central to WEF Nexus strategies. Addressing the needs of vulnerable populations, including refugees, migrant workers, and low-income households, is essential for ensuring that the benefits of sustainable resource management are shared equitably. This requires targeted policies, such as social safety nets, affordable access to essential services, and programs that promote education and awareness about resource conservation. Moreover, promoting gender equality in WEF Nexus planning and implementation can enhance the effectiveness and sustainability of interventions.

Migration, both within and into the region, has significant implications for the WEF Nexus. The case of Jordan highlights the impact of large-scale refugee influxes on water resources, while Kuwait's experience with a substantial migrant worker population demonstrates the challenges and opportunities associated with managing resource demand in a diverse and growing population. National development plans and WEF Nexus strategies must explicitly consider the implications of migration for resource management and incorporate measures to address the needs of both host communities and migrant populations.

The importance of data, monitoring, and evaluation cannot be overstated. Robust data collection and analysis are essential for understanding the complex dynamics of the WEF Nexus, informing evidence-based policymaking, and tracking progress towards sustainability goals. Establishing comprehensive monitoring and evaluation frameworks, both at the national and regional levels, will be crucial for ensuring accountability, adapting strategies as needed, and demonstrating the impact of WEF Nexus interventions.

In conclusion, the challenges facing the Middle East's water, energy, and food systems are substantial, but they are not insurmountable. By embracing the WEF Nexus approach, countries in the region can move towards a more integrated, sustainable, and resilient future. The policy recommendations outlined in this chapter provide a comprehensive framework for action, encompassing governance reforms, technological innovations, behavioral changes, and a commitment to social equity and regional cooperation. While the specific context and priorities of each country will vary, the overarching principles of the WEF Nexus provide a common foundation for addressing the shared challenges of the region. By working together, leveraging their unique strengths, and investing in a sustainable future, the countries of the Middle East can transform their resource challenges into opportunities for growth, development, and enhanced human security. The journey will undoubtedly be complex and require sustained commitment, but the potential rewards—a more secure, prosperous, and resilient future for all.

**Acknowledgements** This work would not be possible without the support of the NATO Science for Peace and Security Programme and the results of the Advanced Research Workshop ARW G6276 on Security Enhancement for Climate Changes Impacting Urban Resources (SECCURE) held in Montelibretti (Rome) in July 2024.

## References

1. Bazilian M et al (2011) Considering the energy, water and food nexus: towards an integrated modelling approach. Energy Policy 39(12):7896–7906
2. Simpson GB, Jewitt GP (2019) The development of the water-energy-food nexus as a framework for achieving resource security: a review. Front Environ Sci 7:8
3. Garcia DJ, You F (2016) The water-energy-food nexus and process systems engineering: a new focus. Comput Chem Eng 91:49–67
4. Hoff H (2011) Understanding the nexus
5. Zhang C et al (2018) Water-energy-food nexus: concepts, questions and methodologies. J Clean Prod 195:625–639
6. Howells M et al (2013) Integrated analysis of climate change, land-use, energy and water strategies. Nat Clim Chang 3(7):621–626
7. Nationen V (2015) Transforming our world: the 2030 agenda for sustainable development: A/Res/70/1. United Nations, Division for Sustainable Development
8. Gain AK, Giupponi C, Benson D (2018) The water–energy–food (WEF) security nexus: the policy perspective of Bangladesh. Sustainability in the water energy food nexus. Routledge, pp 183–198
9. Kim H et al (2024) Understanding the role of biodiversity in the climate, food, water, energy, transport and health nexus in Europe. Sci Total Environ 171692
10. Mabhaudhi T, Chibarabada T, Modi A (2016) Water-food-nutrition-health nexus: linking water to improving food, nutrition and health in Sub-Saharan Africa. Int J Environ Res Public Health 13(1):107
11. Schmeller DS, Bridgewater P (2016) The intergovernmental platform on biodiversity and ecosystem services (IPBES): progress and next steps. Springer, pp 801–805
12. Benson D, Gain AK, Rouillard JJ (2015) Water governance in a comparative perspective: from IWRM to a'nexus' approach? Water Altern 8(1):756–773
13. Cairns R, Krzywoszynska A (2016) Anatomy of a buzzword: the emergence of 'the water-energy-food nexus' in UK natural resource debates. Environ Sci Policy 64:164–170
14. Allouche J, Middleton C, Gyawali D (2015) Technical veil, hidden politics: interrogating the power linkages behind the nexus. Water Altern 8(1)
15. Liu J et al (2017) Challenges in operationalizing the water–energy–food nexus. Hydrol Sci J 62(11):1714–1720

16. Salam PA et al(2017) Water-energy-food nexus: principles and practices. Wiley
17. Waughray D (2011) Water security, the water-food-energy-climate nexus: the World Economic Forum Water Initiative. Island Press Washington, DC
18. Mancosu N et al (2015) Water scarcity and future challenges for food production. Water 7(3):975–992
19. Water U (2024) Progress on the level of water stress
20. IRENA F (2021) Renewable energy for agri-food systems: towards the sustainable development goals and the Paris Agreement
21. Siddiqi A, Anadon LD (2011) The water–energy nexus in Middle East and North Africa. Energy Policy 39(8):4529–4540
22. Rogers P (2017) The triangle: energy, water & food nexus for sustainable security in the Arab Middle East. Water, energy & food sustainability in the Middle East: the sustainability triangle. Springer, pp 21–43
23. Initiative WEFW (2011) Initiative, innovative water partnerships. In: Water security: the water-food-energy-climate nexus, pp 225–233
24. BMU B (2012) The water, energy and food security nexus–solutions for a green economy. Policy Recommendations, Bonn
25. Long G, Clough E, Rietig K (2022) Global partnerships for the SDGs. Partnerships and the sustainable development goals. Springer, pp 27–39
26. Avellán T, Roidt M (2022) The nexus: concepts and frameworks. In: Handbook on the water-energy-food nexus. Edward Elgar Publishing, pp 16–35
27. Mohtar RH, Daher B (2016) Water-energy-food nexus framework for facilitating multi-stakeholder dialogue. Water Int 41(5):655–661
28. Rasul G, Sharma B (2016) The nexus approach to water–energy–food security: an option for adaptation to climate change. Clim Policy 16(6):682–702
29. Hogeboom RJ et al (2021) Resilience meets the water–energy–food nexus: mapping the research landscape. Front Environ Sci 9:630395
30. Mohtar RH, Daher B (2017) Beyond zero sum game allocations: expanding resources potentials through reduced interdependencies and increased resource nexus synergies. Curr Opin Chem Eng 18:84–89
31. Daher B et al (2021) Toward resilient water-energy-food systems under shocks: understanding the impact of migration, pandemics, and natural disasters. Sustainability 13(16):9402
32. Carmona-Moreno C, Dondeynaz C, Biedler M (2019) Position paper on water, energy, food and ecosystems (WEFE) nexus and sustainable development goals (SDGs). Publications Office of the European Union Luxembourg.
33. Peña-Torres D, Boix M, Montastruc L (2022) Optimization approaches to design water-energy-food nexus: a litterature review. Comput Chem Eng 167:108025
34. Albrecht TR, Crootof A, Scott CA (2018) The water-energy-food nexus: a systematic review of methods for nexus assessment. Environ Res Lett 13(4):043002
35. Daher BT, Mohtar RH (2018) Water–energy–food (WEF) nexus tool 2.0: guiding integrative resource planning and decision-making. In: Sustainability in the water energy food nexus. Routledge, pp 36–59
36. Solano-Pereira P, García-González A, Miguel González LJ (2025) Economic representation in water–energy–food nexus models: a systematic review of system dynamics approaches. Energies 18(4):966
37. Bakhshianlamouki E et al (2020) A system dynamics model to quantify the impacts of restoration measures on the water-energy-food nexus in the Urmia lake Basin, Iran. Sci Total Environ 708:134874
38. Deng H-M et al (2020) Managing the water-energy-food nexus in China by adjusting critical final demands and supply chains: an input-output analysis. Sci Total Environ 720:137635
39. White DJ et al (2018) The water-energy-food nexus in East Asia: a tele-connected value chain analysis using inter-regional input-output analysis. Appl Energy 210:550–567
40. Liu Z et al (2021) Water-energy nexus within urban agglomeration: an assessment framework combining the multiregional input-output model, virtual water, and embodied energy. Resour Conserv Recycl 164:105113

41. Magliocca NR (2020) Agent-based modeling for integrating human behavior into the food–energy–water nexus. Land 9(12):519
42. Schreinemachers P, Berger T (2011) An agent-based simulation model of human–environment interactions in agricultural systems. Environ Model Softw 26(7):845–859
43. Castelli C et al (2022) A review of macroeconomic models for the WEFE nexus assessment
44. Castelli C et al (2024) Exploring macroeconomic models in the water, energy, food, and ecosystem (WEFE) field: a comprehensive review. Environ Res Lett
45. Lazaro LLB et al (2022) Ten years of research on the water-energy-food nexus: an analysis of topics evolution. Front Water 4:859891
46. Saltelli A et al (2020) Five ways to ensure that models serve society: a manifesto. Nature 582(7813):482–484
47. Pianosi F et al (2016) Sensitivity analysis of environmental models: a systematic review with practical workflow. Environ Model Softw 79:214–232
48. Voinov A, Bousquet F (2010) Modelling with stakeholders. Environ Model Softw 25(11):1268–1281
49. Soleimanian E, Afshar A, Molajou A (2022) A review on water simulation models for the WEF Nexus: development perspective. Environ Sci Pollut Res 29(53):79769–79785
50. Borge-Diez D, García-Moya FJ, Rosales-Asensio E (2022) Water energy food nexus analysis and management tools: a review. Energies 15(3):1146
51. Kaddoura S, El Khatib S (2017) Review of water-energy-food nexus tools to improve the nexus modelling approach for integrated policy making. Environ Sci Policy 77:114–121
52. Baas S, Trujillo M, Lombardi N (2015) Impact of disasters on agriculture and food security
53. Rising KL et al (2020) Development of the uncertainty communication checklist: a patient-centered approach to patient discharge from the emergency department. Acad Med 95(7):1026–1034
54. Cai X et al (2018) Understanding and managing the food-energy-water nexus–opportunities for water resources research. Adv Water Resour 111:259–273
55. Munshi K (2003) Networks in the modern economy: Mexican migrants in the US labor market. Q J Econ 118(2):549–599
56. Programme WF (2017) At the root of exodus: food security, conflict and international migration. World Food Programme Rome
57. Scott A, Worrall L, Pickard S (2022) Energy, migration and the 2030 Agenda for Sustainable Development. ODI
58. Sušnik J, Staddon C (2022) Evaluation of water-energy-food (WEF) nexus research: perspectives, challenges, and directions for future research. JAWRA J Am Water Resour Assoc 58(6):1189–1198
59. Maftouh A et al (2022) The application of water–energy nexus in the Middle East and North Africa (MENA) region: a structured review. Appl Water Sci 12(5):83
60. Van der Helm AW et al (2017) Developing water and sanitation services in refugee settings from emergency to sustainability—the case of Zaatari Camp in Jordan. J Water Sanit Hyg Dev 7(3):521–527
61. (MWI) M.o.W.a.I (2023) National water strategy 2023–2040
62. Rajsekhar D, Gorelick SM (2017) Increasing drought in Jordan: climate change and cascading Syrian land-use impacts on reducing transboundary flow. Sci Adv 3(8):e1700581
63. Yoon J (2017) Evaluation of water security in jordan using a multi-agent hydroeconomic model. Stanford University
64. Al-Ansari N et al (2014) Water supply network losses in Jordan. J Water Resour Prot 6(2):83–96
65. Aburamadan R, Cotella V, Darwish R (2024) Sanitation infrastructure challenges in global south poverty pockets: towards a multidisciplinary approach to sustainable health—the Jordanian case. J Infrastr Policy Dev 8(8):5309
66. Pappas K et al (2022) National and international funding for financing sustainable development goal 6: case of Jordan. SDGs in Africa and the Middle East Region. Springer, pp 1–29

67. Karnib A (2019) Water as crosscutting factor in the SDGs under review at the high-level panel forum for sustainable development (HLPF) 2019 in the Arab States. UNESCO WWAP consultant. In: Proceedings of the UNESCO world water assessment…
68. UNHCR (2020) Jordan: Zaatari Camp Factsheet. https://data.unhcr.org/en/documents/details/73845
69. UNHCR (2019) Jordan: Azraq refugee camp fact sheet. https://reliefweb.int/sites/reliefweb.int/files/resources/70193.pdf
70. UNHCR (2019) Jordan—Zaatari refugee camp fact sheet. https://data2.unhcr.org/en/documents/download/70183.
71. ISG (2019) Evaluation of UNICEF's response to the water, sanitation, and hygiene needs in Jordan as a result of the Syrian refugee crisis (July 2012 to July 2017). https://gdc.unicef.org/media/4626/download
72. Refugee R (2018) 3RP regional refugee & resilience plan 2019–2020 in response to the Syria crisis. https://www.unhcr.org/sites/default/files/legacy-pdf/589497237.pdf
73. Adam-Bradford A et al (2024) Climate and water-related vulnerabilities in refugee hosting communities in northern Jordan: Irbid and Ramtha municipalities
74. Badran A (2017) Climate change and water science policy in management. In: Water, energy & food sustainability in the Middle East: the sustainability triangle, pp 3–19
75. Falkenmark M, Lundqvist J (1998) Towards water security: political determination and human adaptation crucial. In: Natural resources forum. Wiley Online Library
76. Al-Addous M et al (2023) Water resources in Jordan: a review of current challenges and future opportunities. Water 15(21):3729
77. Verdeil É (2014) The contested energy future of Amman, Jordan: between promises of alternative energies and a nuclear venture. Urban Studies 51(7):1520–1536
78. Alshwawra A, Almuhtady A, Sakhrieh A (2023) Electricity system security in Jordan: a response for Arab uprising. Heliyon 9(5)
79. Starz A, Van Sickle M (2014) IAEA reviews Jordan's progress in developing nuclear power programme. Int Atomic Energy Agency
80. Bashour I (2017) Soils and food security in the Arab world. In: Water, energy & food sustainability in the Middle East: the sustainability triangle, pp 283–298
81. Kumaraswamy P, Singh M (2018) Jordan's food security challenges. Mediterr Q 29(1):70–95
82. Leck H et al (2015) Tracing the water–energy–food nexus: description, theory and practice. Geogr Compass 9(8):445–460
83. W.F.P. Jordan (n.d.) https://www.wfp.org/countries/jordan
84. Welthungerhilfe CW (2020) 2020 Global Hunger Index: one decade to zero hunger: linking health and sustainable food systems. https://www.globalhungerindex.org/pdf/en/2020.pdf
85. UNCC (2002) Report and recommendations made by the panel of commissioners concerning the third instalment of "F4": Claims. United Nations Compensation Commission Governing Council Kuwait, UNCC, Kuwait
86. Al-Sarawi M, Massoud M, Al-Abdali F (1998) Preliminary assessment of oil contamination levels in soils contaminated with oil lakes in the greater Burgan oil fields, Kuwait. Water Air Soil Pollut 106(3–4):493–504
87. Hirschmann K (2009) The Kuwaiti oil fires. Infobase Publishing
88. Brown G, Porembski S (1998) Flora and vegetational aspects of miniature dunes in a sand-depleted Haloxylon salicornicum community in the Kuwait desert. Flora 193(2):133–140
89. Aleisa E (2024) Navigating Kuwait's water scarcity challenges: a holistic analysis of water resources, environmental impact, cost, and policy implications. Desalination 586:117827
90. Alkandari AJ (2024) Climate and climate change aspects of Kuwait. Terrestrial environment and ecosystems of Kuwait: assessment and restoration. Springer, pp 57–91
91. Aliewi A, Alomirah H (2020) Assessment of the significance of water-energy-food nexus for Kuwait. Resilience, response, and risk in water systems: shifting management and natural forcings paradigms. Springer, pp 357–367
92. Tariq MAUR et al (2022) A detailed perspective of water resource management in a dry and water scarce country: the case in Kuwait. Front Environ Sci 10:1073834

93. Gelan AU, Atkinson G (2022) Climate change and food security: assessing the prospect for Kuwait using an economy-wide model. Kuwait J Sci 49(4)
94. Alsayegh OA (2021) Barriers facing the transition toward sustainable energy system in Kuwait. Energ Strat Rev 38:100779
95. Olver-Ellis S (2020) Building the new Kuwait: vision 2035 and the challenge of diversification
96. Endo A et al (2015) Methods of the water-energy-food nexus. Water 7(10):5806–5830
97. Siderius C et al (2019) Characterising the water-energy-food nexus in Kuwait and the Gulf region
98. Alzankawi A et al (2025) 7 Kuwait at the crossroads. In: Geoeconomics of the sustainable development goals, p 104
99. Alotaibi S (2011) Energy consumption in Kuwait: prospects and future approaches. Energy Policy 39(2):637–643
100. Al-Saidi M, Saliba S (2019) Water, energy and food supply security in the Gulf Cooperation Council (GCC) countries—a risk perspective. Water 11(3):455
101. Baldwin-Edwards M (2011) Labour immigration and labour markets in the GCC countries: national patterns and trends
102. Palier B, Azoulay R, Louer L (2021) Kuwait's welfare system. Laboratory for Interdisciplinary Evaluation of Public Policies
103. Forstenlechner I, Rutledge EJ (2011) The GCC's 'demographic imbalance': perceptions, realities and policy options. Middle East Policy 18(4):25–43
104. Alghais N, Pullar D (2018) Projection for new city future scenarios—a case study for Kuwait. Heliyon 4(3)
105. Sharp D, Sadliwala B, Al-Shammari A (2024) Recognising the right to urban climate justice in Kuwait. Geoforum 155:104099
106. ElMassah S, Hassanein EA (2023) Economic development and environmental sustainability in the GCC countries: new insights based on the economic complexity. Sustainability 15(10):7987
107. Moossa B et al (2022) Desalination in the GCC countries-a review. J Clean Prod 357:131717
108. Ali J, Moser S (2024) New cities for a 'new Kuwait': planning for national continuity and stability. Int Plan Stud 29(3):252–267
109. Alamri Y et al (2023) The economic nexus between energy, water consumption, and food production in the Kingdom of Saudi Arabia. Economies 11(4):113
110. Tan T, Al-Khalaqi A, Al-Khulaifi N (2014) Qatar national vision 2030. Sustain Dev Apprais Gulf Reg 19(1):65–81
111. Aggarwal G et al (2024) Sustainable investment strategies for Oman vision 2040—contribution towards socioeconomic development and environmental conservation
112. Khalifa M, Woods NE, Eltahir EA (2023) Estimates of Sudan's historical water withdrawals from the Nile. J Hydrol 624:129858
113. Ersoy SR et al (2022) Sustainable transformation of Lebanon's energy system: development of a phase model
114. Juaidi A et al (2016) An overview of energy balance compared to sustainable energy in United Arab Emirates. Renew Sustain Energy Rev 55:1195–1209
115. National Center for Agricultural Research and Extension (Jordan) (NCARE) (2023)
116. Baljon M (2023) Revolutionizing Saudi Arabia's agriculture: the IoT transformation of water management. J Adv Res Appl Sci Eng Technol 36(1):217–240
117. Geglio T et al (2021) Water-and energy-efficient appliances for circular water economy: conceptual framework development and analysis of greenhouse gas emissions and water consumption. ACS ES&T Eng 2(3):409–422
118. Memon AM et al (2022) Aquaponics in Saudi Arabia: initial steps towards addressing food security in the arid region. Agriculture 12(12):2094
119. Saleh E (2020) Can irrigation water saving options cope with water scarcity in Egypt. Environ Sci Agric Food Sci 6:167–176
120. Al Bannay S, Takizawa S (2022) Decoupling of water production and electricity generation from GDP and population in the Gulf Cooperation Council (GCC) countries. Sustainability 14(9):5386

121. Abdelhafez AA, Metwalley SM, Abbas H (2020) Irrigation: Water resources, types and common problems in Egypt. In: Technological and modern irrigation environment in Egypt: best management practices & evaluation, pp 15–34
122. Maksoud A et al (2021) Generating optimum form for vertical farms skyscrapers in UAE. Int J Eng Res Technol 10(3):689–700
123. Aleisa E (2015) Water consumption and conservation behavior in Kuwaiti households. Kuwait J Sci 42(3):91–113
124. Kazemian M, Behzadmehr A, Sarvari S (2010) Thermodynamic optimization of multi-effect desalination plant using the DoE method. Desalination 257(1–3):195–205
125. Tarsheed (2023) Qatar general electricity & water corporation. https://www.km.qa/tarsheed/

# Author Index





A. Angelini et al. (eds.), *Security Enhancement for Climate Change Impacting Urban Resources - SECCURe*, NATO Science for Peace and Security Series C: Environmental Security, https://doi.org/10.1007/978-94-024-2345-7

# Index





A. Angelini et al. (eds.), *Security Enhancement for Climate Change Impacting Urban Resources - SECCURe*, NATO Science for Peace and Security Series C: Environmental Security, https://doi.org/10.1007/978-94-024-2345-7

GPSR Compliance

*The European Union's (EU) General Product Safety Regulation (GPSR) is a set of rules that requires consumer products to be safe and our obligations to ensure this.*

*If you have any concerns about our products, you can contact us on ProductSafety@springernature.com*

In case Publisher is established outside the EU, the EU authorized representative is:

Springer Nature Customer Service Center GmbH
Europaplatz 3
69115 Heidelberg, Germany

**Batch number: 10755329**

Printed by Printforce, the Netherlands